NATURE, NURTURE AND CHANCE

THE LIVES OF FRANK AND CHARLES FENNER

NATURE, NURTURE AND CHANCE

THE LIVES OF FRANK AND CHARLES FENNER

FRANK FENNER

Visiting Fellow, John Curtin School of Medical Research
The Australian National University

THE AUSTRALIAN NATIONAL UNIVERSITY
E PRESS

E PRESS
Published by ANU E Press
The Australian National University
Canberra ACT 0200, Australia
Email: anuepress@anu.edu.au
Web: http://epress.anu.edu.au

National Library of Australia
Cataloguing-in-Publication entry

Fenner, Frank, 1914- .
Nature, nurture and chance : the lives of Frank and Charles Fenner.

ISBN 1 920942 62 9
ISBN 1 920942 63 7 (online)

1. Fenner, Frank, 1914- . 2. Fenner, Charles, 1884-1955. 3. Microbiologists - Australia - Biography. 4. Virologists - Australia - Biography. 5. Geographers - Australia - Biography. 6. Educators - Australia - Biography. I. Title.

579.092

Indexed by Frank Fenner.
Cover design by ANU E Press .
Frank Fenner, by Mathew Lynn, 1999, courtesy of the John Curtin School of Medical Research.
Charles Fenner, by Ivor Hele, 1935, courtesy of Education Centre, Adelaide.

Table of Contents

List of Figures v

Preface vii

Acknowledgements xiii

Acronyms xv

Part I. The Life of Frank Fenner

Introduction 3

Chapter 1. Childhood, 1914 to 1932 9

Chapter 2. The University Years, 1933 to 1940 21

Chapter 3. The War Years, May 1940 to February 1946 27

Chapter 4. Walter and Eliza Hall Institute, 1946 to 1948; Rockefeller Institute, 1948 to 1949 47

Chapter 5. Professor of Microbiology, John Curtin School of Medical Research, 1949 to 1967: Administrative and Domestic Arrangements 55

Chapter 6. Professor of Microbiology, John Curtin School of Medical Research, 1949 to 1967: Research 81

Chapter 7. Director of the John Curtin School of Medical Research, 1967 to 1973 97

Chapter 8. Activities Associated with the Australian Academy of Science 113

Chapter 9. Director of the Centre for Resource and Environmental Studies, 1973 to 1979 121

Chapter 10. Smallpox and its Eradication, 1969 to 1980 137

Chapter 11. Visiting Fellow, John Curtin School of Medical Research, From 1980 159

Part II. The Life of Charles Fenner

Chapter 12. The Fenner Lineage 197

Chapter 13. Childhood, University, Marriage and Family 209

Chapter 14. Charles Fenner, the Educational Administrator 217

Chapter 15. Overseas Trips, Diary Extracts on Education 239

Chapter 16. The Scientist and Science Communicator 265

Chapter 17. Overseas Trips, Diary Extracts on Science 281

Chapter 18. Reflections, Frank Fenner 327

Index of Names 341

Subject Index, Part I, Frank Fenner, Chapters 1–11, 18 351

Subject Index, Part II, Charles Fenner, Chapters 12–17, 18 355

List of Figures

1.1. Iramoo, the house at 42 Alexander Avenue, Rose Park in 1931, after adding the second storey as the boys' shared bedroom 11

1.2. The family in 1937 12

2.1. Adelaide University Hockey Club, Inter-varsity team, Brisbane 1937 23

3.1. Officers' Mess, 2/6 Field Ambulance, Woodside, September 1940 28

3.2. Frank Fenner at the site of the 2/2 Australian General Hospital at Hughenden a day after the cyclone 35

3.3. Photograph of Bobbie Roberts in 1942, by Julian Smith 41

5.1. The temporary buildings of the John Curtin School of Medical Research 60

5.2. The Fenner house at 8 Monaro Crescent 61

5.3. Viewing the Canberra Meritorious Award for Architecture, 1956 62

5.4. Staff of Department of Microbiology in the temporary laboratories, June 1955 64

5.5. Team assembled in Indonesia at request of Ambassador Walter Crocker to investigate the establishment of a medical school in Sumatra, July–August 1956 67

5.6. Academic staff, visiting fellows and students of the Department of Microbiology in 1962, on the roof of Infected Animal House, with Black Mountain in the background 72

6.1. European and Brazilian Rabbit 84

6.2. Frank Fenner at the bench, inoculating chick-developing embryos with a virus suspension 91

10.1. The Smallpox Recognition Card, showing a child with smallpox at the pustular stage 138

10.2. Map showing my travel by air and road around South Africa and Namibia in 1978 150

10.3. Frank Fenner addressing the World Health Assembly on 8 May 1980 157

11.1. Frank Fenner receiving the Prime Minister's Prize for Science, from Prime Minister John Howard, August 2002 187

12.1. Map of Hesse and the Schwalm valley 198

12.2. Schwälmer men's and women's attire 199

12.3. The Fenner coat-of-arms 206

12.4. Fenner Hall, The Australian National University, Canberra 207

13.1. Students and teachers at Dunach School in 1890 210

13.2. Charles Fenner with his father Johannes Fenner, in 1903, when Charles was 'principal' of two one-teacher schools 211

13.3. Photograph of Miss Emma L. Hirt 213

13.4. Photograph of the married couple 213

14.1. Cartoons of Charles Fenner published in *The Bulletin*, 1929, 1930 and 1935 225

14.2. Charles Fenner, Director of Education, at his desk 226

16.1. Charles Fenner with Sir Edgeworth David, examining a piece of fossil wood 271

16.2. Map of Australia showing the 'strewnfield' of australites 273

16.3. Australites 274

16.4. Medals 275

Preface

Frank Fenner AC, CMG, MBE, FRS, FAA, MD, DTM, is undoubtedly the most highly decorated and awarded Australian scientist of the 20th and 21st century. Beginning with the David Syme Research Prize for his work on mousepox in 1949, through his diverse and pioneering work in virology and microbiology, he has picked up a glittering array of scientific honours which include the Mueller and Matthew Flinders Medals (1964 and 1967), the Britannica Australia Award for Medicine (1967), the Burnet Medal (1985), the Prime Minister's Prize for Science (2002), and, internationally, the highly prestigious Japan Prize (1988), the Copley Medal of the Royal Society of London (1995), and the Albert Einstein World Award for Science (2000). Focusing his high place in university education, Fenner Hall, a college of The Australian National University (ANU), was named for him at its establishment in 1992.

For a researcher who never had time to gather a PhD, it is a formidable record. But at 91, Frank, with characteristic modesty, observes, 'You just have to live a long time' However, as the President of the Australian Academy of Science recently summed up, Professor Fenner is `the doyen of virology and one of the greatest scientists Australia has produced'.

Frank Fenner's life, as it unfolds in these pages, marks a significant piece of the history of Australian science. His chapters are studded with the names of many of the major players in pathology, microbiology and the rising field of virology with whose lives he interconnected, and his mode of inserting boxed information on these participants throughout his text provides valuable biographical portraiture.

It was World War II that shaped Frank's course. A graduate in Medicine of the University of Adelaide in 1938, he enlisted in the Army Medical Corps after his resident year, but, noting that unusual infectious diseases existed in the Middle East—a likely future theatre of war—he went first to Sydney University to do a three-month diploma course in tropical medicine. Thus, through chance and foresight—a combination that would mark his life—he seized on a path that would determine his future research career. Well prepared, he served as a malariologist with the Australian forces in Palestine where he met the influential Hamilton Fairley. In 1942, he returned to Australia to serve as pathologist at the Australian General Hospital at Hughenden, Queensland, where he treated patients from the New Guinea campaign suffering from malaria and dysentery. Subsequently, he was transferred to New Guinea to become a key participant in the control of malaria and other insect-transmitted diseases in Papua and New Guinea, a crucial contribution in these theatres of war. His report on the 'wastage' through sickness and disease in the Australian capture of Lae and Finschhafen put him on the scientific map.

It was at Hughenden that Frank encountered that great 'enabler', Bill Keogh, and most importantly and for him most enduringly, met the charming, highly skilled nursing sister, 'Bobbie' Roberts, with whom he worked on malaria diagnosis and who later became his wife.

His research career, begun with his report and papers on disease among the Australian troops, took him in 1944 to an initial spell at the Walter and Eliza Hall Institute under Macfarlane Burnet, where he investigated North Queensland tick typhus, and, with his discharge from the Army early in 1946, he moved to a position there as Senior Haley Research Fellow. It was at the Hall Institute that he began his work on the experimental epidemiology of the ectromelia virus (mousepox). His baptism in virology, as he puts it, had begun, and it presaged his lifetime interest in the poxviruses and his ultimate involvement in the World Health Organization's smallpox eradication campaign.

It was in 1949, during a postdoctoral period in America at the Rockefeller Institute for Medical Research (an opportunity set in train by Bill Keogh for former wartime workers in the Directorate of Hygiene and Pathology) working under the mentorship of René Dubos, that he received news of his appointment as first Professor of Microbiology in the new John Curtin School of Medical Research at the ANU. After travelling to London to make plans for the School with Howard Florey, he would hold this pioneering post until 1967, when he became Director of the School.

It was while Professor of Microbiology that Frank carried out his monumental work on myxomatosis. Once more, 'fortune favoured the prepared mind'. Myxomatosis, a virus disease long known to kill European rabbits, which had become a major pest in Australia; in the absence of country boys from the rural scene in the war years, the population of this imported species had multiplied spectacularly. From 1948, spurred by the strong advocacy of Jean Macnamara, the Commonwealth Scientific and Industrial Research Organization (CSIRO) was trialing the impact of the introduced virus in the field and during the Christmas—New Year period of 1950-51, the disease escaped from one of its trial sites and killed millions of rabbits. It was then that Frank decided to embark on the study of the virology of the disease. With Gwen Woodroofe and Ian Marshall as his assistants, he worked on the evolution of virulence of the virus, the genetic resistance of the rabbit, and conducted collaborative studies with entomologists and ecologists at the CSIRO on the ecology of the disease. He also began what became a lifetime of journeying when he travelled to conduct comparative studies of the disease in Britain, Europe and America. His definitive book, *Myxomatosis,* co-authored with Francis Ratcliffe, came out in 1965.

Under his leadership, the John Curtin School of Medical Research became a famous institution in virology and expanded through the appointment of diverse and illustrious talent into new disciplinary fields. In 1973, he moved to another

Directorship at the ANU's new Centre for Resource and Environmental Studies (CRES) where he turned his ranging mind to environmental problems and international collaborative ventures. His travel mileages soared.

Yet Frank Fenner is undoubtedly best known to wide audiences in Australia and overseas for his part in the eradication of smallpox. 'For almost the whole of my career at the laboratory bench', he records mildly in an oral interview at the National Library of Australia, 'I worked on pox viruses'. Through his expert research and his growing involvement in the burgeoning international committees and commissions of the Smallpox Eradication Program of the World Health Organization, initiated in 1967, he became a major figure in the development of strategies to eliminate the disease and offer a system of 'certification' of the eradication of the disease across the world. In May 1980, he addressed the World Health Assembly, which accepted the global eradication of smallpox. Never entirely complete, research continues on vaccination, vaccine stocks and monkeypox in which Frank emerges again as the authoritative public spokesman.

By any reckoning, Frank Fenner's career evokes the loud exclamation 'What a life!' Yet, as the Chinese saying goes, 'his nets never dried'. At his retirement from CRES 26 years ago, he moved across the ANU campus to a room in the John Curtin School of Medical Research as a perpetual Visiting Fellow. Large scientific volumes continued to flow from his desk, embracing every form of virology; microbiology; myxomatosis; smallpox eradication; an historical overview on the John Curtin School of Medical Research itself; and, most recently, a major historical compilation, *The First Fifty Years* of the Australian Academy of Science, a society in which Frank has, since its inception, been a notably active Fellow, ideas man, and benefactor. For the most part uncritical, and given to factual chronicle and reportage, Frank's historical volumes draw on contributions from other players. But it is, undoubtedly, his missionary zeal and energy that bring them into print. There has, indeed, always been something about the quiet persistence of the man, a steady mode of dealing with ideas, organizations, and local and world scientific initiatives, that, coupled with a deep intention to record, has made him a unique phenomenon in Australia. Rich in information, his writings provide invaluable archival resources for historians of medicine, the biological sciences, and the steady rise of 20th-century science itself.

Appropriately, his book closes with his reflections on something at which he has always been singularly gifted—friendship and special friends. It leads him too to the *raison d'être* of the accompanying biography of his father, educationalist, scholar, and science communicator, Charles Fenner.

Frank once observed that he had always believed that his father was a greater intellectual than himself but did not have the same opportunities. Nature, nurture and chance all played their decisive part. Frank's grandfather—Charles Fenner's father—was born in the village of Niedergrenzebach, in the province of Hesse.

Emigrating to Victoria attracted by the prospect of gold, he brought to Australia a German lineage that dated back to the 15th century. (The family's coat of arms is now the emblem at Fenner Hall).

Born at Dunach near Ballarat, where his father was the licensee of the Dunach Hotel, Charles left school early, in circumstances very different from those of his son, and was apprenticed at a printing office, becoming in turn pupil-teacher in local primary schools and then `principal' simultaneously of two one-teacher bush schools. Ambitious, with a lucid mind, he took a two-year course at Melbourne Teacher's College, taught briefly, but revealed his real ability when he gained a Kernot Research Scholarship at Melbourne University, where he majored in geology and biology under two great professors, E.W. Skeats and Baldwin Spencer, and graduated BSc with first class honours in 1912. He took a Diploma of Education the following year.

Geology and its fieldwork became his *métier* and, serving as Headmaster at Mansfield Agricultural High School, he published his first scientific paper. In 1914, he was appointed Principal of the Science Departments in the Ballarat School of Mines where he again concentrated on geology and conducted intensive field work in nearby Werribee Gorge, Bacchus Marsh and the Glenelg River. His substantial papers on these areas won him a DSc degree at Melbourne University. Two years later, he was appointed to the new post of Superintendent of Technical Education in the South Australian Department of Education, and moved his family to Adelaide. He was a highly cultured man who, during his 23 years in the post, worked against the odds to raise the profile of technical education as a necessary industrial underpinning in the State. He also emerged as a contributor and liaison officer of the ABC's Educational Broadcasts in South Australia, a part-time lecturer in the Geography Department of the University of Adelaide (1927-39) and a highly productive and distinguished science communicator.

Frank shows an evident pride in his father's accomplishments and his scholarship in science. But unfavourable chance (absent from his own experience) delayed Charles Fenner's elevation to the Directorship of Education in South Australia, which came late in 1939, a decade later than he and his family had privately hoped. Yet his father's interest in writing books—five came from his pen, along with chapter contributions and numerous articles—influenced Frank's own addiction to producing works with a wider reach than the mainstream of his scientific papers. 'Always generalize', his communicator father advised him. It was a skill, he remembers, that shaped his writing of his most long-enduring paper, 'The pathogenesis of the acute exanthems', published in *The Lancet* in 1948 and reprinted as a classic paper in *Reviews in Medical Virology* in 1996.

Binding their lives in nature, nurture and chance, Frank Fenner has brought us into close acquaintance with two generations of men, bonded yet different, each making a distinctive contribution—one widely international, the other, State-oriented—to the knowledge and education of this country.

Ann Moyal
Canberra, March 2006.

Acknowledgements

In Chapter 14, I have made extensive use of the article written by Bernard K. Hyams, published in *Biography, an interdisciplinary quarterly,* 13,1, (Winter 1990) 57–75. I am grateful to Dr Hyams and to Stanley Schab, managing editor of *Biography*, Centre for Biographical Research, University of Hawai'i, for permission to reproduce this material. In the same chapter, I have also made use of the entry of his name in the *Australian Dictionary of Biography*, written by Lynne Trethewey; I thank both the publishers, ANU E Press and The Australian National University, and the author. I thank the World Health Organisation for permission to reproduce Figure 26.4 on page 1205 of *Smallpox and its Eradication* and the Smallpox Recognition Card.

Brian Swan helped with information on the Wongana Circle (Chapter 16), and my sister Winn, my brother Bill, and nephew Max Fenner have also kindly read over the drafts of sections relating to our family and producing the brief biographies in Chapter 1. Erica Jolly provided useful information about my father in *A Broader Vision,* her monumental book on vocational education in South Australia.

I owe special thanks to Dr Ann Moyal for writing the Preface to the book.

A number of institutions in South Australia provided useful information and/or photos relating to my father's career: the State Library of South Australia, the University of Adelaide, and the Royal Society of South Australia, the Royal Geographical Society of South Australia and the Field Naturalists' Society of South Australia. The source of most data in Part I was in MS143 of the Basser Library Archives of the Australian Academy of Science, located in Canberra. Much of the material used in Part II was in MS178, in the Basser Library Archives. Barbara Holloway provided invaluable help as editor and in final organising of the chapters on the ANU E Press template. Compilation of all this data and, indeed, my long post-retirement career would have been impossible were it not for the generosity of the John Curtin School of Medical Research for providing me with a large study, a succession of computers, and access to such facilities as the Eccles Library, the School Photography Service and the IT Help Desk.

Acronyms

AAAS	Australasian Association for the Advancement of Science
ABC	Australian Broadcasting Commission (until 1983), Australian Broadcasting Corporation (thereafter)
AC	Companion of the Order of Australia
ACIAR	Australian Centre for International Agricultural Research
ACIH	Agency for Cooperation in International Health
ACT	Australian Capital Territory
ADAB	Australian Development Assistance Bureau
AIF	Australian Imperial Force
AGH	Australian General Hospital
ANU	The Australian National University
ANZAAS	Australia and New Zealand Association for the Advancement of Science
ANZAC	Australia and New Zealand Army Corps
ASID	Australasian Society for Infectious Diseases
ASM	Australian Society for Microbiology
AUC	Australian Universities Commission
BAAS	British Association for the Advancement of Science
BBC	British Broadcasting Corporation
BE	Bachelor of Engineering
BSc(Eng)	Bachelor of Science (Engineering)
CCC	Civilian Conservation Corps (USA = Youth Employment Scheme)
CCS	Casualty Clearing Station
CDC	Centers for Disease Control (USA)
CMG	Companion of the Order of St Michael and St George
CNR	Centre for Natural Resources
CRA	Conzinc Riotinto Australia
CRES	Centre for Resource and Environmental Studies
CSIR	Council for Scientific and Industrial Research
CSIRO	Commonwealth Scientific and Industrial Research Organization
CUMC	Chinese Union Medical College
CUP	Cambridge University Press
CVI	Children's Vaccine Initiative
DCM	Distinguished Conduct Medal
DSc	Doctor of Science
DTM	Diploma of Tropical Medicine
ESCAP	Economic and Social Commission for Asia and the Pacific

FNS	Field Naturalists Society
FRS	Fellow of the Royal Society
GP	general practitioner
GTC	Government Training Centre (UK)
IC	Instructional Centre (UK)
ICI	Imperial Chemical Industries
ICMR	Indian Council for Medical Research
ICSU	International Council of Scientific Unions
ICTV	International Committee on Taxonomy of Viruses
IIASA	International Institute of Applied Systems Analysis
JCSMR	John Curtin School of Medical Research
LHQMRU	Land Headquarters Medical Research Unit
MB BS	Bachelor of Medicine, Bachelor of Surgery
MD	Doctor of Medicine
MM	Military Medal
MSU	Moscow State University
NGO	Non-Governmental Organization
NHMRC	National Health and Medical Research Council
NIH	National Institutes of Health (USA)
NOC	National Occupational Conference
NSF	Nature and Society Forum
NZ	New Zealand
OAM	Order of Australia Medal
OBE	Officer of the Order of the British Empire
RAAF	Royal Australian Air Force
RV	Recreational Vehicle
SBS	Special Broadcasting Service (Australia)
SCOPE	Scientific Committee on Problems of the Environment
SPF	Schools Patriotic Fund
TFM	Training of Fit Men (UK)
TQM	Total Quality Management
UCLA	University of California Los Angeles
UNDP	United Nations Development Programme
UNEP	United Nations Environment Programme
UNESCO	United Nations Educational, Science and Cultural Organization
UNHCR	United Nations High Commission for Refugees
UNICEF	United Nations Emergency Fund
USAID	United States Agency for International Development
WAAF	Women's Australian Air Force

WHO	World Health Organization
WPA	Works Progress Administration (USA)

Part I. The Life of Frank Fenner

Introduction

I have chosen the title *Nature, Nurture and Chance: The Lives of Frank and Charles Fenner,* because I think that the first three words encapsulate the three elements that determine the lives of all humans. In one way, quite apart from the changes in the world around us, my father and I have had very different careers, yet my life, of which I know much more than I could hope to find out about his, clearly demonstrates the great importance of all three.

My father, Charles Fenner, was born in 1884 and died in 1955, I was born in 1914, and this book was published in 2006; a period of 122 years, during which the world has changed more than at any other time in human history. This book covers the period from 1895, when Father first got a job, until the present. He grew up in a poor family, in the country; I grew up in a middle class family, in the city. Each of us had the same two great loves in our intellectual lives: a love of science and a love of writing. The first 11 chapters (Part I) deal with my career, the next six (Part II) with my father's life, and the last chapter deals mainly with my life but also contains reflections on the contrasting lives of father and son. The detailed information is located in two large files in the Basser Library, Australian Academy of Science, Canberra; File 178 for Charles Fenner's life and File 143 for mine.

Chapter 1 describes my childhood as the second of five children and contains a description of family life. Chapter 2 describes my life at the University of Adelaide, where I initially enrolled in Science and transferred to Medicine in my second year. I was introduced into research by involvement with the Board for Anthropological Studies' expeditions into Central Australia and my research on Aboriginal skulls in the South Australian Museum, which formed the basis for the award of the degree of Doctor of Medicine in 1942.

Chapter 3 describes my career in the Australian Army in World War II, an experience that largely determined my subsequent career. Before enlisting, I obtained a Diploma of Tropical Medicine at the University of Sydney, which, after 13 months in the Middle East, led to appointments as pathologist to a 1,200-bed hospital in north Queensland for nine months and then as a malariologist in New Guinea. My future wife, Bobbie Roberts, was a transfusion expert in the hospital; we married when we were both in Melbourne during a short break I had from New Guinea. As a malariologist, my research interests switched from physical anthropology to infectious diseases and led to my acceptance as a research worker with Australia's leading virologist, Macfarlane Burnet, on discharge in February 1946.

Chapter 4 describes my immediate post-war career at the Walter and Eliza Hall Institute in Melbourne, where Burnet suggested that I should work on the experimental epidemiology of ectromelia, a disease of mice caused by a virus which he had shown was related to vaccinia virus. I spent two and half years doing very interesting work with this disease, especially investigations about what happened during the week-long incubation period. Although the Hall Institute was a very lively place, during the days before air travel we saw very few international visitors, so Burnet arranged a year's work for me at the world's leading medical research institution at the time, the Rockefeller Institute of Medical Research in New York. Here I worked on mycobacteria with a charismatic French-American, René Dubos, who later became an environmental guru. While there, I received an offer of the Chair of Microbiology in The Australian National University (ANU), which I accepted.

The rest of Part I describes my life in the ANU. I was Professor of Microbiology in the John Curtin School of Medical Research (JCSMR) from July 1949 to September 1967. Chapter 5 describes the administrative and domestic arrangements during those 18 years. Bobbie and I had gone to England to meet fellow professors and Sir Howard Florey in July 1949 and, after six months travelling around Europe, we arrived back in Melbourne in February 1950. There were no laboratories in Canberra, so initially I worked again in the Hall Institute, then, from November 1952, in temporary laboratories which had been built in Canberra, and, from 1957, in the permanent JCSMR building. Chapter 5 also includes descriptions of the house we built and interesting trips that I made to Indonesia in 1956, to China in 1957, to India in 1960–61, to Churchill College in Cambridge in 1961–62, and to Moscow State University in 1964.

Chapter 6 summarizes my research during that period. Initially, I continued work on mycobacteria, principally on *Mycobacterium ulcerans,* but from February 1951, with two assistants and extensive collaboration with ecologists in the Commonwealth Scientific and Industrial Research Organisation (CSIRO), the principal theme was the newly introduced myxoma virus and the evolutionary changes in its lethality and the genetic resistance of rabbits. This work culminated in my second book, written in collaboration with the senior CSIRO ecologist, Francis Ratcliffe, entitled *Myxomatosis* and published by Cambridge University Press. From 1957, I also carried out laboratory experiments on the genetics of vaccinia virus and I spent most of my time in 1965–66 writing a 900-page book, *The Biology of Animal Viruses.*

In 1967, the Head of JCSMR, Sir Hugh Ennor, resigned from the ANU to take a senior post in the Commonwealth Public Service. As described in Chapter 7, I applied for and was appointed Director of JCSMR. I came to the position just as the governance had shifted from a School Committee (the professors) to a Faculty/Faculty Board structure. My period in the post, 1967–73, was a time of considerable expansion, with the establishment of several new departments and the replacement of retiring senior staff, all of which demanded much of my time.

I found that I could not supervise research without being involved in bench work myself, so I filled in my spare time by writing review articles and books. The first, with the collaboration of a former PhD student, David White, was a textbook for medical students, *Medical Virology*, published in 1970, and the next a second edition of *The Biology of Animal Viruses*, with four co-authors, published in 1974.

The other institution that has played a major part in my life in Canberra is the Australian Academy of Science (AAS), and Chapter 8 describes the many aspects of its activities with which I have been involved. I am the only surviving Fellow of the initial elections, in 1954. From 1958–61 I was Secretary, Biological Sciences, which greatly broadened my interest in biological and environmental sciences. From 1969–82 I was involved with Academy committees dealing with environmental problems. In 1978, with Lloyd Rees as co-editor, I prepared a history of the first 25 years of the AAS. In 1984, I initiated donations to establish annual conferences on environmental problems, which have been held annually in the Dome since 1988. I was also instrumental in introducing video histories of Fellows, a concept which has been promoted by Council since 2001 for as many Fellows as possible.

In 1973, approaching the end of my appointment as Director of JCSMR, I was appointed Director of the newly created Centre for Resource and Environmental Studies (CRES), as described in Chapter 9. I resigned from all medical science committees except those concerned with smallpox, but having been appointed to the major international environmental committee, the Scientific Committee on Problems of the Environment (SCOPE) in 1971, I was appointed Editor-in-Chief of SCOPE publications, 1976–80, and attended all meetings of the SCOPE Executive Committee. Chapter 9 contains a summary of the activities of CRES during its first six years (1973–79).

Chapter 10 describes my participation in the World Health Organisation (WHO) Intensified Smallpox Eradication Program. Initially, I served on its major research committee, which consisted of poxvirus experts, then on several of the International Commissions for the Certification of Smallpox Eradication, in Africa, India and China, and from 1977 to 1979 I was Chairman of the large Global Commission for the Certification of Smallpox Eradication. The 1979 meeting of the Global Commission agreed that smallpox had been eradicated globally and set out recommendations for steps to be taken in the post-smallpox world. As Chairman of the Global Commission, I presented this report to the World Health Assembly in May 1980, where it was unanimously approved.

Chapter 11 describes my career as a Visiting Fellow in JCSMR after retirement in December 1979. It includes an account of the committees set up to see that the recommendations of the Global Commission on the Certification of Smallpox Eradication were carried out and subsequent national committees concerned with the use of smallpox as a bioterrorist weapon. The recommendation that occupied most of my time was the production of a book describing the campaign; I was senior editor of a massive book, *Smallpox and its Eradication*, which was published by WHO in 1988.

I still work at JCSMR every weekday, usually from about 7 am to 3.30 pm, for most of that time sitting in front of a word-processor and, during my 'retirement', writing some 140 papers and book chapters and 14 books (including new editions). The books include the textbooks on medical and veterinary virology, a book on the orthopoxviruses and another on monkeypox, another book on myxomatosis, a history of microbiology in Australia, a history of the JCSMR, and the 50-year history of the Australian Academy of Science. This chapter also records the receipt of several prestigious awards, including the Japan Prize, the Copley Medal and the Prime Minister's Award for Science.

Bobbie and I were also able to take some more leisurely trips overseas, including six months in the US National Institutes of Health as a Fogarty Scholar. In 1989 Bobbie had a colonectomy for carcinoma of the colon; five years later she had secondaries that did not respond to treatment and she died in December 1995.

Part II of *Nature, Nurture and Chance: The Lives of Frank and Charles Fenner* consists of six chapters describing the life and career of my father, and a final chapter, 'Reflections'.

In contrast to my numerous flights overseas (see Part I), my parents travelled overseas, by ship, on only two occasions. In 1931, when my father was 47 years old, he was chosen as one of the Australian delegates to the Centenary Meeting of the British Association for the Advancement of Science. He and my mother travelled to England, via the Gulf of Suez, on the *Balranald,* and back around South Africa, on the *Jervis Bay*. The second trip was in 1937, when they went to North America and Europe, returning to Australia via South Africa, across the Pacific Ocean on the *Monterey*, the Atlantic on the *Queen Mary*, to South Africa on the *Stirling Castle*, and across the Indian Ocean to Adelaide on the *Anchises*. This trip was funded by a grant from the Carnegie Foundation and was focused on educational matters.

On both of these trips, Father kept a diary, in which he recorded as much as he could of the activities each day, often inserting drawings and photographs of items of interest. In 2002, I had the diaries from both trips typed up. They are quite long, the 1931 diary containing 131,244 words and the 1937 diary containing 208,403 words. They are so revealing of the character, interests and relationship between husband and wife, and reveal so much of his ideas on both education and science, that I have included selected and edited passages from them in this book. On both trips, Father's interests in geology and human geography appear on many pages. There are some comments on educational matters in the 1931 diary, but many more in the 1937 diary. Another feature that the reader will gather from the diaries is that Father suffered from gastric problems and could not have made the trips without the support of his wife Peggy, who carried out all the repeated packing and unpacking, and took care of him when he was ill.

Chapter 12 describes briefly the lineage of the Fenner family, which goes back for many generations in Germany and three generations in Australia; extracts from the diaries on visits to Germany constitute 80 per cent of this chapter.

Chapter 13 describes Father's childhood, his education at the University of Melbourne, which he entered as a student teacher, received a scholarship and graduated BSc with First Class Honours in Geology and Biology. He married another student teacher in 1911. She gave up teaching when her first child was born in 1912. Father rose rapidly as a teacher and educational administrator, being appointed Principal of the Ballarat School of Mines in 1914 and carrying out research on physiography in his spare time.

Chapter 14 describes Father's career in the Education Department of South Australia, in which he was Superintendent of Technical Education from 1916 to 1939 and Director of Education from 1939 until 1946, when he retired because of ill health. He suffered from a stroke in 1954 and died in 1955. In 1990, Bernard Hyams wrote a long article about my father in the journal *Biography*, entitled 'Charles Fenner: Scientist Who Would Be Administrator'. The abstract of this article reads: 'The career of Dr Charles Fenner in Australia tells us much about the history of education and educational administration in that country. The struggles undertaken by this gentle scholar to advance the cause of technical education emphasizes two of the dominant characteristics of Australian public education: resistance to vocational training and the tradition of recruitment of administrators from the rank and file of teachers.' Hyams' essay is published almost in full, followed by contributions from several South Australian educationalists. This formal account is supplemented by Chapter 15, which consists of extracts dealing with educational matters taken from his 1931 and 1937 diaries.

As described in Chapter 16, there is no question that Father lived a double life, indeed, a triple life, for he was the loving and life-long husband of Peggy (née Hirt) and father of four boys and one girl, as well as being an educational administrator, and, at the same time, he was an important figure in Australian science, both in original research and as a science communicator. While at the Ballarat School of Mines, he published papers on the geomorphology of Werribee Gorge, for which he was awarded the degree of DSc in 1917. In 1920, he received the first award of the Sachse Gold Medal of Royal Geographical Society (Victorian Branch) and, in 1929, the Syme Prize of the University of Melbourne for work on the geography of South Australia.

While in Adelaide, he wrote numerous papers on the geomorphology of the Mount Lofty Ranges and books for both university and school students on the geography of South Australia, and, from 1920 to 1940, he was the Lecturer in Geography at the University of Adelaide. In addition, from 1916 until 1939, he wrote fortnightly 'Science Notes' for the Melbourne weekly magazine, *The Australasian*. He was a valued member of the Royal Society of South Australia and the Royal Geographical Society of South Australia, serving various offices, including President, in both. He was the first recipient of the John Lewis Gold Medal of the Royal Geographical Society of South Australia.

Chapter 17, the longest chapter in the book, supplements this formal account of his life as a scientist with enthralling extracts from his diaries, which reveal a truly remarkable breadth and depth of knowledge, considering that there were few radio and no television programs in those days to keep Australians in daily touch with the rest of the world.

Finally, in Chapter 18, I recall close personal friendships from all periods of my life and reflect on 'chance', the contrasting opportunities in my life and my father's.

Chapter 1. Childhood, 1914 to 1932

Birth and Housing

My father, Charles Fenner, had been appointed to the Ballarat School of Mines in November 1914, and, with his wife, Peggy, and Lyell, the only child at that stage, had moved into a house at 2 Doveton Street, Ballarat. I was born there on December 21, 1914, the second of a family of five. I was given the name 'Frank Johannes', the second name being that of my grandfather. The house was right next to 101 Eyre Street (which we used to pronounce 'Ay-er'), where Mother's two unmarried sisters and a widowed sister lived. Later, when we had moved to Adelaide, one or other of the children would be sent to stay with their aunts for the Christmas holidays. On my father's appointment as Superintendent of Technical Education in South Australia, in November 1916, the family (now three children, with the birth of my sister Winifred in Ballarat on 28 August, 1916) moved to Adelaide and initially lived in a rented house in Barton Terrace, North Adelaide. In 1918, the family moved to a house in 42 Alexandra Avenue, Rose Park, which was very close to the Rose Park Primary School, where all five children received their primary education.

In the latter part of 1938, when I was in final year Medicine at the University of Adelaide, I must have been getting worried about Hitler's antics and thought that I should change my second name from 'Johannes' to 'John', a change which Father endorsed before a Justice of the Peace on 11 October, 1938. Several German place names in South Australia were also changed at that time. In retrospect, it must have been hurtful to my father, since it was the only acknowledgement of his parents in the children's names, but I don't remember knowing of any misgivings he may have had. For many years, I have rarely used the name 'John' or the initial 'J'—only 'Frank'.

Family Life

'Nature, nurture and chance.' The next few chapters will reveal two examples of 'nature'—i.e., the effects of my genes—namely, my ability to work hard and my lack of mathematical skills. What about the effects of 'nurture', essentially my family life, on my career? This was undoubtedly substantial. As described below, we were a very happy family. Our parents were always loving, to each other and to all of us. We played happily together at home, and we had occasional wonderful motoring holidays. The one I remember best was during the summer holidays in 1930, on what we called the 'Africa Speaks' expedition, after a film of the time. This was a trip from Adelaide through the Coorong and around much of country Victoria, where our uncles, aunts and cousins lived. On this trip, and whenever we went into the country, as on trips organised by the Field Naturalists Society ('Field Nats'), of which he was an active member, Father would explain features of the countryside to us, geological, botanical, historical, in a fascinating way. I was attracted to geology from a very early age; my parents kept a drawing I had made of the section of a volcano at the age of four years, and, while I was still at secondary school, I had accumulated quite a good collection of fossils during our trips around Victoria and South Australia and by exchange, including a Triassic fossil of *Ginkgo* leaves (I now have the best *Ginkgo* tree in Canberra in my garden). I am sure that this childhood experience played a large part in my later interest in environmental problems.

My mother's influence was important but less obvious. As was usual in that generation, although she had trained as a school-teacher, after she had children she devoted her life to her family. One has only to read the diaries Father wrote on his two overseas trips in the 1930s to see how much he depended on her (see Chapters 15 and 17). As well as offering all possible support for the children to study at home (there was not even a wireless to distract us then), she made the home a haven of affection and support during the sometimes turbulent times of adolescence, and she skimped and saved as much as possible to ensure that each of the five children had as good an education as possible, within our interests and capacity.

Finding a Home in Adelaide

Father, Mother and the three children, Lyell, Frank and Winn, moved from Ballarat to Adelaide in November 1916. Two more boys, Tom and Bill, were born there. For about two years, Father rented a house in Barton Terrace, North Adelaide. After investigating the quality of the primary schools in various suburbs he decided that the best primary school was at Rose Park. After looking at houses near that school, the family bought the house at 42 Alexandra Avenue, Rose Park. Father called it 'Iramoo', after the Aboriginal name for the plains where Melbourne now stands. The name was first used by John Batman on 6 June, 1835, when he 'bought' Melbourne from the local Aboriginal tribes. When I was at Thebarton Technical High School (see below) I made an appropriate copper name-plate, which was put on that house. Some years after I moved to Canberra I placed it on the front of our holiday house at Dalmeny, on the South Coast of New South Wales.

Alexandra Avenue was one of the very few streets in Adelaide that consisted of two roads, one on either side of a wide strip of lawn containing two rows of elms. There was also a row of oak trees on each footpath. It was close to the Victoria Park Racecourse, which was situated in the parklands that surround the city of Adelaide. Some years later, as the children grew up, we used to join the local children to play football on the lawn, and I played cricket and learned to play tennis on the open spaces and public tennis courts near the Racecourse. The house itself was five houses down the street from the Rose Park Primary School. There were two tram-lines within easy walking distance and the house itself was only about a five kilometre walk from North Terrace and the Museum, Art Gallery, University, Botanic and Zoological Gardens and the Royal Adelaide Hospital.

Figure 1.1. Iramoo, the house at 42 Alexander Avenue, Rose Park in 1931, after adding the second storey as the boys' shared bedroom

On the verandah, from the left: Winn, Tom, Lyell, Mother, Bill, Frank.

The house was relatively large. There was a lane at the back, which provided access to a garage, and reasonably large lawns at both back and front. Around the back lawn there were several quite productive fruit trees: grapevines, two fig-trees, and apricot, peach, lemon and orange trees. Father was no gardener, but I became interested in the garden from the time that I undertook first-year Botany at the university. In the early 1930s, Father had a second storey built on the house (Figure 1.1); this was a large single room with fly-wire around three sides, accessed by stairs at the back. The four boys slept there and, at the side of the stairs, there was a series of shelves that I used for my collection of fossils.

The Fenner Children

At the time of the move, the family comprised three children: Charles Lyell—born in Melbourne, 17 August, 1912; Frank Johannes—born in Ballarat, 21 December, 1914; and Winifred Joyce—born in Ballarat, 26 August, 1916. Two more children were born in Adelaide: Thomas Richard—born on 18 June, 1918; and William Greenock—born on 11 March, 1922. Brief biographies of my siblings follow.

Figure 1.2. The family in 1937

From the left: Thomas (in naval uniform), Lyell, Father, Winifred, Mother, Frank, William.

Charles Lyell Fenner

Born in Fitzroy, in Melbourne, on 17 August, 1912, the eldest child was named after the famous English geologist, Charles Lyell (1797–1875), and always went by his second name. Like the rest of the children, he received his early education at Rose Park Primary School. For secondary education, he went to Adelaide High School, where he progressed as far as Leaving (Fourth Year), which was the matriculation year at the time. He then went to the Teachers College and the South Australian School of Arts and Crafts, and, after training, he waited for a teaching post to become available. In 1933 he made an overseas trip to England and several European countries with the Australian Scout contingent, attending the World Jamboree at Gödöllö in Hungary in August. After his return he was appointed to a number of small, one-teacher country schools.

On 10 September, 1938, he married Theodosia (Thea) Kleinig, who he had met while teaching at Dutton. They went to live at Watchman, north of Balaklava, and the eldest of their three children, Theodore (Ted) was born in Balaklava Hospital in July 1939. When the Area School system was being planned, Lyell decided to apply for a position as a Manual Training Instructor in 'Boys Craft'—or, in more recent parlance, a Technical Studies teacher. This involved a year of training in the performance and teaching of woodwork, sheet metalwork, saddlery and boot repairing. The woodwork and metalwork courses were conducted at various Boys Technical High Schools, chiefly Goodwood and Thebarton, but the only place for saddlery and boot repair training was the Magill Reformatory. The last named institution was a grim and forbidding place, and the conditions that prevailed there made a long lasting impression on Lyell. Goodwood Boys Technical High School was a much more convivial place, especially since one of the young art teachers there was Jeff Smart, now the famous artist Jeffrey Smart, who used to hone his skills by drawing 'heads' of staff and students. An early Jeffrey Smart, in the form of a pencil 'head' of Lyell, made in 1942, serves as a reminder of the training period.

After gaining the requisite certificates, Lyell was appointed as a Boys Craft teacher at Maitland Area School in 1943. Two further sons, Max and Christopher, were born in Maitland in 1945 and 1948 respectively. Lyell taught there until May 1951, when he was transferred to the Wudinna Area School, in Central Eyre Peninsula (usually known as the 'West Coast'). This was one of the most enjoyable periods of Lyell's teaching career. As well as having many talented craft pupils, there was new scenery to draw and paint, and he could indulge his love of natural science to the full in the Nature Study courses, with a huge surrounding area of diverse and frequently ill-researched source material. It was the time of booming agricultural prices, and much land was cleared for cropping, with the consequent destruction of wildlife habitat, a practice that brought forth relatively little comment in those days. Fortunately the Nature Study lessons had some impact upon the community, so farmers and their children kept an eye out for all sorts of animals and plants that 'the Boys Craft chalkie and his missus might be interested in'. As a result, the family acquired a considerable variety of native animal pets, including a tawny frogmouth, numerous finches of several different species, 'mountain devils' (small, spiky lizards), 'sleepy lizards' (stump-tailed skinks), honey possums, a brush-tailed possum, fat-tailed marsupial mice and a family of Mitchell's hopping mice, along with transient wombats and echidnas. In addition, several boxes of insect and plant specimens, together with maps and collection data, were sent to the Entomology and Botany Departments of Adelaide University, or delivered personally by Lyell during school holiday visits. The resultant enumeration of new species and revelations of unexpected distributions prompted a number of visits by research staff from those Departments and also some zoologists, who were equally surprised.

In December 1957, Lyell was transferred to Adelaide, where he returned to primary school teaching at Challa Gardens and Campbelltown schools, becoming the librarian at the latter for the last few years before retirement in 1977. He also prepared and presented several nature science broadcasts for the ABC. After retirement, he spent much time with the Retired Teachers Association, Art Gallery and Museum visits, reading and, to a more limited extent, travelling. He fulfilled part of a lifelong ambition when he took one of the first Antarctic flights offered by the airlines and finally saw Antarctica, even though no landings were possible. His wife, Thea, died on 15 August, 1993, in her 80th year, and Lyell died on 25 May, 1997, aged 84.

Winifred Joyce Fenner

Born in Ballarat on 26 August, 1916, Winn took primary education at Rose Park Primary School and then went to a private school, Wilderness School, located in Medindie, for her secondary education from 1930 to 1933. She then learnt 'retouching'—i.e., removing blemishes from negatives using a very sharp pencil—and then practised this in photography studios as a career. Then came the war, and she contemplated joining the Women's Australian Air Force (WAAF), but the Education Department was crying out for teachers, since so many young men were joining the armed forces. So she entered the Adelaide Teachers College in 1941, and, after one year's training, went to a one-teacher school near Lock on the West Coast. After a year there she returned to Adelaide and did an 'Area School' course and then went to Area Schools at Loxton, on the River Murray, for 1944–45 and then to Penola, in the South East, in 1946. Father resigned from the Education Department later that year and Winn moved to Walford Girls School, a large private school in Hyde Park. She stayed there from 1947 to 1976, always cycling to work. She was Sports Mistress and taught art and design. After retiring, she continued to go to Walford as a volunteer, doing calligraphy and helping in other ways until 2001. From 1951, she lived with a somewhat older friend, Phyllis (PM) Stoward, initially at Colonel Light Gardens, and then at her current residence in Myrtle Bank. PM died in 1967. Since then, Winn has always had a dog and, at the age of 89, still plays tennis twice a week.

Thomas Richard Fenner

Born in Adelaide on 18 June, 1918, Thomas (Thos) was named after his uncle, who died (from 'friendly fire') while fighting on the Western Front in 1916. After completing his primary education at Rose Park Primary School, he went to Thebarton Technical High School for one year and then to Unley High School, to study subjects required for entry to the Naval College. At the age of 14, on 1 September, 1932, he was admitted as a Cadet Midshipman to the Royal Australian Naval College, then located on Westernport Bay in Victoria. He graduated as a midshipman on 1 May, 1936, was promoted to Sub-Lieutenant on 16 November,

1938, and was seconded to the Royal Navy in Britain from 30 January, 1937,until 27 January, 1939. He was promoted to Lieutenant on 16 March, 1940, and Lieutenant-Commander on 16 March, 1948. When Mother and Father were in London in 1937, they met with him while he was working as a lieutenant with the Royal Navy. He served with the Royal Australian Navy right through World War II, being discharged on 26 January, 1950, with the rank of Lieutenant-Commander.

He married Beverley Slaney on 12 February, 1942, while on leave. They had one daughter, Vicki, born on 1 March, 1943. A few months later, because of a fire in the house, Beverley died of burns, but Vicki survived. Some years later, Tom married Margaret Legge. They had two children by the second marriage. Later, when Vicki was about seven years old, my wife Bobbie and I were advised by a friend and paediatrician, Dr Stanley Williams, that Vicki was so badly treated by Margaret that we should adopt her, which we did, with Tom's approval.

After discharge from the Navy, Tom served for a few years with the Australian Security and Intelligence Organisation, but he did not enjoy that and got a managerial job with the firm that was involved with the construction of the Tullamarine airport. Early in 1964, he was diagnosed with lymphoma, which progressed rapidly in spite of treatment. He died on 21 September, 1964, aged 46.

William Greenock Fenner

Bill was born on 11 March, 1922. His second name, Greenock, is the name of the volcano behind Dunach, where Father was born. Like the rest of the children, Bill had his primary education at Rose Park Primary School. In 1935, he went to The barton Technical High School. However, in 1937, when Father and Mother went overseas, he was sent to St Peters College as a boarder for one year. He stayed there as a day boy until 1939, when he matriculated, enrolling in Mechanical Engineering at the University of Adelaide in 1940 and graduating BSc(Eng) in 1943. He went back in 1945 to do some extra subjects so that he graduated BE at the end of that year. From 1947 to 1950, he worked in the South Australian Department of Mines, as a geophysicist, then from 1951 with Imperial Chemical Industries (ICI). This involved a number of different jobs and included his first trip abroad, to Europe in 1958. In 1946, he was married to Monica Lewis, and while still in Adelaide they had three children, Murray (b. 1947), Peter (b.1949) and Patricia (b. 1955). In 1961, Bill decided to leave ICI and got a job with Comalco (part of Conzinc Rio tinto Australia (CRA)), which meant moving to Melbourne. Over the next 14 years, he had a leading role in four major projects: bauxite mining at Weipa (north Queensland), a salt field at Dampier (Western Australia), an alumina plant at Gladstone (Queensland) and an aluminium smelter in New Zealand. These projects took Bill all over Australia and the world, especially to Japan. Then came a major change, a senior management position.

in Victoria Railways. While in this last job he had become very interested in Total Quality Management (TQM) and, after retirement in 1985, this, together with a hobby farm at Heathcote, became his major activity, involving him in lectures, conferences, much overseas travel and the production of a massive 588-page book, *Quality and Productivity for the 21st Century.*

Family Life

Our life as a family was very happy, even though much of it occurred during the Great Depression. In our house, this led, symbolically, to the use of lard instead of butter and torn-up newspaper as toilet paper. Mother was the 'Rock of Gibraltar', loving to all of us, very supportive of Father at home and indispensable on his overseas trips. She was helped in the house from time to time by her unmarried sisters, especially Christina (Crin), who came over from 101 Eyre Street and lived with us for months at a time. Another Aunt, Anna, died in our house in 1927. For all of the children, it was our first experience of death.

In his otherwise perceptive essay (see Chapter 14), Hyams comments that Charles Fenner was 'a rather stern father to his children', presumably because he 'never took them to sports games; instead he delighted to conduct them on nature study walks and expeditions'. The latter statements are correct, but in the collective memory of the three children currently alive, Frank, Winn and Bill, he was anything but a stern father; the children had little interest in watching sports games but considerable interest in playing them, and we all delighted in the nature study walks and drives. As noted in the testimonial written at the time Father left the Ballarat School of Mines, Charles Fenner had 'the power of inspiring in his students a love for the subjects he teaches', and this applied to his children as well.

As well as the usual bedrooms, bathroom (water heated as required by burning paper), 'front' room and dining room (both used mainly for visitors, although we used to sit in front of the dining room fire in winter), kitchen and bedrooms, there was a large room at the back of the house, the 'den', which was sacrosanct. It was here that Father wrote his scientific papers and fortnightly articles for *The Australasian,* and the shelves and desks were filled with his papers. Besides his own extensive library, Father bought us Arthur Mee's *The Children's Encyclopedia* and we were encouraged to use it whenever we asked a question, and among the other books was a series on the *Myths and Legends of Greece and Rome,* which I remember reading from end to end. We had plenty of space to play in the back garden and the lawn on Alexandra Avenue, and for much of the time we had a much-loved dog, a fox terrier named 'Flinders', after the great explorer. We had fowls and different children had responsibility for looking after them as we grew up.

Associated with his government work, in the early years Father had use of a car and chauffeur, and later we had our own car. As well as his use of the car for driving to work, the family were often taken for drives to Golden Grove and Teatree Gully, in the Mt Lofty foothills, and during the Christmas holidays, trips to various parts of South Australia (notably the tip of Yorke Peninsula) and Victoria, where the majority of our relatives, on both sides of the family, lived. These were memorable because Father knew so much about the geology, wildlife, plants and human history of the countryside and explained it all to us.

Other experiences that remain in my memory were the summer holidays on a wheat farm at Ebenezer, near Kapunda, which was owned by German relatives on my mother's side, the Kleinigs. It was wonderful for city kids to be able to wander in the countryside, see horses and cows at first hand, participate in the reaping, have a chance to try milking the cows, and enjoy the family meals, always preceded by a prayer in German (which I could not understand).

Primary School

One of my few memories of my early days was when my mother took me to the kindergarten at the beginning of 1919. I remember crying bitterly when she left me, but I soon adjusted and enjoyed my school days. Once past kindergarten, the classes consisted of 50 to 60 students, teaching was largely by rote, the schoolyard was small and paved with asphalt, and woodwork was taught across the road. My best friend there was John Dowie, who was almost exactly the same age as me. He became an outstanding sculptor. I enjoyed school and was, by nature, a hard worker. When I did the Qualifying Certificate examination in 1926, I received top marks in the State and got my photograph in the local newspaper. In the accompanying article, I attribute much of my success to teachers Miss Vera Dawe and Mr N. Carmichael; there is also the comment that I 'had a keen desire to be a farmer', possibly because of the enjoyment of school holidays at the farm of the Kleinig family at Ebenezer.

Secondary Schools

In 1927, when I was due to go to secondary school, my father was promoting an experiment in teaching at the Thebarton Technical High School on the use of the Dalton Plan for secondary education (see Chapter 14). The essence of this plan was that there was only one lesson a week in each subject and work for the rest of the week was set out in assignments and individual study at the student's own pace. I enjoyed this and, as well as doing woodwork, sheet-metalwork and architectural drawing, I did several additional 'academic' subjects. Thebarton Technical High School (later called Thebarton Technical School) was a boys' school situated on the far side of town from our house, and I had to take two trams daily each way. I enjoyed travelling through the city and I liked school. The mode of teaching, involving considerable individual initiative, was a very good introduction for the type of teaching offered at the university, and I had none of the problems of school–university adjustment that were common among my contemporaries.

I passed the Leaving Certificate at the end of fourth year, with more subjects that were usually taken and several credits. I then went to Adelaide High School (then a co-educational school in the Grote Street, in the centre of Adelaide), in the hope and expectation that I might win one of the 12 University bursaries offered each year at the Leaving Honours examination. My experience was mortifying. The course concentrated on six subjects: Physics, Chemistry, Mathematics I, Mathematics II, English and French. I failed in mathematics but, like many other students, went back next year to try again for a bursary. The Great Depression was under way, my father's salary had been cut, and a bursary would have been a boon. Next year I passed all subjects with some credits, but again failed to win a bursary. I did not enjoy Adelaide High as well as 'Thebby Teck'. The competition at the Leaving Honours level was substantial, since the class comprised the bulk of the students in the State whose parents could not afford to send them to one of the two élite private schools (St Peters and Prince Alfred Colleges) but who sought to gain a University bursary. The teaching consisted of a continuous series of lessons and practical work. Mathematics loomed large, in physics and chemistry as well as in maths itself, and although the maths teacher, 'Doggy' Nietz, was a good teacher, I had little intrinsic mathematical ability.

Religion

From the time I knew them, neither of my parents was a practising Christian, although each had taught at Sunday Schools before their marriage. However, because both were the children of German rather than British parents, they believed that an understanding of the Christian faith as practised in an Anglican church was a desirable part of our upbringing for life in what was then an Anglo-Celtic, Christian community. On 26 September, 1920, I was baptized at the Anglican Church of St Theodore, Rose Park, where I subsequently went to Sunday School and was confirmed at the age of 12. I was even then sceptical about the truth of Christian teachings, but occasionally with boy friends (and later with the hope of meeting some girls) went on Sunday evenings to one of the local churches, St Theodore's or the Congregational church across the road, until I was 17 or 18 years old. Since then, I have attended churches only for official events: marriages, occasionally baptisms, and deaths. I am still interested in religion as a widespread human activity (an 'anthropological' interest) but I am without religious beliefs myself—I accept the 'unknowable'. At times I was intolerant of what I saw and still see as the evils of some Christian teaching, especially that of evangelical Christians, and the Roman Catholic prohibition of contraception.

Sport

Father was no athlete, although he had represented Melbourne University at rifle shooting during his student days there. Although the Victoria Park Racecourse was very close, none of us took any interest in racing, nor in football (Australian Rules was the only game in Adelaide at that time). However, until I went to the University I was an enthusiastic cricketer, and participated in an early morning 'school' for young hopefuls, given at 6 am one day a week by former Test player George Giffen in the Parklands, some five kilometres away from our house. I used to walk over and back before breakfast on Saturdays. I was a passable slow bowler, and at one school match notched up eight wickets for 36 runs. However, when I came to play at the University my bowling was too slow and after the first season I gave up the game, although until after the War I went to all Test matches played in Adelaide and was a great admirer of Clarrie Grimmett as a bowler and Bradman and Jackson as batsmen.

At Thebarton Technical High School, I tried playing Australian Rules football, but found that I ran up and down the field and never seemed to touch the ball. Influenced by Norman Dowdy, a teacher who one evening each week conducted gymnasium classes that I attended (along with Lindsay Pryor) at a church building across the road, I tried playing hockey for a team called 'The Wanderers'. It turned out that I was quite good at that, and quickly gained entry to grade A hockey as soon as I went to the University.

The other game that I learned to play locally, with a friend on the public courts at Victoria Park, was tennis. Although I dropped this during and just after the War, I took it up again as soon as I came to Canberra, and I still play doubles at a court near our house in Canberra every Saturday morning.

Chapter 2. The University Years, 1933 to 1940

Selection of Course

Although I had failed to get a bursary, my parents were determined that I should go to the university; naturally, to the University of Adelaide, then the only university in South Australia. With the exposure that I had had to geology, it was not surprising that I wanted to do science, majoring in geology. However, my father dissuaded me. My year of entry was 1933, before the mineral boom of the 1940s. He pointed out that there were very few jobs in geology; besides the Government Geologist (one job, held by Dr L. Keith Ward, brother of the Professor of Bacteriology in Sydney, whom I later came to know very well), the number of positions at the university had just been expanded by 33 per cent by the appointment of Kleeman as a lecturer. Further, Father had observed that when he was studying biology and geology at the University of Melbourne in 1910–12 there were many 'duffers' there studying medicine who were now getting salaries three or four times as high as his, and, more importantly, that there was a much wider range of choice in professional courses such as medicine and engineering than in a specialized field of science. If I wanted to, he suggested, I could become a physician, a surgeon (both with many potential specialties), a pathologist, a general practitioner or even a research worker. I accepted his argument, but then had to spend a good deal of time studying Latin to Intermediate level, since this was a required subject for entry to medicine. Ever since, I have found it useful in understanding the origins of many English words, but I did not do enough Latin to get a feel for Roman culture.

In 1933, since the courses in Physics and Chemistry for First Year Science and Medicine at the university covered much the same ground as those in Leaving Honours, and medical students did only one term of Botany and two terms of Zoology, I initially enrolled in the Faculty of Science and did a full year course in Botany and Zoology. I won the John Bagot Scholarship for Botany, but was unable to accept it because I had transferred to Second Year Medicine. I also received the top credit in Zoology and a credit in Chemistry.

Medicine, 1934 to 1938

Since there was no examination at the end of second year, which was largely devoted to anatomy (four students to a human cadaver), I took First Year Geology as an evening student, with Sir Douglas Mawson and Dr Cecil Madigan as teachers; I received a credit at the end-of-year examination. I had joined the Science Students Association during my first year, maintained my association with it throughout my time at Adelaide University, and was elected its President in 1937.

I did quite well in medical studies in the succeeding years, gaining top place and Dr Davies Thomas Scholarships in 1935 and 1936, and the Lister Prize for Clinical Surgery in 1937. However, to my disgust (which I can still vividly remember), I obtained only third place in the final year examinations. It was a small compensation to gain the Dr Charles Gosse Medal for Ophthalmology.

In the summer of 1937–38 there was a severe outbreak of poliomyelitis in Australia. The Northfield Infectious Disease Hospital appealed to fifth year students for help, and a friend and somewhat older colleague, David Shepherd (who was already a qualified pharmacist) and I volunteered and spent the long vacation working there, mostly on poliomyelitis and diphtheria. I was not worried about 'infantile paralysis', as poliomyelitis was then called, but my confidence was undermined when, soon after my arrival, I admitted a young man with paralysis who had been in the same class as me at Rose Park Primary School. The Superintendent of Northfield Hospital at the time, Dr Alan Finger, was an active and sincere Communist. Although I read, and indeed still have, some of the books about communism that he gave me, I refused his invitations to attend meetings, and have in fact never attended a political meeting of any kind. However, for some time after enlistment in the Australian Army, all my outgoing letters were opened and censored (the convention was that officers censored their own letters). I was unaware of this until much later. I enjoyed my time at Northfield, doing laboratory work, mainly diagnosis of diphtheria, and helping with patients of all kinds. In addition, I had my first experiments with sex there, with an older and wiser nurse, who was very tolerant of my 'falling in love' with her.

Activities in Student Affairs

I took an active part in student affairs at the university and was a member of various students' committees: Union, Science Students, Medical Students and Sports Association. I gained a University 'Blue' for hockey in 1936, and was captain of the Adelaide University hockey team in 1937 and 1938. As I relate below, in 1936 I became associated with Professor Frederic Wood Jones, FRS, Professor of Anatomy in the University of Melbourne. In 1935, he had established the McCoy Society for Field Investigation and Research (named after Professor Sir Frederick McCoy, the first Professor of Natural Science at Melbourne University), which arranged for staff and students to carry out biological surveys of interesting places in Victoria during vacations (Ashton, 2001). Stimulated by this, I persuaded Cecil Madigan (Senior Lecturer in Geology) to establish a 'Tate' Society (named after Ralph Tate, an early South Australian naturalist) and

Figure 2.1. Adelaide University Hockey Club, Inter-varsity team, Brisbane 1937

Front row: J. E. Kelly, N. C. Hargrave, A. W. Cocks, F. J. Fenner (captain), R. Motteram, W. M. Rolland, M. C. Newland.
Rear row: J. T. Hutton, G. M. Turnbull, M. C. Knight, B. L. G. Johns, J. McPhie.

participated in its first trip, to caves at Swan Reach, in the lower Murray River, in December 1937. Others in the party included Pat Mawson (daughter of Sir Douglas), Roy Sprigg (hon. DSc, ANU, 1980) and Leigh Parkin (later Director of the South Australian Geological Survey). Although another excursion was arranged, to Jankalilla Beach, in December 1938, the Tate Society was an early war casualty.

Studies in Physical Anthropology

As important for the development of my subsequent career as any of the ordinary course-work, was an event that occurred in my second year, 1934, a year when the students spent most of their time dissecting human cadavers. Wood Jones had been Professor of Anatomy in Adelaide from 1919 until 1926. One of his last official acts, before he resigned in 1926 to take up the Rockefeller Chair of Physical Anthropology at the University of Hawai'i, was to successfully recommend that the University Council establish the Board for Anthropological Research, for which Professor J. B. Cleland, the Professor of Pathology, had long lobbied, as a permanent committee of the University (Jones, 1987). The Board organized regular trips to different parts of Central Australia during the August vacations to study Australian Aborigines. Norman Tindale, of the South

Australian Museum, always participated in these expeditions. Because of Wood Jones's interests, anthropometric observations were an important component of their work. In 1933, the Lecturer in Anatomy at the University, Hugo Gray, had made these measurements, but he taken a position in England and no-one was available for the August 1934 expedition to Pandi Pandi, on the Diamantina River, near Birdsville. Probably because of my father's acquaintance with some of the academics involved, notably Dr T. D. Campbell, the head of the Dental School and organizer of the expeditions, I was offered the job.

This was the beginning of a long association with physical anthropology and university and museum staff interested in various aspects of Australian Aborigines and Aboriginal life: Norman B. Tindale, in the Museum, C. P. Mountford, Frederic Wood Jones, Professors T. D. Campbell (Dental School), T. H. Johnston (Zoology) and J. B. Cleland (Pathology), and later an American anthropologist, J. B. Birdsell. The trip to Pandi Pandi was a revelation to me: travelling along the 'Birdsville Track' from Maree to Birdsville; contact with Aborigines of several Central Australian tribes; and the production (with Tindale) of a movie film (16 mm, black and white), an article—'Sandhills and Gibber Plains'—in the Adelaide newspaper, *The Advertiser*, to raise funds for the Board, and my first substantial scientific paper (Fenner, 1934). I went on two subsequent trips: to Nepabunna in the northern Flinders ranges in 1937, and to Eucla in 1939. In 1938, Dr Grenfell Price (then Master of St Mark's College, University of Adelaide), with the support of the State government, organized an expedition to central Australia to examine what were reported to be 'Leichhardt remains'. I was asked to go in case there were any skeletal remains and, although it was August of my final year, I very much looked forward to the prospect of travelling by camel to a remote part of Australia. However, three days before the expedition left I was playing inter-varsity hockey and suffered from a fractured patella, an event that evoked commiserations from *The Bulletin*, August 24, 1938. Nothing of significance was found.

Frederic Wood Jones

Born in London in 1879, Frederic Wood Jones came to Australia as Professor of Anatomy at the University of Adelaide in 1919. He was a highly original scientist, interested in many aspects of biology, and carried out ground-breaking research summarized in his *Mammals of South Australia*, illustrated with his excellent drawings. He was elected a Fellow of The Royal Society in 1925. In 1927 he moved to Hawai'i as Rockefeller Professor of Physical Anthropology and while there wrote his first paper on the non-metrical morphological features of human skulls. He moved to the University of Melbourne in 1929, and from 1936 until 1942, he and I conducted a prolonged correspondence. He attempted, unsuccessfully, to have my long paper (Fenner, 1939) published in the *Philosophical Proceedings of The Royal Society*. He was a charismatic man, of great mental energy, an outstanding lecturer, with a very facile pen; he published 15 books and over 100 scientific papers (see W. E. Le Gros Clark, 1955).

Contact with Tindale led to other activities associated with the Museum. In December 1934, I was asked to join him on an expedition to Flinders Chase, the major nature reserve on Kangaroo Island, to make insect collections. While there we also discovered and dug out some diprotodon bones, and Tindale included me as a co-author of a short paper on these. During 1935, I developed an interest in Aboriginal skulls, of which the South Australian Museum had an excellent collection. Rather than making formal measurements, as specified by an international agreement (Hrdlicka, 1920), I was attracted to a paper written by Wood Jones on 'non-metrical morphological characters' of skulls (Wood Jones, 1931), and I wrote to him in March 1936 suggesting that I should make a similar study of Australian Aboriginal skulls. He wrote back supporting the idea. Over the period 1936 to 1938, I spent most lunchtimes in the basement of the South Australian Museum, and in the long vacations obtained funds from the David Murray Scholarship Fund of the University of Adelaide to visit museums in Melbourne, Sydney and Canberra to examine Aboriginal skulls in their collections. In 1937, I stayed in Beauchamp House in Canberra, opposite the Institute of Anatomy, where the skulls were located. Beauchamp House is now the Ian Potter House of the Australian Academy of Science. I was appointed 'Honorary Craniologist' at the Museum (I believe a unique designation, not politically correct these days) and published one major paper (Fenner, 1939) and several minor papers on these skulls. When I was in Palestine in 1941 (see Chapter 3), I arranged to have these papers submitted for the degree of Doctor of Medicine at the University of Adelaide. After I had submitted evidence from Brigadier Neil Hamilton Fairley, the Director of Medicine in the Australian Army, to the Dean of the Medical School, Sir Trent de Crespigny, that I had 'an advanced knowledge in the principles and practice of medicine', I was awarded the degree of Doctor of Medicine (MD) (in absentia) in 1942. The examiners were Professors F. Goldby, of the University of Adelaide, and A. N. Burkitt, of the University of Sydney.

Resident Medical Officer, Adelaide Hospital

I spent the year 1939 as a resident medical officer at the Adelaide Hospital, my first experience of living away from home for a prolonged period. During this period I spent four months as resident physician for Dr S. R. (Ginger) Burston, who was later to become Director-General of Medical Services for the Second Australian Imperial Force (AIF), and as surgical clerk for Mr (later Sir) Ivan Jose.

Although I enjoyed this work, I never intended to practice medicine; I wanted to do research. Hugo Gray had written to me early in 1939, suggesting that I should come to the United Kingdom and get a job in anatomy: 'With its two big lines, human evolution and experimental anatomy, and its temporary dearth of good men, it offers better scope than anything else in medical science.'

By this time, Wood Jones had left Melbourne for Manchester and, in a long letter in April 1939, he told me that anatomy in Britain was at a low ebb. He suggested that I should investigate the possibility of working with Dr E. Weston Hurst, the Director of the newly-established Institute of Medical and Veterinary Science at the Adelaide Hospital. I explored this possibility, and applied for a grant from the National Health and Medical Research Council (NH&MRC) to enable me to undertake research on viruses with him. During the discussions he told me that the two leading virologists in the English-speaking world were C. H. Andrewes, of the National Institute for Medical Research in London, and F. M. Burnet, who, he said, had everything in high degree, especially originality. Much later, I learned that my application to NH&MRC had been unsuccessful. But, on 3 September, 1939, Australia declared war on Germany and plans for the future, for all who had been students in my year, were scrapped.

References

Ashton, D. H. 2001, The history of the McCoy Society, *The Victorian Naturalist*, vol. 118, pp. 321–7.

Fenner, F. J. 1936, Anthropometric Observations on South Australian Aborigines of the Diamantina and Cooper Creek Regions, *Transactions of the Royal Society of South Australia,* vol. 60, pp. 46–54.

Fenner, F. J. 1939, The Australian Aboriginal Skull: its Non-Metrical Morphological Features, *Transactions of the Royal Society of South Australia,* vol. 63, pp. 248–306.

Hrdlicka, A. 1920, *Anthropometry*, Wistar Institute of Anatomy and Biology, Philadelphia.

Jones, P. G. 1987, South Australian Anthropological History; the Board for Anthropological Research and its Early Expeditions, *Records of the South Australian Museum*, vol. 20, pp.71–92.

Le Gros Clarke, W. E. 1955, Frederic Wood Jones, *Biographical Memoirs of Fellows of The Royal Society*, vol. 1, pp. 119–134.

Wood Jones, F. 1931, The non-metrical morphological characters of the skull as criteria for racial diagnosis. Part I, General discussion of the morphological characters employed in racial diagnosis, *Journal of Anatomy*, vol. 65, pp. 179–95.

Chapter 3. The War Years, May 1940 to February 1946

Enlistment in the Australian Army Medical Corps

There was only one woman among the 17 students who graduated in medicine in 1938 and became resident medical officers in the Adelaide Hospital in February 1939. All of us, except for her and two of the men, enlisted in the Army or Air Force when they finished their year of residence. I was not a pacifist and I thought that there was no alternative to war with Hitler. With hindsight, I realise that I would have faced serious problems of conscience if I had been in any service other than the medical corps and had had to kill another person. Of course, I would not have been alone in that, it was just so much easier to serve as a medico.

As early as December 1939, it had been decided that Australian troops would initially be sent to Palestine, and the advance party had arrived there in early January. I knew that a number of unusual infectious diseases that occurred in tropical and semitropical countries were found in the Middle East, and I wanted to have the chance to be something other than a regimental or field ambulance medical officer. In 1939, resident medical officers at Adelaide Hospital received board and lodging and about £5 a week, plus a bonus of £200 if they stayed on for the full year, i.e., until February. I used the bonus to go to Sydney and study for the Diploma of Tropical Medicine (DTM), a three-month course available in Australia only at the School of Public Health and Tropical Medicine at the University of Sydney. I stayed at St Andrew's College, which was on the University grounds. My best friend there was Edgar Mercer, who had been at Adelaide High School a couple of years ahead of me. He had a flat in King's Cross and we used to go there most weekends. Coming from Adelaide, which has hot but dry nights, I found Sydney's humid nights during February hard to take. I worked for three months at the School of Public Health and Tropical Medicine, where, amongst others, I met Ted (later Sir Edward) Ford, who was lecturer in bacteriology. He had been a close personal friend and admirer of my mentor in physical anthropology, Wood Jones, and Ted and I were later to work as colleagues in malaria control in New Guinea. While there I also played for the Sydney University hockey team. Altogether, I greatly enjoyed my time there. Returning home to Adelaide in April, I finally enlisted as a captain in the Australian Army Medical Corps on 9 May, 1940, commencing duty with the 2/6 Australian Field Ambulance on 12 June, 1940.

Military Training at Woodside

During the six-month training period at Woodside, in the Adelaide Hills, as well as all the other training, we used to go for long marches every few days, during which I used to read a book as soon as we got going, an uncommon but not illegal action. The longest march we had was from Woodside to Mannum, where we officers met with the local doctor, whose name was Alpers. I remember meeting his two sons, one of whom, Michael, eventually became the Director of the Papua New Guinea Institute of Medical Research. In recent communications with Michael, he also remembers my arrival there. We usually had weekends off and most of us would go back to Adelaide. I continued playing hockey with the Adelaide University team, and one day was hit in the eye by the ball and got a black eye. As might be expected, this aroused a lot of derisive comment from the troops at drill on Monday.

In our spare time, Noel Bonnin (a surgeon six years older than me and also an officer in the 2/6 Field Ambulance) and I set up a small laboratory and carried out a number of experiments on the treatment of gas gangrene in guinea pigs by the local application of sulphanilamide, the only antibacterial drug then known. We published an article describing our results (Bonnin and Fenner, 1941). An article in the Adelaide newspaper, *The News*, in January 1942, mentioned that research workers at Tulane University, in the United States, had published a paper suggesting the use of another sulphonamide, sulphathiazole, based on our results.

Figure 3.1. Officers' Mess, 2/6 Field Ambulance, Woodside, September 1940

At head of table: Colonel R. Southwood (ADMS, Southern Command), on his left, Lieut-Col. E. H. Beare, Commander of the Unit. First on left of table, Frank Fenner; others include R. S. Wilkinson, R. Sands, N. J. Bonnin, F. K. Mugford, W. M. Irwin, J. R. Magarey, H. M. Fisher, R. A. Higginson, D. W. Sands, and C. G. Rankin.

Service in Palestine

Along with many other troops, the Unit embarked for Palestine in December 1940, disembarking at Suez. We then moved up to Gaza, where the majority of Australian troops were located immediately after they arrived. A field exercise there a few weeks after disembarkation included an exchange of officers and other ranks between the 2/4 and the 2/6 Field Ambulances, during which I earned the wrath of the brigadier in charge of the operation. As far as I could ascertain, he was upset because I had followed the advice of the Officer-in-Charge of the 2/6 Field Ambulance, Lieut-Col. E. (Teddy) Beare, that officers should always 'march with the men'. Thus, I left the decision about where to place an Advanced Dressing Station to the Staff Sergeant, who had gone ahead with the equipment on a truck. The brigadier arrived at the same time as I did, regarded the site that had been selected as very dangerous, and rightly blamed me for it. During that exercise we had the opportunity to visit many Palestinian villages, and see the threshing of wheat, camels at work, and so on. I still feel sympathy for the Palestinians ousted by the state of Israel.

A few weeks after this exercise I was transferred to Headquarters, First Australian Corps, where I worked closely with the Deputy Director of Medical Services, Brigadier W. W. S. Johnson, a fine physician and a fine gentleman. There may have been other considerations in my transfer, possibly related to my Diploma of Tropical Medicine, but for me it was a most fortunate change. Johnson took me with him when he visited Jerusalem and there I met Dr Saul Adler, FRS, an outstanding parasitologist and an expert on malaria in the region. I met Ted Ford again, he was responsible for a Mobile Bacteriological Unit attached to Corps headquarters. I also made the acquaintance of Colonel (later Brigadier Sir) Hamilton Fairley, Director of Medicine for the Second AIF, and Colonel J. S. K. Boyd, Director of Pathology for the British forces in the Middle East, both outstanding experts in tropical diseases.

Neil Hamilton Fairley, 1891–1966

Born in Melbourne in 1891, Fairley graduated MB BS with first class honours from Melbourne University in 1915. In 1916, he enlisted in the Australian Army Medical Service and sailed to Egypt as pathologist to the 11th Australian General Hospital. He was subsequently promoted Major and, in 1918, at the age of 27, was appointed Senior Physician to the Hospital, with the rank of Lieutenant-Colonel. He returned to Melbourne in 1920 as first assistant to the Director of the Walter and Eliza Hall Institute, but in 1922 went to Bombay, where he worked on schistosomiasis, dracontiasis and sprue. From 1925 to 1929 he worked again at the Walter and Eliza Hall Institute, studying snakebites and snake venoms.

In 1929 he went back to London as lecturer at the London School of Hygiene and Tropical Medicine, where he worked on filariasis, leptospirosis and malaria. With the outbreak of World War II, Fairley enlisted again in the Australian Army Medical Service, with the rank of Colonel, and joined the Headquarters of the Australian Forces as Consulting Physician. I met him when he visited 2/1 Casualty Clearing Station in Nazareth. Returning to Australia in 1942, he was promoted to Brigadier and became Director of Medicine in the Australian Military Forces and Chairman of the Combined Advisory Committee on Tropical Medicine, South Pacific Area, and as such, directly responsible to General MacArthur. His major contributions in this role were ensuring that all available sulphaguanidine was made available to troops on the Kokoda Track, where dysentery was undermining their fighting capacity, and the setting up of the Land Headquarters Medical Research Unit (LHQMRU) in Cairns in June 1943. The latter consisted of an entomological section, which was sent 20,000 anopheline larvae from New Guinea weekly, a pathology section, and a clinical section, which supervised the use of human volunteers subjected to infection with falciparum malaria. Boyd (1966) notes that the keynote of success in all these experiments was the subinoculation test, which was carried out by my wife, Bobbie.

After the War, Fairley became Wellcome Professor of Tropical Medicine at the London School of Hygiene and Tropical Medicine and followed up the experiments on the pathogenesis of malaria, but in 1948 he had a serious illness from which he never fully recovered. He continued to serve on many committees, but in 1962 retired to the country and died in 1966 (see Boyd, 1966).

Fairley, who had worked on malaria in Macedonia before the War, advised the Commander-in-Chief, General Wavell, not to send British and Australian Troops to Macedonia, a highly malarious area, to bolster Greek resistance to a German attack, as he had planned, but rather to send seasoned Greek troops. Wavell reacted violently, but after an interview he withdrew his criticisms. Knowing from our conversations in Nazareth that I had a Diploma of Tropical Medicine, he dispatched me to a staging camp in Alexandria to await shipment to Greece as a malariologist for the Australian forces. I was there, in a very dusty staging camp, for about five weeks. During this time I explored Alexandria, including the catacombs, where I met a South Australian anthropologist who was studying the burials there, and made friends with some very interesting local people. However, after about five weeks it became clear that the German stukas (dive bombers) were overwhelming the Allied forces, and my only participation in the Greek campaign was to go to Crete as medical officer on a ship evacuating civilians, some weeks before the German parachutists invaded the island. We returned to Alexandria with an imposing naval escort.

2/1 Casualty Clearing Station

Shortly after I had returned to Corps Headquarters, I was transferred as a physician to the 2/1 Casualty Clearing Station (2/1 CCS), which was located in Nazareth as a field hospital for Australian soldiers involved in the Syrian campaign, which was developed to oust the Vichy French, who then occupied Syria and Lebanon. Since many of the patients suffered from malaria or dysentery, I set up a small laboratory and carried out malaria diagnosis by examination of thick films. While there, my colleagues called me 'Noffie' (an abbreviation of *Anopheles*, the malaria mosquito), a nickname that stuck until I moved to New Guinea as a malariologist. At the conclusion of the Syrian campaign, the Unit moved to Beirut, where it set up in what was said to have been the only mental hospital in the Middle East, at Asfurieh, about 20 km to the east of Beirut. We were there for six months. I made friends with an American microbiologist who worked in the American University in Beirut, and also with a well-to-do Lebanese family named Hitti, with whom I spent several very pleasant weekends at their country home in the mountains.

During this period Professor Sydney Sunderland, who had succeeded Wood Jones as Professor of Anatomy at the University of Melbourne, entered into negotiations with me and the Army authorities in Australia to secure my release, so that I could return to Australia and take up the position of Senior Lecturer in his department. I replied saying that my current commitments were such that I could not accept his offer, but that I looked forward to joining him after the War. However, after my experience with malaria in New Guinea (see below), I had decided that research in infectious diseases, not anatomy or physical anthropology, was to be my post-war activity. I therefore did not respond to a later invitation from the University of Adelaide to apply for the vacant chair of anatomy.

My friend, Noel Bonnin, who was by then a surgeon in 2/1 CCS, and I also arranged two very interesting trips from Beirut during two separate periods of leave. For the first, we hired a taxi with a driver who spoke French but no English, and our French was very primitive. Noel acquired two four-gallon tins of petrol from friends in a nearby British Field Ambulance, and we travelled south through Haifa, Jerusalem and Jericho, bathed in the Dead Sea and went on into Jordan. We stopped at Jerash, a wonderfully well-preserved Roman city, then proceeded through Amman, the capital of Jordan, and south through the desert until we came to a narrow gorge to Petra, the 'rose-red city half as old as time', and spent a day exploring the wonderful buildings carved out of the red sandstone. Unaccustomed to travelling over corrugated desert roads, which we had to use on our trip down, our driver drove slowly, while with our experience of travelling on corrugated roads in outback Australia we tried to encourage him to drive faster. We came back along the mountains that form the western shore of the Dead Sea. Here our driver felt at home (from his experience in the mountains of Lebanon) and drove so fast around the winding roads that we shouted *'Lentement! Lentement!'*, to no avail. We visited some impressive castles on the mountains, and eventually stopped off in Damascus, then visited Homs, and back to Asfurieh.

My next adventure, with two medical orderlies and in a field ambulance, was to accompany a regiment of the 9th Division from Palestine through Egypt to Mersa Matruh, in Libya. This passed without incident, and I then came back at my own pace, and stopped off near the pyramids in Cairo for a couple of days to explore that part of the world. This proved a useful preparation for my second week's holiday with Noel Bonnin. We had purchased and read a Penguin book on the wonders of ancient Egypt and, a few months later, spent a week's leave travelling by train from Beirut through Cairo and along the Nile Valley to Aswan, where a major dam was under construction, then back to Luxor, where there were an amazing number of famous ruins. Usually a very popular tourist destination, it was deserted because of the War. 'Knowledgeable', as we thought we were, having read the Penguin book, we interviewed several guides at Luxor before selecting the one we judged to be the best. He turned out to be a splendid guide. We stayed in a luxurious but almost empty hotel and we went to all the local sites.

Apart from the marvels of the ancient Egyptian ruins, there were a couple of incidents on this trip that I vividly remember, 65 years later. The first occurred on the way to Cairo. There had been a severe sandstorm in the Sinai Desert and the train ran off the line just before it reached Kantara, on the Suez Canal. The 2/2 Australian General Hospital (AGH) was located at Kantara and Noel's brother was a physician there, so we spent the night with him before proceeding to Cairo the next day. We had planned to spend a day in Cairo and, importantly, to visit the British Pay Office to get some Egyptian currency. However, we arrived a day late, in the early evening, in a blacked-out city, and the train to Aswan on which we were booked was due to leave in an hour. I thought that I could remember where the Pay Office was from my earlier trip (to Mersa Matruh), so we ran through the blacked-out city streets, arriving just as the paymaster was packing up, and got our money. Then a run back to the station and we threw ourselves into the train just before it left. The other memorable incident was our departure from Luxor. The train to Cairo was filled with British troops coming down from Khartoum. Our excellent guide ran back and forth, speaking with the railway officials, and an extra carriage was hooked onto the train for our convenience.

Australia, April 1942 to March 1943

With the entry of Japan into the War, Prime Minister Curtin insisted that all Australian troops except the 9th Division, which was part of Montgomery's force at El Alamein fighting against Rommel, should immediately return to Australia. I came back as the medical officer for a transport battalion on a small and very old ship, the *Pundit*, leaving from Suez on 8 February, 1942. We stopped for a week in Colombo, while many passenger ships and a protective fleet of warships was assembled. By the end of the first day at sea after leaving Colombo the fleet was almost out of sight; *Pundit* could not keep up. The fleet commander signalled, 'Goodbye, Good luck' and steamed away. There were some scares about Japanese submarines, and we steamed ahead at full speed (12 knots an hour!) but, fortunately, these were false alarms. A couple of other memories of that trip were that, on the fortnightly payday, all the troops of the transport battalion would play two-up until by five o'clock all the money was redistributed, and that I read several quite substantial books, including H. A. L Fisher's 1,300-page *A History of Europe*. We lived on bully beef and biscuits, and I lost about a stone and a half on the trip. Finally, as I remember it, it took us about seven days to cover the last 350 nautical miles, until we disembarked at Fremantle. Looking over the side, the water seemed to be moving ahead of the ship. *Pundit* never left Fremantle; it was not considered to be seaworthy. From Perth troops from the eastern states took the train across the Nullabor, arriving in Adelaide on 6 April, 1942, where I took leave with the family at 42 Alexandra Avenue.

Pathologist, 2/2 Australian General Hospital

The 2/1 CCS set up a small hospital at Ipswich, near Brisbane, and, soon after that, when my leave was over, I rejoined them. One day, Brigadier Fairley visited the Unit and asked me whether I would like to become a hospital pathologist. With visions of six months in Sydney for a training course, I had no hesitation in accepting. Instead, a few days later I found myself on the narrow-gauge train steaming north to Hughenden, in Central Queensland. I was the only male on the train, but there were a couple of hundred women, nurses moving up to the tented 2/2 AGH. I did not talk to any of them, but I remember seeing a particularly attractive nurse combing her long hair; she was later to become my wife. I was replacing Major (later Colonel) E. V. (Bill) Keogh, who had been pathologist there when the hospital was at Kantara, in Palestine. When the troops returned to Australia he was appointed Director of Hygiene and Pathology at Land Headquarters in Melbourne and, as such, was my boss until the end of the War. One of the surgeons at 2/2 AGH was Major Edgar King, who was appointed Professor of Pathology at the University of Melbourne at the end of the War, so my principal worry, that I did not have skills in histology, was relieved. Promoted to Major on 10 November, 1942, I worked at the 2/2 AGH for about nine months, during which time there was a constant stream of patients from New Guinea, most with malaria or dysentery. I published three papers (of mediocre quality) in the *Medical Journal of Australia* as the result of my work at 2/2 AGH.

It turned out that the young woman with the long hair who I had noticed on the train was Sister Ellen Margaret (Bobbie) Roberts, who, in June 1945, was to receive the honour of Associate of the Royal Red Cross for her work in the 2/2 AGH blood bank (see below). There was not much demand for blood transfusions in Hughenden, and she was assigned to help me with haematology and malaria diagnosis. Major Patrick de Burgh, who was to succeed Professor Hugh Ward as Professor of Bacteriology at the University of Sydney, ran a Mobile Bacteriological Laboratory near the hospital, and he and I examined Bobbie for her skill in thick film diagnosis of malaria; she passed with flying colours. Thereafter, she worked for a few hours each day in my laboratory.

On 3 December, 1942, the tented hospital at Hughenden was hit by a cyclone. Every tent was blown over. My laboratory, one of the very few wooden buildings, was tipped sideways but held up by the water pipe leading to the laboratory tap. After a few days, the tents were re-erected, but early in January the hospital was moved to Rocky Creek, inland from Cairns on the Atherton Tableland, which was high enough to be free of *Anopheles* mosquitoes *(Anopheles punctulatus,* an effective vector, was common in Cairns).

Malariologist in New Guinea

Ted Ford had carried out malaria surveys in New Guinea before the War. To find out exactly what was happening in the field in New Guinea, Colonel Keogh posted him as Deputy Director of Hygiene and Pathology at Port Moresby, with the rank of Lieutenant Colonel. In the campaign at Milne Bay, in September 1942, relatively untrained Australian troops had achieved the first defeat of the Japanese on land; they later suffered severe casualties from malaria (quinine was ineffective as a suppressive drug against New Guinea strains of malignant tertian malaria). Early in December, Ford sought and obtained an interview with the Commander-in-Chief of the Australian Army, General (later Field-Marshal Sir) Thomas Blamey. In his quiet persuasive way, Ford convinced Blamey that, unless malaria was controlled, the army in New Guinea would be totally destroyed by the disease. Blamey acted immediately. New Routine Orders dealing with malaria, prepared by Keogh with Ford's assistance, were promulgated and enforced. To provide expert advice and dramatise the importance of malaria, three new posts of malariologist were established to supplement the work of the Assistant Directors of Hygiene. In March 1943, Ford was appointed senior malariologist, based in Port Moresby, and two other medical officers, Major J. C. English and myself, were appointed malariologists. I moved up to Port Moresby in April and initially shared an office there with Ford. In July, I moved to Buna, on the north coast, where troops were preparing for the Lae-Finschhafen campaign.

Figure 3.2. Frank Fenner at the site of the 2/2 Australian General Hospital at Hughenden a day after the cyclone

Sir Edward (Ted) Ford

Brian Gandevia noted, and I can confirm that Ted Ford had several personal qualities rarely seen in one man: an ever-present gentleness, a great depth of kindness and understanding, a wonderful generosity and a sincere humility, and a keen sense of humour and wit, none of which precluded determination and firmness when the occasion demanded (as was evident in his conversation with General Blamey). He was a late starter in medicine, working in the Postal Department at night and doing medicine by day. Graduating in 1932 at the age of 30, he entered academic medicine by becoming a lecturer in anatomy, where he came under the charismatic influence of Frederic Wood Jones. After obtaining a Diploma of Tropical Medicine in 1938, he investigated venereal disease and malaria in New Guinea. Back in Australia, he became a lecturer in bacteriology at the Sydney School of Hygiene and Tropical Medicine (which is where I first met him in 1940; in 1947 he became Director of the School) and in 1941 he went to Palestine as commanding officer of a Mobile Bacteriological Laboratory. Back in Australia, Bill Keogh organized his transfer to New Guinea as Bill's deputy, and in 1943 he was appointed to the new post of Senior Malariologist, where, of course, I had very close relations with him (see Gandevia, 1994).

Jim English and I learned what our job was as we went along. By this time, quinine had been replaced by atebrin. Our first mission was to convince all those in charge of troops in the field, from colonels to platoon commanders, that they must ensure that all troops under their control swallowed one tablet of atebrin daily. I and all troops who did this became greenish-yellow in appearance, and some suffered from skin and neurological problems as complications (but not impotence, a rumour that was widely circulated among the troops). As malariologists, we also supervised the work of the Malaria Control Units and Entomology Research Units associated with the campaigns to which we had been allocated. As malaria came under better control, we took responsibility for the prevention of other insect-transmitted diseases, mainly dengue and scrub typhus, as well as malaria. Besides keeping in close touch with Keogh via 'demi-official' letters, I had frequent contact with Ian Mackerras (then Director of Entomology and later the first Director of the Queensland Institute of Medical Research) and his deputy, Francis Ratcliffe, with whom I was later to collaborate in studies of myxomatosis. I also met our equivalents in the US Army in New Guinea, Majors McGhee Harvey and Fred Bang, both professors at Johns Hopkins University, who remained good friends in my post-war days.

During this period I was instructed by Colonel Keogh to apply for the position of Director of the Institute of Medical and Veterinary Science in Adelaide, where Dr Weston Hurst had been Director during my intern days. Major R. J. Walsh, officer-in-charge of the Blood Transfusion Unit in Sydney, who later became a very good friend of mine, had received similar instructions; Keogh insisted that possible candidates for such positions who were in the services should not be overlooked. Fortunately, in terms of our subsequent careers, neither of us was appointed.

The major campaign with which I was closely associated in New Guinea was the capture of Lae and Finschhafen by the troops of Second Australian Corps, comprising the 7th and 9th Divisions, over the period September 1943 to March 1944. I wrote a long technical paper on malaria control during this campaign, which is summarized in A. S. Walker's history of medical aspects of the War (Walker, 1952). I set out below an abbreviated version of Walker's summary:

> The terrain was highly malarious; the military operations were highly successful. This campaign began with better prospects than others previously fought. Equipment was better, protective clothing was worn, mosquito repellent and atebrin were available, and mosquito control was applied at an early stage in the operations, the control units moving along with the troops Notwithstanding all this nearly 10,000 men were evacuated with malaria.
>
> There was no question that the malaria risk was high in Lae and Finschhafen; the Japanese suffered heavily, 308 died out of 708 admitted to one of their field hospitals in the Huon Gulf area. Conditions were favourable for survival of adult mosquitoes long enough to enhance the risk of a rising infection rate. It was most important to realise that poor antimalarial discipline increased the risk of a gametocyte reservoir among the Allied troops. In Lae and Finschhafen control of adult mosquitoes was ineffective; in Lae gametocyte carriers were promptly segregated, with the result that larval control was rapid and effective, whereas in Finschhafen this segregation was ineffective and larval control was consequently slowed.
>
> The malarial risk was high during the first month after the landing, and it was only after this that control reduced the risk. Fenner made an analysis of the capture of Lae, the capture of Finschhafen and the enemy counter-attack, with the following offensives on Sattelberg and Wareo, and the final capture of the Gusika-Wareo line. Though allowances must be made for the different nature and intensity of these actions, the sick wastage figures given in Fenner's report are most significant, as seen in the table below.

	4–17 Sep	22 Sep–10 Dec	3 Dec–1 Mar
Date and place	Lae	Finschhafen	Gusika to Saidor
Killed & missing	150	291	83
Wounded	397	1,037	186
Malaria	62	3,400	4,300

Walker goes on to analyse the relative importance of other factors, such as the length of service in New Guinea, the physical condition of the troops on arrival there, the severity of the fighting, and the provision of reinforcements.

Award, Member of the British Empire

One (to me) unexpected outcome of my report was decoration with the Member of the British Empire (MBE) award in 19 July, 1945, on the recommendation of Colonel G. W. G. Maitland, DDMS, 2 Aust Corps. The citation read:

> Major Fenner has been Malariologist attached to NEWGUINEAFORCE, 1943–44 and has been responsible for the coordinated control of malaria throughout those areas of PAPUA and NEW GUINEA occupied by Australian troops.
>
> In the Technical Administration of the Malaria Control Units he has exhibited a devotion to duty exceeding that normally required of an officer and has contributed to the scientific knowledge of malaria control in the Army.
>
> Through the coordinated functioning of the Malaria Control Units under Major Fenner's administration and the improved anti-malaria discipline in the Force, the incidence of malaria in the troops is now reduced to a minimum.

Other Research in New Guinea

While I was in New Guinea and later in Morotai, but not involved in campaigns, I became interested in severe enteric fever that occurred amongst New Guinea natives, Japanese prisoners and Australian troops in New Guinea. With the cooperation of Dr Nancy Atkinson, an expert on *Salmonella* bacteria working in Adelaide University, who identified the causal organism as *S. blegdam*, I wrote a paper on the disease among Australian troops (Fenner and Jackson, 1946) and another on cases among the New Guinea natives (Jones and Fenner, 1947).

Bill Keogh

Esmond Venner (Bill) Keogh was a remarkable man. I am only one of many whose lives have been greatly influenced by his actions behind the scenes. Initially enlisting in November 1914 as a non-combatant soldier in the Third Light Horse Field Ambulance and serving in Gallipoli and Egypt, in 1916 he transferred to the Third Australian Machinegun Battalion and served in France with such courage that he was awarded a Military Medal (MM) and a Distinguished Conduct Medal (DCM). During the World War II, when he had to wear his decorations on his jacket, he would insert that part beneath his lapel. After World War I, he initially worked as a dairy farmer, then in 1922 he embarked on a medical course, graduating in 1927 with first-class honours in medicine. He wanted a laboratory job, and joined the Commonwealth Serum Laboratories in May 1928. After travelling about Australia a good deal, in 1935 he established a small research unit at the Commonwealth Serum Laboratories and also worked for a time with Burnet at the Hall Institute. Bill was in the United States in September 1939, and returned immediately to Australia and was gazetted as a major in the Royal Australian Medical Corps on 13 October, 1939. On 14 February, 1940, he sailed as pathologist with the 2/2 Australian General Hospital, initially to Gaza, moving in November to Kantara, where he remained until the hospital moved back to Australia in February 1942. He was immediately moved to Army headquarters at Victoria Barracks as a full colonel and Director of Hygiene, Pathology and Entomology. As well as all his official duties, which included appointing me as pathologist and then as a malariologist to 2/2 Australian General Hospital, he played a major role, with Brigadier Fairley and Ted Ford, in establishing the LHQ Medical Research Unit in Cairns, with the specific mission of developing effective antimalarial drugs (see Gardiner, 1990).

Wartime Work at the Walter and Eliza Hall Institute

At the conclusion of the Lae-Finschhafen campaign, the Seventh Division was withdrawn to the Atherton Tablelands and I also went back, by air. Because there were very few people aboard the plane, the pilot let me have a short spell at the 'wheel'. My memory of that was a realisation that even slight movements of the controls would result in surprising changes in altitude. I arrived back in Townsville in August 1944, staying at the 2/2 AGH but pursuing my own work. Colonel Keogh had arranged that I should come down to Melbourne for six weeks from October and work at the Walter and Eliza Hall Institute, where Macfarlane Burnet was now the Director. I thought it best if I could bring a small problem with me, and took advantage of the fact that an old Adelaide University colleague and friend of mine, John Funder, was working there on a newly discovered variety of typhus, a rickettsial disease called North Queensland tick typhus. This had been discovered by doctors working at 2/2 AGH while I was in New Guinea (Andrew et al., 1946). Before going down to Melbourne, I therefore worked in the field with staff of a Malaria Control Unit and an Entomology Research Unit to collect sera and ectoparasites from a range of wild animals in an effort to discover the reservoir of this newly-discovered zoonotic disease, and I took this material to Melbourne for testing. At the Hall Institute I carried out complement fixation tests on these sera, using an antigen prepared from yolk sacs infected with the North Queensland tick typhus rickettsiae. Eight animals, of five different species, gave positive results; five of the positives came from a localized area of rain forest, which was also the site of infection of several human cases (Fenner, 1946). I later found that Keogh had suggested to Burnet that this would be an opportunity for him to decide whether he wanted to recruit me as a research worker at the end of the War (see below).

Marriage

Captain Bobbie Roberts, as she was then, had been sent from the 2/2 AGH to the Heidelberg Military Hospital in Melbourne for the period 8 September to 8 December, 1944, to give classes on blood transfusion. This overlapped with my spell at the Hall Institute. Almost immediately after I arrived in Melbourne, I proposed to her in a little room at the Heidelberg Hospital. Since she was a Roman Catholic and I was an atheist, I was given a lecture by a Catholic priest . Three days later, we were married in a side chapel of the Catholic Cathedral, with a former laboratory assistant of mine, Lieutenant Mavis Freeman, and Bobbie's closest friend, Nurse Reuben Warner, as witnesses. Major Kevin Brennan, an officer in Keogh's section at Victoria Barracks, kindly let us live for three weeks in his house, which was vacant at the time. Reflecting the religious intolerance then common in Adelaide, my mother was initially very upset that I should marry a Catholic but, after they became acquainted, they became very close friends. My father, on the other hand, wrote her a warm letter, saying that 'Life is too short, and too full of pitfalls, to waste any opportunity of happiness'. I returned to Atherton on 24 November, Bobbie went to the LHQ Medical Research Unit in Cairns on 8 December.

This Unit had been set up by Brigadier Fairley to test antimalarial drugs on human volunteers and Bobbie had already worked there, as had several physicians at 2/2 AGH. She was the only person locally available who could carry out direct blood transfusions from artificially infected volunteers to other, uninfected volunteers. She did this by using a device invented and given to her by Dr Julian Smith, who was father of the senior surgeon of the 2/2 AGH. She also assisted Major Josephine Mackerras with entomological work. I used to drive from the Tablelands to Cairns every weekend and stay with her in one of the local hotels. We studied *Ideal Marriage* by van de Velde (one of the few sources of such information at the time) to learn more about sexual pleasures. Colonel Talbot, the Commanding Officer of 2/2 AGH, very kindly offered us use of his holiday home at Yeppoon, then a beautiful and almost uninhabited settlement on the coast, for a honeymoon. We spent a blissful few days there, and then I received a message that I had to be in Townsville by 6 April to embark

Figure 3.3. Photograph of Bobbie Roberts in 1942, by Julian Smith

on the *General Butner* to go to Morotai, in the Halmaheras. Just before I embarked I met with Bill Keogh and he told me that Burnet intended to offer me a job when I got out of the Army.

In June 1945, Bobbie received the award of Associate of the Royal Red Cross. Her citation reads:

> WFX1536 Captain Ellen Margaret Fenner
>
> Sister Fenner commenced duty with 2/2 Aust Gen Hosp and embarked for overseas service on 30 Apr 40...At El Kantara she did outstanding work as Sister-in-charge of acute surgical wards and blood bank centre. She assisted the medical staff in carrying out research work and her work at all times has been brilliant and untiring.
>
> On return to Australia she carried on with the work in the blood bank when the hospital was established in Queensland. In early 1944 she was detached for duty with a Malaria Control Unit. She showed exceptional ability for this kind of work and her work was outstandingly good. Because of her work with this unit it was requested that she should be posted to the LHQ Medical Research Unit where she is at present serving.

Service in Morotai and Borneo

I disembarked in Morotai on 16 April, 1945. While I was on the ship and overseas, Bobbie and I wrote to each other every day. I was not able to keep her letters but, unbeknownst to me, she kept all of mine, organized by the month and tied together in neat bundles kept in a drawer next to her bed. I became aware of them when she was in bed with advanced pulmonary secondaries from a colon cancer. When I found her reading one, in October 1995, she looked at me with an expression of such deep love that I could not bear to read the letters, until October 2001, when I undertook to update my Basser Library archives. They are pretty torrid love letters, but mention facets of my scientific career not available elsewhere. The catalogue of my Basser Library archives contains explanatory notes as well as the letters.

For the rest of the War, until 27 August, 1945, I was based at 1 Aust Corps Headquarters on Morotai, and was involved with malaria control in the attacks on Brunei Bay, Tarakan and Labuan. Throughout these campaigns, the malaria rates were very low, with only 97 cases being admitted to medical units between April 6 and September 7, from a force of 17,000. After hostilities ceased, I visited Brunei, Labuan, Tarakan, Balikpapan and Sarawak. With Francis Ratcliffe, I organized some trials of mosquito control by dispersion of DDT from aircraft. As reflected in my letters to Bobbie, there were long periods of boredom and anxiety about the job with Macfarlane Burnet at the Walter and Eliza Hall Institute.

Appointment as Francis Haley Research Fellow

Eventually, on 10 July, 1945, I received a letter from Burnet. The relevant part read:

> The establishment of the Chair of Experimental Medicine [at the University of Melbourne, specifically for Burnet] has made it possible for me to look forward to having a full time senior man [on a Francis Haley Fellowship] in that department. The primary interest is epidemiology in the broad sense...it must be concerned with cancer, T.B. or other widespread and important human diseases.
>
> Would you be interested in a preliminary offer of such an appointment at a salary of £1000 p.a. to start as soon as your release from the Army? You would have a very considerable latitude in regard to choice of a particular field...There is a specially good opportunity in experimental epidemiology for the study of a virus disease (ectromelia of mice). In January last we found that this disease was antigenically almost identical with smallpox and vaccinia. The disease is spontaneously infective for mice...Any appointment would, of course, be subject to University approval.

I had no problem in replying immediately, noting that, as far as release from the Army was concerned, I should be in the first group, having had five years service. On 30 July, I received a reply from Burnet and a request to should send him a *curriculum vitae* and list of published work, hoping to satisfy the University that advertisement was unnecessary. On 27 August, I received a letter from Burnet confirming my appointment as Senior Haley Research Fellow in the Department of Experimental Medicine of the University of Melbourne, on a salary of £1000 p.a.

Bobbie's mother had been quite ill and, on 1 August, 1945, Bobbie was transferred to 110 Perth Military Hospital, so that she could be near her. She stopped in Adelaide for a few days on the way over and stayed with my family.

Leave Prior to Discharge

With the war over, and because my father was also ill, on 27 August, 1945 I was granted seven days compassionate leave and 32 days annual service leave. On 28 August, I flew from Morotai to Brisbane and, on my way by troop train to Adelaide, I met up with Ted Ford in Sydney, and had dinner with Mac Burnet and his wife Linda in Melbourne. He said everything was going well as far my appointment was concerned, but that the University had not yet decided whether the position should be advertised (in fact it never was). He also mentioned that Ian Wood, who had been senior physician at 2/2 AGH and in charge of the blood bank there, was to take charge of the Clinical Research Unit at the Hall Institute and that Mavis Freeman, who had been my lab assistant at 2/2 AGH and was one of the witnesses at our marriage, was to be biochemist in Mac's team. From 2 to 21 September, 1945, I stayed with my parents in Adelaide, ringing Bobbie in Perth most days. On 21 September, I flew to Perth on a military plane, a Liberator, and Bobbie and I stayed together at the Adelphi Hotel, in the centre of Perth. I hired a small car and spent a wonderful few days exploring the beauties of Southwest Western Australia in spring.

Discharge from the Army

Bobbie stayed in Perth until her discharge from the Army on 1 November, 1945, after 2,122 days service, outside Australia for 696 days. I had been posted to the 115th Heidelberg Military Hospital on 19 October, so at the conclusion of my leave I went back to Victoria. I spent most of my spare time reading the most comprehensive book on viral diseases of humans, van Rooyen and Rhodes (1948), and anything that I could get on ectromelia virus, since this was to be what I would work on with Burnet. Eventually, on 31 January, 1946, I was discharged from the Army, with 2,059 days service, outside Australia for 1,086 days. I started work at the Hall Institute on 1 February. We had found a suitable flat in Milswyn Street, South Yarra, just opposite the southern entrance to the Royal Victorian Botanic Gardens. I developed the habit of walking through the Gardens and along the bank of the River Yarra to Elizabeth Street, where I would catch a tram to the Hall Institute, which was located in a wing of the Royal Melbourne Hospital. Initially Bobbie worked part-time in the Alfred Hospital, just to the south of our flat, where her best friend, another Western Australian nurse, Jean Freeman, was working. Later, Bobbie joined me as an unpaid technical assistant.

References

Andrew, R. R., Bonnin, J. M. and Williams, S. E. 1946, Tick typhus in Northern Queesland. *Medical Journal of Australia,* vol. 2, pp. 253–8.

Bonnin, N. J. and Fenner, F. 1941, Local implantation of sulphanilamide for the prevention of gas gangrene in heavily contaminated wounds: a suggested treatment for war wounds, *The Medical Journal of Australia,* vol. 1, pp. 134–40.

Boyd, J. 1966 Neil Hamilton Fairley, 1891–1966, *Biographical Memoirs of Fellows of The Royal Society*, vol, 12, pp. 123–41.

Fenner, F. 1946, The epidemiology of North Queensland tick typhus: Natural mammalian hosts, *Medical Journal of Australia,* vol. 2, pp. 666–8.

Fenner, F. and Jackson, A. V. 1946, Enteric fever due to *Salmonella enteritidis* var. *blegdam (Salmonella blegdam):* a series of fifty cases in Australian soldiers from New Guinea. *Medical. Journal of Australia,* vol. 1, pp. 313–26.

Jones, H. I. and Fenner, F. 1947, Infection with *Salmonella blegdam* amongst natives of New Guinea: An account of fourteen cases with post-mortem reports of four fatal cases, *Medical Journal of Australia,* vol. 2, pp. 356–62.

van Rooyen, C. E. and Rhodes, A. J. 1948, *Virus Diseases of Man,* Second edition, Thomas Nelson and Sons, New York.

Walker, A. S. 1952, *Australia in the War of 1939–45, Series 5, III. The Island Campaigns,* Canberra, Australian War Memorial.

Chapter 4. Walter and Eliza Hall Institute, 1946 to 1948; Rockefeller Institute, 1948 to 1949

Research on Ectromelia, February 1946 to August 1948

When I arrived at the Walter and Eliza Hall Institute on 1 February, 1946, almost all the staff there were working on influenza virus, which, remembering the disastrous outbreak of influenza just after the World War I, Burnet had undertaken as his contribution to the war effort. As he had suggested, I undertook studies of various aspects of the epidemiology of infectious ectromelia virus. As I found when I had access to the library there, Burnet's suggestion that I should work on the experimental epidemiology of this virus stemmed from work carried out with it in England (Greenwood et al., 1936). Selection of this virus was made more attractive because of Burnet's discovery (Burnet, 1945) that it was an Orthopoxvirus, i.e., it belonged to the same group of viruses as smallpox and vaccinia viruses.

I used the same animal room as the other research workers to carry out experiments, but had a separate room to house infected mice—a wise precaution in view of the disastrous outbreaks of this disease that occurred in laboratory mouse colonies in Europe and the United States. Throughout this work, Bobbie acted as my part-time and unpaid technical assistant. We started by using cylindrical cages that could be attached together according to the number of mice in the cages, as developed by the British team. The first experiments looked at the effect of vaccination with vaccinia virus, the next two on the portal of entry and the sites of elimination of the virus in naturally infected mice.

At the end of the first year, I learned that the University of Melbourne had just introduced the PhD degree (previously, scientists went to the United Kingdom for their PhD). Since as an ex-serviceman I would not have to pay any fees, I applied to the University to be admitted as a PhD student, only to receive the reply that as a Senior Haley Research Fellow of the University of Melbourne I held too senior a post to do a PhD.

During these early experiments we handled all the mice in the cages every day, and made the observation, not easy because of the hair all over the mouse's body, that those mice which did not die of acute hepatitis usually developed skin lesions all over their body—a generalized rash. This had never been described before and was of particular interest because it suggested that ectromelia would be a good model for studying the pathogenesis of smallpox and other generalized diseases that produced a rash. We immediately undertook experiments to investigate the pathogenesis of this generalized infection, in particular to find out what happened during the incubation period. We had to use the crude techniques available in those days, namely infecting mice by inoculation of a small dose, in a small volume, in the footpad, and then titrating virus from the internal organs (blood, lymph nodes, spleen and liver) and, when a rash developed, from the lesions and from seemingly uninfected skin. The titrations were supplemented by histological studies, since most ectromelia-infected cells produced very characteristic inclusion bodies. Most of my papers were published in the 'Adelaide journal', the *Australian Journal of Experimental Biology and Medical Science*, but I published the detailed experimental results of the pathogenesis experiments in the *British Journal of Experimental Pathology* and a paper drawing generalized conclusions in *The Lancet*. Nearly 50 years later, the latter paper was republished in 1996, as a 'Classic paper', in *Reviews in Medical Virology*. Over the next two years I published five more papers dealing with various aspects of the experimental epidemiology of mousepox, as we now called the disease and the virus, concluding with a long and comprehensive review in the *Journal of Immunology* (Fenner, 1949). My baptism in virology, the study of ectromelia virus, led to a lifetime's interest in the poxviruses and ultimately to my involvement in the smallpox eradication campaign of the World Health Organization.

Frank Macfarlane Burnet

My relations with Burnet, who was the most creative and imaginative scientist that I have known, were very cordial. At the time I arrived there in 1946, he and all other staff were working on influenza virus. Burnet kept tight control over their investigations, for in those days of non-existent overseas travel, he thought that he had to compete with large teams in the United States. In contrast, he allowed me complete freedom to do as I wished within my topic, the experimental epidemiology of ectromelia. At that time he worked at the laboratory bench from 9.30 am until 4 pm each week-day, and although we met at the tearoom, he was a reserved man and talked little. However, when I had completed an investigation and written it up I would give the manuscript to Burnet. He would read it that evening, and at 4 pm next day we would meet in his office to discuss its publication, and he would then ask about my current and ongoing work. In contrast to common practice in many laboratories, then and now, Burnet never put his name on a paper involving experimental work unless he had done some of the bench work, and all 11 of the papers on mousepox were published in my own name, or linked with that of my wife.

Helping Burnet Write a Review Article and a Book

Unlike most biomedical scientists, Burnet wrote many books as well as scientific papers. Early in 1948, after he had an opportunity to evaluate my writing, Burnet asked me to collaborate with him in an article he had been asked to write for a new international journal, *Heredity* (Burnet and Fenner, 1948). He must have been satisfied with my performance, because he then asked me to collaborate with him in producing a second edition of *The Production of Antibodies*, the first edition of which he had published in 1941 (Burnet et al., 1941). I helped chase up some of the work done since then, notably Medawar's studies of transplantation immunity. Burnet was responsible for all the interpretation and speculation. The second edition is notable because it contains the first mention of the concept of immunological tolerance, the topic cited in the award of the Nobel Prize to Burnet and Medawar in 1960.

Overseas Study at the Rockefeller Institute for Medical Research, August 1948 to July 1949

During the last few months of the war, Bill Keogh had travelled overseas and made arrangements with several funding bodies, notably the Nuffield Foundation, the Carnegie Foundation and the Rockefeller Foundation, to enable men who had been in the Directorate of Hygiene and Pathology during the War to gain overseas experience. He had visited and been impressed by René Dubos at the Rockefeller Institute, and after discussion with Burnet they had agreed that I should go to work with him, in what was then the leading medical research institute in the world. I received a Rockefeller Foundation Fellowship of £3,000 and a grant of £300 from Burnet to help cover travel costs. Before I left, Burnet told me that, when I had spent a year at the Rockefeller, he would provide me with a job at the Hall Institute, perhaps as Deputy Director, or, he said, there may be a job in 'Florey's new institute in Canberra'.

Before we left, I wrote to the senior virologist at the Rockefeller Institute, Frank Horsfall, suggesting that I would like to bring some ectromelia virus with me, so that I could make use of some of the more sophisticated equipment there. This was greeted with horror, because of the bad reputation that virus had of decimating colonies of laboratory mice, a matter on which I spoke at a conference in New York as late as 1980 (Fenner, 1981). Instead, knowing that Dubos was working on mycobacteria, I took some strains of the newly-discovered 'Bairnsdale bacillus' with me.

After shipping our heavy goods to New York, Bobbie and I went by the *Aorangi* to Vancouver. Travel was arranged so that the ship would stop for at least one day at several islands: Auckland in New Zealand, Suva in Fiji and Honolulu in Hawai'i. At each of these places we spent an interesting day travelling around the city and the country near it, all quite different from Australia or the countries of the Middle East, where Bobbie and I had been during the War. Burnet had written to Dr C. E. Dolman, Professor of Bacteriology in Vancouver, and he met us and took us all around Vancouver and its environs before we boarded the Canadian Pacific Railway to cross the continent. We had hand luggage only, and stopped off at Lake Louise for a day to see the mountain scenery, which was breathtaking. Then through the Rockies to the prairies of Saskatchewan, Manitoba and Winnipeg to the Great Lakes. We got off in Toronto, where we met up with van Rooyen and Rhodes, authors of the book on medical virology that I have mentioned earlier (see Chapter 3). After an interesting day there we went south, spending a day at the Niagara Falls, then by train to New York.

Accommodation in New York

Arriving at New York Central Station at 9 am on 20 September, we booked in at a hotel near the Rockefeller Institute, which was located on East River between 39th to 42nd Street. After meeting with Dubos and his team, one of them drove us around looking for a suitable apartment. In 1948–49, New York was an exciting city, clean and still reasonably safe, if one didn't wander into some parts of Harlem (as both Bobbie and I were warned by bus drivers, on different occasions). We eventually finished up sharing a flat with a 70-year-old Georgian Jew, Mr de Kika, who had a large apartment at 141 West Street, on the opposite side of town from the Institute, which we arranged to share at a cost of $80 monthly. That evening we met up with my old friend Noel Bonnin, who was in New York for a few days on his way back to Australia from England. After getting our heavy baggage out of Customs we set up at the flat. Early in our stay in New York, we visited scientists who had been working with Burnet when we were there, Fred Nagler in New Brunswick and Bernard Briody at Yale University. We also visited some of the marvellous museums in New York, the Guggenheim, the Metropolitan Museum, the Museum of Modern Art and the American Museum of Natural History. I enjoyed the bus trip across the city, which with delays and walking took about an hour each way, because it was a great chance to see something of this great city and its varied inhabitants. We also went to the Metropolitan Opera several times; I remember particularly Wagner's *Tristan and Isolde* and *Siegfried*.

It would have been difficult to live and travel as we did on the £3,000 of the Fellowship if Bobbie had not got employment to look after the young child of a local doctor, Dr Grogan O'Connell. On 19 December, we both moved into his large apartment at 438 East 88 Street, rent-free and within easy walking distance of the Rockefeller Institute.

My Research at the Rockefeller Institute

Dubos had laboratories on the third floor of New North Wing, just below the laboratories of the Rockefeller Foundation, where Nobel Prize winner Max

Theiler worked, with a number of other distinguished arbovirologists. Just along the passage were three other distinguished scientists: Peyton Rous, who was to receive the Nobel prize in 1966, 50 years after his demonstration that Rous sarcoma virus was responsible for malignancy; Rollin Hotchkiss, who was patiently demonstrating that pneumococcal 'transformation', discovered by Oswald Avery, was indeed due to DNA and not a protein; and Merrill Chase, who had just escaped from the teutonic supervision of Karl Landsteiner to carry out important work on tuberculosis, independently of supervision.

Being new to the field, I carried out research suggested by Dubos and some experiments, more or less in parallel, with the 'Bairnsdale bacillus'. Five papers emerged from my research there, the most important being a method of counting viable tubercle bacilli (Fenner, 1951).

René Jules Dubos

Dubos was a fascinating character, very different from Burnet. He had five post-doctoral fellows (as I was classified) working with him. He himself did little laboratory work at that time, but planned the experiments and often wrote them up. At the end of each day he would assemble all of us in his office, sit with his feet up on the table, and ask each of us to describe any interesting results. Then he would pick out an 'interesting result' and erect what I called an 'inverted pyramid' of speculation, which usually fell down but occasionally led to novel experiments. He invited Bobbie and me to his house on the upper Hudson River for weekends on several occasions. I will never forget my first experience of spring in a deciduous forest. He and I developed very close relationships. I always visited him when I went to New York, we maintained an extensive correspondence for many years, and in 1968 I persuaded him and his wife Jean to come out to Canberra, an experience that he greatly enjoyed. He later became an environmental guru who certainly influenced my later life as Director of the ANU's Centre for Resource and Environmental Studies (see Moberg, 2005).

Dubos as a Mentor

In those days, all the academic staff gathered together for lunch (a substantial meal supplied at a cost of 25 cents) in the Institute dining room. Dubos took special care to see that all of his team met the personalities of the Institute. On different days, he would move with his group from one table to another and initiate discussions with whoever he had chosen to sit with that day—Gasser, Goebels, van Slyke, Stein, Rivers and many others— so I met a wonderful group of medical scientists, which of course was just what Burnet and Keogh had planned. Periodically, Dubos organized dinners at various 'ethnic' restaurants near the Institute, to which all his postdocs would be invited, along with scientists from Cornell University and the Public Health Research Institute, to discuss tuberculosis. I remember particularly Walsh McDermott, Jules Freund and Bernie Davis.

Other Personal Experiences While at the Rockefeller Institute

A number of other important matters occurred while I was at the Rockefeller Institute. The most important, personally, was that Bobbie was found to be suffering from carcinoma of the uterus and had to have a total hysterectomy. The operation was performed by surgeon H. C. Taylor and our friend, gynaecologist Grogan O'Connell, on 17 June, at the Sloane Hospital for Women, in 168th Street. She was in hospital for four weeks, with daily X-ray treatment, which continued for two weeks after leaving hospital. I commuted each afternoon from the Institute to 168th Street to see her. We remained good friends of Grogan O'Connell and his wife, and over the next 20 years I made a point of visiting him whenever I went to New York.

The other two matters were more pleasant. The first was a letter from Sir Howard Florey, dated 19 December, 1948, stating that he had been authorized by the Interim Council of The Australian National University to offer me the Chair of Microbiology in the John Curtin School of Medical Research. He also mentioned that he had enjoyed reading my paper in *The Lancet* (Fenner, 1948b). He suggested that while in the United States I should look into the equipment and design of microbiological laboratories, with a view to the design of the building in Canberra. I accepted the offer immediately, and said that during the next three months I intended to spend about four days each three weeks visiting microbiological centres in New York, New Haven, Boston, Philadelphia, Baltimore and Washington. He knew that I was leaving America at the end of July, and later wrote to say that the other two men who had been appointed as professors in the John Curtin School, Adrien Albert and Hugh Ennor, would be visiting him in Oxford during the first week in August.

The other very pleasant surprise, in June 1949, was the news that I had been awarded the 1949 David Syme Prize for Scientific Research, given for 'the best original research work in biology, natural philosophy (physics), chemistry or geology during the preceding two years'. It was founded in 1905, was open to all persons resident in Australia for at least five years, and was at the time one of the most prestigious awards in Australia. I was doubly delighted because my father had been awarded the same prize 20 years earlier, in 1929, for his studies in physiography. Since both father and son had had such a close association with Ballarat, in 1978 I presented both medals and accompanying photographs of the recipients to the Ballarat Historical Park Association. Much later, in 2001, when I was assembling my various medals for presentation to the ANU, the

University of Melbourne kindly provided me with a duplicate of my 1949 Syme Prize medal.

Other Meetings in the United States

As well as the regular seminars at the Institute, many of which I attended, especially those run by virologists Tom Rivers and Frank Horsfall, I went to international and national scientific meetings held in New York, Boston, Denver and Cincinnati. Bobbie and I also travelled extensively, by Greyhound bus, to many places on the East Coast and as far west as Kentucky, with René Dubos' introductions, always being welcomed by the many scientists that I met. Among the most memorable was a drive from Elizabethtown up the Adirondack Mountains to Lake Saranac in mid-May. When we started, the trees were in full leaf. At Saranac Lake there was still snow on the ground and the trees were just coming into leaf. My visa expired at the end of July, so by arrangement with the ANU, Bobbie and I embarked on the *Queen Mary*, to return to Australia via Europe.

References

Burnet, F. M., Freeman, M., Jackson, A. V. and Lush, D. 1941, *The Production of Antibodies; a Review and Theoretical Study,* Monograph of the Walter and Eliza Hall Institute in Pathology and Medicine, No. 1. Macmillan, Melbourne.

Burnet, F. M. 1945, An unexpected relationship between the viruses of vaccinia and infectious ectromelia of mice, *Nature*, vol. 155, p.543.

Burnet, F.M. and Fenner, F. 1948, Genetics and immunology, *Heredity*, vol. 2, pp.289–324.

Burnet, F.M. and Fenner, F. 1949, *The Production of Antibodies. Second Edition,* Monograph of the Walter and Eliza Hall Institute, Macmillan and Company Limited, Melbourne.

Fenner, F. 1948a, The clinical features and pathogenesis of mousepox (infectious ectromelia of mice), *Journal of Pathology and Bacteriology,* vol. 60, pp.529–52

Fenner, F. 1948b, The pathogenesis of the acute exanthems, *Lancet,* vol. 2, pp.915–20

Fenner, F. 1949, Mousepox (infectious ectromelia of mice): A review, *Journal of Immunology,* vol. 63, pp.341–73.

Fenner, F. 1951, The enumeration of viable tubercle bacilli by surface plate counts, *American Review of Tuberculosis,* vol. 64, pp.353–80.

Fenner, F. 1987, Frank MacFarlane Burnet, 1899–1985, *Historical Records of Australian Science,* vol. 7(1), pp.39–77.

Fenner, F. 1981, Mousepox (infectious extromelia). Past, present and future, *Laboratory Animal Science,* vol. 31, pp.553–9.

Greenwood, M., Bradford Hill, A., Topley, W. W. C. and Wilson, J. 1936, *Experimental Epidemiology*, Medical Research Council Special Report Series No. 209, His Majesty's Stationery Office, London.

Moberg, C. L. 2005, *René Dubos, Friend of the Good Earth; Microbiologist, Medical Scientist, Environmentalist*, ASM Press, New York.

Chapter 5. Professor of Microbiology, John Curtin School of Medical Research, 1949 to 1967: Administrative and Domestic Arrangements

Europe, July 1949 to February 1950

As described in the previous chapter, Sir Howard Florey had made arrangements for the first three professors appointed to the John Curtin School of Medical Research to meet him in Oxford early in August 1949. Adrien Albert (Medical Chemistry) was working in the Wellcome Laboratories in London and Hugh Ennor (Biochemistry) had come over from Melbourne. Bobbie and I arrived in England on 2 August. She stayed with my friend Cecil Hackett and his wife Beattie at Northwood, just out of London. I went up to Oxford and spent a very busy four days talking about the future of the School with my new colleagues (it was the first time that we had met) and with Florey. With the help of Florey's colleague, Gordon Sanders (who later came out to Canberra for a few months to help us with the planning), among other things we decided on an H-shape for the JCSMR building, with the main laboratories on the south side of each wing, to avoid direct sunlight, and rooms for special facilities on the north side, with narrow passages, to make cluttering with equipment difficult. The spine of the H was reserved for School requirements, with the library on the top floor, a lecture theatre and seminar rooms, administrator's offices and a tea-room on the ground floor, and stores on the bottom floor. Since there were no laboratories in Canberra at the time, the ANU had arranged with Burnet to make available two laboratories at the Walter and Eliza Hall Institute until we were able to move into laboratories in Canberra.

Sir Howard Florey, 1898–1968

Howard Walter Florey, an Adelaide Rhodes Scholar and since 1935 Professor of Pathology at the University of Oxford, was invited to Australia by Prime Minister Curtin in 1943 to advise on the production of penicillin in Australia. He came out in August 1944 and spent six months here, during which he visited the mainland capitals. After commenting on the lack of adequate facilities for medical research in Australia, Curtin invited him to develop a plan for a national medical research institute. This was ultimately realized by the inclusion of the John Curtin School of Medical Research as one of the initial four research institutes in the ANU (Foster and Varghese, 1996). In 1947, three Australians and one New Zealand expatriate were persuaded to meet as the Academic Advisory Committee for the new Australian National University, Florey being Advisor to the John Curtin School of Medical Research. He continued in this role until 1956, visiting Canberra several times. In the early days I, and I presume other heads of the first four departments, corresponded regularly with him. I have a file containing 35 such letters in my Basser Library archives, the great majority written in 1951 and 1952, before Hugh Ennor had been installed as Dean. From 1960–65, Florey was President of The Royal Society of London and, from 1965 until his death in 1968, he was Chancellor of the ANU. He was not an imaginative scientist of the calibre of Burnet or Dubos, but he 'got things done' in relation to such matters, for example, as penicillin production, the John Curtin School of Medical Research and new buildings for The Royal Society. He had the English habit of addressing everyone except close personal friends by their surname. I found him very helpful whenever I looked for advice (see Fenner and Curtis, 2001).

The Act setting up the ANU was passed in August 1946, and with it came a statutory grant of £325,000. We discussed with Florey how best to spend this money, and decided to use our portion of it to buy back-sets of important journals, get equipment for our laboratories, and provide ANU scholarships to enable young Australian scientists to do a PhD degree in England (none of us wanted to take on PhD students until we had settled into our labs in Canberra). I recommended three such students: J. H. Bennett to work with R. A. Fisher (he later became Professor of Genetics at the University of Adelaide), F. W. E Gibson to work with D. D. Woods on chemical microbiology (he became Professor of Biochemistry here in 1967 and Director of the JCSMR, 1977–79), and W. K. Joklik, who worked for 8 years in my department and then became Professor of Microbiology and Immunology at Duke University, in North Carolina.

Cecil John Hackett, 1905–95

I insert this note on Cecil Hackett because almost every time that I went to London, from 1949 to 1993, I stayed with Cecil and his wife Beattie, initially at his home in Northwood, then in Geneva, when he was working at the World Health Organization, and most often at their flat in Moscow Road, London, just north of Hyde Park. Born in Adelaide in 1905, Cecil graduated MB BS at the University of Adelaide in 1927 and MD in 1936. I met him when he lectured to me on physiology in 1936 and subsequently because of his involvement in the expeditions into Central Australia organized by the Board of Anthropological Studies (see Chapter

2). From 1937 to 1940, he worked in Uganda, investigating the bone lesions associated with the tropical disease yaws. During the War, he served as a medical officer in the Royal Air Force in West Africa, India and Burma, and from 1945 to 1954 he was Director of the Wellcome Museum of Medical Science in London. During the next decade he worked as a Senior Medical Officer at WHO Headquarters in Geneva, concentrating on yaws and its elimination from Indonesia. Back in London in 1965, he continued studies on lesions of syphilis and yaws in bones. In 1980–81 he was the prime mover in arranging for a stone recognizing the contributions of Lord Florey to Great Britain to be placed on the floor of Westminster Cathedral, where his wife, an avowed non-believer, worked so hard as a voluntary helper that she was awarded an MBE and a carving of her head was used as a gargoyle on the Cathedral (see Fenner, 1995).

From August to November Bobbie and I stayed with the Hacketts at Northwood, some 20 km west of London. Since no laboratories were available in Canberra, we took the opportunity to travel around Europe, in the process seeing as many microbiologists as I could. We bought a Ford Prefect car and drove first around England and Scotland (a total of some 3,000 miles) then through France, Belgium, Holland, Germany and Italy (a total of about 4,200 miles). In Germany we went through the Ruhr, still bearing most of the wounds of the War, to the peaceful countryside of Neidergrenzebach, where we met our closest relatives on the Fenner side in the house where my paternal grandfather was born and I saw my great-grandfather's grave. I also attended several national and international scientific conferences in England. Bobbie and I enjoyed learning about the countryside and looking through the great art galleries and museums of Europe. It was an unforgettable experience.

On 8 November, we moved into Nuffield House, in London, where we stayed until we left for home, embarking on the *Orontes* on 12 January, 1950, and arriving in Melbourne on 18 February. Such was the interest in The Australia National University in those days that there were items in the newspapers recording my arrival in Perth, in Adelaide on 15 February, and in Melbourne a journalist from *The Age* met the ship and interviewed me, with an article and photo in that newspaper next day.

Accommodation in Melbourne

I thought that it might be as long as 5 years before we would be able to move from Melbourne to Canberra. Initially, therefore, we moved into the house of my wartime friend, pathologist Alan Jackson, who was going with his wife Mavis on his overseas trip, as arranged by Bill Keogh. They lived in a very nice house in 4 Turnbull Avenue, Toorak, close to a railway station. Bobbie looked after their three children while I went to the Hall Institute each day. Later we looked after the three Burnet children in their house at 10 Belmont Avenue, Kew, while Mac and Linda spent nine months in England. Meanwhile, we looked for and bought a house in Blackburn, then on the outskirts of Melbourne but near the railway station, although I usually drove to work.

Our Children

Bobbie, in particular, was anxious to have children, ideally one boy and one girl. We had never used contraceptives but she had never become pregnant and, when we were in Melbourne just after the War, she and I had undergone medical investigations to ascertain whether there was some medical reason that could be corrected, but none could be found. Since she had had a hysterectomy in 1949, when we were in America, we decided to adopt a child. We were assisted in this choice by old army medical friends, notably Dr Stanley Williams. We adopted as a month-old baby, a girl who was born on 27 June, 1950, and named her Marilyn Aldus (Bobbie's mother's maiden name) Fenner. We never enquired about her biological parents, and she has never wished to do so. Some time later, while we were still in Melbourne, Stanley Williams came to see us again and told us that he thought we should adopt my niece, the daughter of my younger brother Tom and his first wife, Beverley (née Slaney). Beverley had died in tragic circumstances, while Tom was away with the Royal Australian Navy, but their only child (Victoria, Vicki, born 1 March, 1943) had been saved from the fire. Tom married Margaret Legge a few years later, but Stan told us that Margaret, who had two children by Tom, was treating Vicki very badly, and that we should adopt her. Tom had no objections to transferring Vicki to our care, so we followed his advice and formally adopted Vicki when she was eight years old.

Staff in Melbourne

With the limited space, only one academic appointment was made while we were in Melbourne. Stephen Fazekas de St Groth completed his term of appointment with Burnet in December 1951. Burnet did not wish to reappoint him, but I was impressed by his intelligence and dedication and offered him a Senior Research Fellowship. I obtained laboratory accommodation for him in the laboratories of the Children's Hospital, where he continued studies on influenza virus. For assistance with my studies on mycobacteria, I appointed Ronald Leach as a research assistant in 1950. He stayed with me until the embryo Department moved to Canberra. In 1951, Gwen Woodroofe and Ian Marshall were appointed as research assistants to help me with work on myxomatosis.

The Move to Canberra, November 1952

The overall plan of the permanent building had been decided when the founding professors met Florey in Oxford in July 1949, and a foundation stone was laid by Prime Minister J. B. Chifley on 24 October, 1949, to honour the memory of John Curtin, Prime Minister of Australia 1941–45. However, in the early 1950s, Canberra was struggling out of war-time restrictions on buildings. University House had been started in 1950 and was opened in 1954, but in late 1951 construction of the permanent laboratories of the John Curtin School seemed so far off that Council feared that the dispersed 'School', with departments located in Melbourne, Dunedin and London, would wither away. With the concurrence of Florey, it therefore authorized the construction of temporary laboratories. The contract was let to a local builder, Karl Schreiner, and the laboratories consisted of prefabricated wooden buildings, two being built adjacent to each other, with the overlapping eaves constituting the roof to the corridor. These were started in 1951; those for Microbiology and Biochemistry were ready for occupation late in 1952 and another double laboratory was built for Eccles and the Physiology Department in 1953. George Mackaness and a small group that had been working in Florey's department in Oxford came out as the embryo Department of Experimental Pathology to another double building in 1954. Two adjacent buildings temporarily housed the School Workshop. The permanent workshop was completed and occupied in September 1953, but it was to take another four years before the permanent laboratories would be completed.

Ian Marshall and Gwen Woodroofe moved from Melbourne to Canberra in early November 1952, to unpack and assemble the laboratory equipment that had been sent up by pantechnicon. One of my technicians, Kathleen Sutton, also moved to Canberra. Shortly after she arrived, she married Ian Marshall. Housing would have been much easier to arrange if they had married while still in Melbourne. Bobbie and I and the children moved up late in November 1952, in a Morris Minor and a Ford Prefect, to a University-owned house located at 3 Torres Street, Red Hill. Later, Bobbie's mother, who was then confined to a wheelchair, came from Perth to live with us.

Our Permanent House, 8 Monaro Crescent, Red Hill

Before we had arrived in Canberra, Eccles, Ennor and I had been provided with a block of land on which to build our houses. In contrast to the very large blocks given to Eccles and Ennor, mine was a rather small block in Hotham Crescent, Deakin. I had asked a Melbourne architect, Robin Boyd, recommended to me by Professor Brian Lewis, the ANU architect, to design my house (Serle, 1995).

A

B

Figure 5.1. The temporary buildings of the John Curtin School of Medical Research

Figure 5.1a. Two prefabricated wooden buildings were juxtaposed and a passage constructed where the adjoining roofs touched. Laboratories or rooms for experimental animals opened on each side of this passage. Figure 5.1b. The laboratories of the Department of Microbiology are on the right; there was a similar double building for Biochemistry behind a double-width coffee and seminar room on the left.

When we called for tenders, the design was so revolutionary that only one builder submitted a proposal, at a price (£25,000) that I could not afford. After living in the house in Torres Street for a few months and getting to know and like our neighbours, I eyed with interest the empty block immediately to the east of our house, on the corner of Torres Street and Monaro Crescent. All land in Canberra was on leasehold, and the law at the time was that the lease-holder had to commence building within six months of being granted the lease. The responsible authority, the Department of the Interior, told me that this block had been leased for six years, and that if I immediately surrendered my existing lease, they would transfer this lease, for Block 1 Section 3, Red Hill, to me. That done, Robin redesigned the plan for the new, much larger block, making it a single storey house, facing slightly east of north, with large windows and wide eaves to make the most of the winter sun while excluding the sun in summer. This time, Karl Schreiner, who was now constructing the permanent John Curtin School building, tendered for the building, without the heating system, at a reasonable £8,500. I signed the contract a few days before I went on my first study leave overseas, in May 1953, leaving Bobbie with any problems that might arise before I was back at the end of October.

A

B

Figure 5.2. The Fenner house at 8 Monaro Crescent

Figure 5.2a. Designed by Robin Boyd, in early 1955; the terrace at the front had been completed but the trees had not started to grow.
Figure 5.2b. The front wing of the house in 2004; the terrace is obscured by the sessile junipers, the trees are now quite large, and the garage of the extension built in 1982 can be seen on the right.

The house (Figure 5.2) was an outstanding success, both architecturally and as a place to live in. It was awarded the first Canberra Meritorious Award for Architecture in 1956 (Figure 5.3), and subsequently declared a Heritage house. There were long illustrated articles about the house in the November numbers of *The Australian Home Beautiful* and *Australian House and Garden*, and there are illustrations of it and a good description in Martin Myles (2002) website. An old school and university friend of mine, Lindsay Pryor, then Keeper of Parks and Gardens and later Professor of Botany in the ANU, designed the garden, which I still maintain, giving special attention to the large vegetable garden.

Figure 5.3. Viewing the Canberra Meritorious Award for Architecture, 1956

From left to right: Karl Schreiner (builder), Vicki Fenner, Mrs Schreiner, E. J. Scollay, (Canberra Chapter of the Royal Australian Institute of Architects), Frank Fenner, Marilyn Fenner and Bobbie Fenner.

Vicki Fenner's Death

The most tragic episode that Bobbie and I ever experienced occurred on 30 March, 1958, just after I had been elected to The Royal Society. Our youngest daughter, Marilyn, had contracted rheumatic fever and, after Dr Lorimer Dods had seen her in Canberra, she was transferred under his care to the Royal Alexandra Hospital for Children at Camperdown, in Sydney. Bobbie had gone down there to be with her. I was left in charge of the house and Vicki. One day I came from the garden into the kitchen and surprised Vicki and Catherine (Kate) Webb, the daughter of Professor L. C. Webb (Professor of Political Science in the Research School of Pacific Studies in ANU), a girl of Vicki's age who lived up Torres Street. They were handling two large kitchen knives. I thought this odd, but dismissed it. Next weekend I was in the garden picking strawberries when Len Webb came running down Torres Street and said that something terrible had happened. We went together up the street, across Mugga Way and up a lane between two houses that led up Red Hill. There lay Vicki, a rifle she had borrowed from Kate Webb beside her. Vicki had placed the tip of the barrel in her mouth and shot herself. She had left a note in her bedroom saying that 'Life is not worth living'. The only possible reason for this statement that I could think of was that she had read Neville Shute's book, *On the Beach*, which tells of the destruction of the world by nuclear war and which I had just read. There was, of course, a coronial enquiry, but they could find no other motive. Kate later became an outstanding journalist, especially during the Vietnam War. Some years later, both Professor Webb and his wife were killed in a motor-car accident.

Development of the Department in the Temporary Laboratories, 1953 to 1957

Completed in late 1952, these buildings are still in use, as offices rather than laboratories. They were quickly got into running order and I proceeded to enrol Ian Marshall as a PhD student and persuade W. K. (Bill) Joklik, whom I had supported with ANU money as a PhD student in Oxford, to join the Department—he was our first expert on the biochemistry of viruses and worked on poxviruses. Bruce Holloway, an Australian who had just got his PhD at the California Institute of Technology, came out as a Research Fellows in 1953 and John Cairns, whom I had met when we were both at Burnet's laboratory and again when I visited Entebbe, in Uganda, in 1953, was appointed as a Senior Research Fellow in 1955. Ian Marshall was appointed a Research Fellow when he graduated PhD in 1956, he and Gwen Woodroofe continued to work with me on myxomatosis. Cedric Mims, an Englishman who, like Cairns, had worked in Entebbe, joined the Department as a Research Fellow in 1956 and commenced his now classical work on the pathogenesis of infectious diseases. Fazekas continued to work on influenza virus, Holloway worked on the genetics of the bacterium *Pseudomonas aeruginosa* and Cairns on various aspects of virus multiplication. Another very important appointment made in 1956 was Alan Logie as my Senior Technician.

Other Activities, 1953 to 1957

Planning the Permanent Building

Besides getting on with our research, the three professors located in Canberra and Adrien Albert, when he came out from London, had to plan the details of their laboratories in the permanent building. In consultation with Florey, each of the existing departments had been allotted space: Microbiology was given three floor levels of the front West wing, plus a special animal house for infected animals; Experimental Pathology (Mackaness) was given the top floor of the front East Wing; Neurophysiology (Eccles) and Biochemistry (Ennor) were given one floor level each in the rear East Wing; and Medical Chemistry (Albert) was given the four floors of the rear West Wing (the lower two as a tall space designed to house production-size facilities). Other floors were left undeveloped until new departments were established and their Heads appointed.

Figure 5.4. Staff of Department of Microbiology in the temporary laboratories, June 1955

Back row (from left to right): G. Hendrick[t], R. Myky towycz[vf], D. O. White[st], B. Comben[st], W. K. Joklik [ac], S. Fazekas de St Groth[ac], B. W. Holloway[ac], I.D. Marshall[st], W. Schaffer[t], A. Winter[t].
Middle row: D. Graham[st], G. Woodroofe[ac], A. Logie (head technician), Mrs W. Joklik (secretary), E. Moore[t], W. Trusedale[t], A. Penkethman[t].
Front row: J. Cairns[ac], F. Fenner[ac], P. Gay[t], B. Renfree[t], B. Konowalow[t], F. Vagg[t], Mrs Scholes[t].
[t]= technician, [ac]= academic staff member, [st]= PhD student,[vf]= visiting fellow

Study Leave, 1953

I took my first study leave in 1953, leaving Canberra on 23 May and returning on 25 October. I travelled extensively in the USA, visiting many laboratories as well as museums and art galleries, until 23 August, when I flew to London. This was my first trip overseas since taking up my appointment in the ANU. On arrival in San Francisco, I stopped for a few days with Bill Reeves, who had visited Melbourne in 1951 to investigate the epidemiology of Murray Valley encephalitis, and got his advice on which labs to visit in the USA. According to my memory and my diary, I visited the labs of almost every virologist in the

United States, learning much and often giving seminars, usually about myxomatosis. I also attended a 4-week long course on bacterial viruses at Cold Spring Harbor, N.Y. My diary for that trip, of some 100 pages, is housed in the Basser Library archives.

I returned to Australia via South Africa, stopping off in Entebbe, in Uganda, on the way. When in Entebbe I met John Cairns again, and we went on a wonderful drive to the Ruenzorei Mountains (the 'Mountains of the Moon') and looked over the scarp of the Great Rift Valley at herds of elephants and buffalo far below. I also persuaded John to come out to Australia and join my department. He came in 1955 and stayed until 1963, when he went to the United States to become Director of the Cold Spring Harbor Laboratories.

Study Leave, 1957

I took my second study leave between 7 September and 14 December, 1957, this time travelling to England first, with a stop-off in Athens for a couple of days to see the ancient sights and the museums. Christopher Andrewes, at Mill Hill, spoke to me about the vacant chair in Virology at St Mary's Medical School, but I said that I was not interested. Besides visiting virologists in London, I had long talk with Florey, who told me that my department was the best one in the School and the only young one with a reputation in England (Eccles, of course, had a great reputation as a neurophysiologist). I spent some time with Allan Downie at Liverpool and met Keith Dumbell, later to be the world expert on variola virus. Then I attended a virology conference at Cambridge and visited the Porton biowarfare laboratories.

In Germany I visited Hannover, Tübingen, where the best virology lab in Germany was located, Paris, to see Lwoff at Institut Pasteur, and Jacotot, Vallée and Virat, all of whom were working on myxomatosis. One night I went to the Folies Bergère, which I thought were worth seeing once, but I wouldn't go again.

Next day I flew to New York and the Rockefeller Institute, where a substantial building program was in progress, with a Dome for lectures and accommodation for visitors (which I often used on subsequent visits). I had a long talk and dinner with René Dubos. He pressed me to keep up the study of myxomatosis since there was no other example of virus and host evolution in action. I then visited George Hirst, who was still working on influenza virus, and his colleagues, who were working on poliovirus and Newcastle disease virus. Over the next couple of days, I talked with colleagues of my early days at the Rockefeller: Rollin Hotchkiss, Cynthia Pierce, Dick Shope and Merrill Chase. I also spent an evening with friends Grogan and Janet O'Connell. The next day, I talked with Frank Horsfall and lunched with Jim Hirsch, then gave an Institute lecture on vaccinia genetics, which was followed by a good discussion. I enjoyed a talk with Max Theiler and his arbovirologists, then went over the road to the Sloan-Kettering.

That evening (17 October 1957), I dressed in a tuxedo to give the Harvey Lecture: 'Myxomatosis in Australian Wild Rabbits: Evolutionary Changes in an Infectious Disease'. This was followed by a pleasant dinner with many friends. During weekends I revisited the New York art museums and the Natural History Museum, and went to Carnegie Hall with the Graces (an Australian doctor long resident in New York).

As in 1953, I saw almost every virologist in laboratories in Philadelphia, Washington, Fort Detrick, Baltimore (where I met wartime malariologist Fred Bang again), Boston, Ottawa, Ann Arbor, Cincinnati (where I spent a day and evening with Albert Sabin), Urbana (where Salvador Luria, who received a Nobel Prize in 1969, advised me to use mutants of one virus for experiments on vaccinia genetics), Chicago (where Gwen Woodroofe, on study leave there, met me), Madison, Denver, Stanford University at Palo Alto (where I met David Regnery, who came out as a Visiting Fellow in 1962), San Francisco (University of California at Berkeley) and Los Angeles (where John Cairns was on study leave at Caltech, and I had long talks with him, Matt Meselson, Howard Temin and Harry Rubin). I stayed on at Caltech to attend an excellent conference on 'Animal Cells and Viruses', which was why I arrived back in Canberra just after staff had moved into the new building. At every place I stopped, I gave a lecture on myxomatosis, similar to the Harvey Lecture, except that at the University of California at Berkeley it lasted for three hours, with a ten-minute break each hour!

Travel to Indonesia, 26 July to 17 August 1956

The ANU was so small in those early days that the heads of departments in the four founding Research Schools got to know each other quite well. In particular, Bobbie and I had come to know Walter Crocker, who was the foundation Professor of International Relations in the Research School of Pacific Studies, partly because he was building a house almost opposite ours in Torres Street. However, early in 1952 he had taken leave of absence to accept an appointment as Australian High Commissioner to India and in 1954 he resigned from the University to join the diplomatic corps. In 1955 he was moved to Indonesia as Australian Ambassador, and between 1958 and 1962 he moved back to India. He published eight books, each replete with critical and penetrating comments about matters on which he had a vast international experience (Fenner, 2002).

Early in 1956, Walter Crocker invited me to lead a group of three Australian academics to examine the possibility of providing Australian assistance to establish a medical school in Sumatra. My colleagues were Sydney Sunderland, Dean of the Faculty of Medicine at the University of Melbourne, and Norrie Robson, Professor of Medicine at the University of Adelaide.

Figure 5.5. Team assembled in Indonesia at request of Ambassador Walter Crocker to investigate the establishment of a medical school in Sumatra, July–August 1956

Left to right: H. A. Nasution (Education Ministry), Prof. S. Sunderland, Prof H. N. Robson, Prof. F. Fenner, Mr S. Dimmick (Australian Embassy) and W. Tooy.

We travelled around Indonesia, visiting five of the six existing medical schools, all in Java, and travelling extensively through Sumatra and to Bali, long before it had been spoilt by hordes of tourists. It was a fascinating trip for each of us. We submitted our report on 6 September, 1956 (Fenner, Sunderland and Robson, 1956), recommending that Australia should assist with the establishment of a medical school in Bukkitingi, inland in Sumatra, but this plan came to nothing when the students there supported an uprising against Soekarno. Eventually, Australia provided aid to establish a medical school in Penang, on the west coast of Sumatra. Sunderland took a leading role in this operation.

Travel to China, 8 April to 7 May 1957

On 31 December, 1956, I was asked by Dr Leonard Cox, a neurologist in Melbourne who had a great knowledge of Chinese ceramics and had recently visited China, to join a group of some 20 Australian doctors who were invited by the Chinese Medical Association. The trip was to last a month and was all expenses paid. Most of the doctors were clinicians from Melbourne and Sydney. The two I knew best were Ted Ford, from my Army days, and Syd Sunderland, from our trip to Indonesia. We left Sydney on 8 April, and spent one day in Hong Kong (still a British colony), where we were entertained by Dr L. T. Ride (father of my Canberra friend David Ride) before taking the train to Canton. There, as everywhere else in China, we saw virtually no motor cars except those we travelled in; there were vast numbers of bicycles and many buses. After three days in Canton, during which I talked at length to the heads of the bacteriology and public health departments of the Canton Medical School, we caught the Shanghai Express but got off in Hangchow, where we went for a ride on the lake, with its many small temples, and in the evening saw the opera, 'The Peacock Flies to the Southeast'. After a day at the Chekiang Medical College in Hangchow we took the train to Shanghai.

Here we spent six days and visited both the First and the Second Shanghai Medical Schools, the Serum and Vaccine Institute (of which there were eight in China), the public health centre and the Penicillin Production Plant. We also visited one of the four schools of traditional medicine, each of which graduated about 120 students annually. In contrast, the 35 Western style medical colleges each graduated about 600 students annually.

I was asked to give a lecture at the Second Shanghai Medical School, where the Professor of Bacteriology, Dr Yu Ho, had known Hugh Ward when both were at Harvard. I planned to talk, off the cuff, about *Mycobacterium ulcerans*. However, Dr Yu had to give the talk in Chinese, so I hurriedly wrote something and he translated it overnight. Next day, I gave my talk to an audience of about 150 (compulsory attendance of all staff) and my 19 colleagues with some difficulty, because the screen for my slides was a cotton sheet that swayed in the breeze. It was followed by a talk in Chinese by Dr Yu, but in contrast to my talk, he had his Chinese audience in fits of laughter, while my colleagues went to sleep.

We were on the train all day Tuesday and arrived in Peking at 8.30 pm. We were taken to the Hotel Peking, a very large and impressive building, where we spent ten days before returning to Hong Kong. I shared a room with Ted Ford and in our spare time we usually wandered around together. We were shown around all the wonderful sights of Peking, the Temple of Heaven, the Imperial Palace, the Forbidden City, the Summer Palace, the Great Wall, the Avenue of Animals and the Ming Tombs. We spent a morning at the Peking Medical College and another at the Chinese Union Medical College (CUMC), where the Rockefeller Foundation had played a major role from 1917 to 1941. The Peking Medical College was 45 years old. It had 403 students in 1948 and 3,285 in 1957. The organization followed the pattern recommended by the Soviet Union, with five faculties: medicine (40 per cent of students), public health (20 per cent of students), dentistry and pharmacy (4-year courses with 15 per cent of the students each) and paediatrics (7 per cent of students). The CUMC had become a postgraduate institution, with 350 postgraduate students and three Professors of Microbiology, specializing in immunology, bacteriology and virology; it had an excellent library.

In parallel with the medical colleges was an Academy of Medical Sciences, which controlled four research institutes and 10 research departments. These were well funded but had difficulties in importing equipment from overseas. We were unable to visit the Academica Sinica.

May 1 was the great day of celebration for the People's Republic of China. Leading up to it, Chou En Lai hosted a cocktail party for all visitors from other countries. He gave an excellent speech and was a most impressive man. Next day there was a great celebratory May Day procession on Tienanmien Square. We had excellent seats in a grandstand just below the balcony on which Chairman Mao and his colleagues stood. For three hours there was a procession: first of children, with variegated costumes and different acts, releasing balloons or doves or waving flowers; then solid blocks of farmers and workers with floats, flowers and flags; then dancers and artists with elaborate dresses, jugglers and trick cyclists, mythical animals and Chinese dragons; and, finally, athletes in solid blocks of 600. They all gathered in Tienanmien Square and there was another great release of balloons, streamers and scrolls, and a barrage of rockets that bore parachutes with scrolls and streamers. In the evening we gathered again in the stand for a splendid display of fireworks. Altogether, this visit to China was a wonderful experience.

Travel to Japan, 8–19 May, 1957

When booking my trip to China, I arranged to take two weeks' study leave so that I could visit Japan to talk with colleagues there who were involved in aspects of virology on which I was working. I therefore stopped off in Hong Kong on the way back from China, left my heavy luggage there and flew to Tokyo. My first stop, for orientation as well as technical information, was the US 106 Medical General Laboratory, where I talked with the virologists and gave the staff a talk on medical education in China. On the evening of 10 May, I had dinner with two of the 106 lab staff and two Japanese scientists with whom I had corresponded, Dr Tamiya and Dr Kitaoka, and arranged a tentative program. The next day, Kitaoka picked me up and I went to the National Institutes of Health, where we discussed rickettsial diseases and several virus diseases, including vaccinia. On Sunday, I took the train up to Nikko, where there are some famous temples, and had a very interesting day there.

The next day, Kitaoka took me to Tokyo University, where I met relevant staff and gave a lecture on myxomatosis, with alternate paragraphs by myself, in English, and Dr Matsumoto, in Japanese. On Wednesday, I travelled to Kyoto on the rocket train. It was crowded, but the service was still very good. Kyoto has some marvellous temples, with superb gardens. Dr Dohi was working on ectromelia virus and had some interesting ideas on A-type inclusion bodies. I stayed in a Japanese style inn, which was quite different from the European style in relation to bed, bath and breakfast. My last working day was at Osaka, where I met Dr Kamahora, who also worked on ectromelia virus, at the Research Institute on Microbial Diseases. On Saturday, Kamahora (whose mother was

English) and Ishigami drove me to Nara, famous for its Great Buddha, temples and a fine museum. My Japanese hosts were very attentive with both entertainment and interesting meals, especially in the evenings. Very different from China, and a very interesting and useful two weeks.

Occupation of the Permanent Building, November 1957

By November 1957 the building at ANU was ready for occupation and each department moved into its allotted space. I had designed laboratories on the top floor with low benches, for 'sit-down' virologists, the middle (ground) floor with higher 'stand-up' benches for biochemical microbiologists, and the western end the lowest floor as a Wash-up and Media Preparation Facility for all departments in the School. Initially there was so much space that Bill Joklik occupied three labs on the ground floor.

A stone matching the Foundation Stone occupies a space on the right hand side of entrance; it commemorates the official opening of the building by then Chancellor Sir Howard Florey, in the presence of Prime Minister Robert Menzies, on 27 March 1958.

Development of the School in the Permanent Building, 1958 to 1967

With the increased space, several new Departments and Units were established during the next decade (Fenner and Curtis, 2001). Adrien Albert moved the Department of Medical Chemistry from London to his four-storey laboratories in Canberra; over the next decade his academic staff increased from seven to 15 and his PhD students from one to 12. A Department of Physical Biochemistry was established in 1959, with Alexander (Sandy) Ogston, from Oxford University, as Head; its academic staff grew from three in 1959 to a maximum of nine in 1965, and PhD students from one in 1959 to five in 1966. In 1964, David Catcheside, an Englishman who had been Professor of Genetics in the University of Adelaide from 1952 to 1954, and then returned to England, accepted the Chair of Human Genetics. In 1967, he became the Foundation Director of the newly established Research School of Biological Sciences in the ANU, but in the three years he was in the John Curtin School, the academic staff increased from two to six and the number of PhD students from one to nine. The establishment of a Department of Clinical Science was accepted by the University Council and the Canberra Hospital Board in 1964, but the Foundation Professor, Malcolm Whyte, was not appointed until 1966.

In addition, two small units were set up. The Biological Inorganic Chemistry Unit, headed by Francis Dwyer, was established in 1959 and continued until 1966, although Dwyer died in 1961. Although electron microscopy had been available in the School since 1955, an Electron Microscope Unit, headed by Edgar Mercer, was established in 1963 and greatly strengthened this service.

The other four existing Departments: Biochemistry, Experimental Pathology, Physiology and Microbiology, also increased their staffing, PhD student numbers and output of publications.

Expansion of the Department of Microbiology, 1958 to 1967

With the increased space and adequate block funding, there was a substantial increase in academic and general staff and PhD students. Bruce Holloway left before we moved, but all the others in the temporary laboratories moved across in November 1957. David Howes was appointed as a Research Fellow, December 1957 to April 1960; Graeme Laver, working on the biochemistry of influenza virus, Research Fellow in 1958, Fellow in 1962 and Senior Fellow in 1964; and Gwen Woodroofe, having graduated PhD, was appointed a Research Fellow in 1958 and promoted to Fellow in 1963. Two overseas scientists, Dennis Lowther and Fritz Lehmann-Grube, were appointed as Research Fellows for three year terms in 1960. Peter Cooper was appointed as a Senior Fellow in 1962 and carried out classical work on poliovirus genetics; on graduation as PhD in 1962, Kevin Lafferty was appointed as a Research Fellow, he transferred to the Department of Immunology in 1964; Alan Bellett, from England, and Ken Easterbrook, when he graduated PhD, were appointed Research Fellows in 1963; and Rob Webster and Joe Sambrook, on getting their PhDs in 1964 and 1965, respectively. In 1965, Stephen Boyden, then a Professorial Fellow in the Department of Experimental Pathology, transferred to my department because I was sympathetic to his wish to broaden his outlook from immunology to human biology. He carried out an important study of the 'metabolism' of Hong Kong and, in 1970, he transferred to the newly-established Department of Human Biology, after I had become Director of the School. He has remained a lifetime friend. In 1966 Nigel Dimmock, from England, was appointed Research Fellow for five years and in 1967 Adrian Gibbs, a plant virologist from England, joined the Department as a Research Fellow.

The other major change possible after moving to the new building was the possibility of bringing established scientists from overseas as Visiting Fellows for periods of a year or more. Some funds for travel costs were available from the ANU, and all departments, especially Microbiology and Physiology, profited from their presence. No fewer that 13 Visiting Fellows came to our Department for at least one year during the period 1958 to 1967, seven from the United States, two from Czechoslovakia, and one each from England, Finland and India. In addition, Alfred Gottschalk, FAA, a biochemist whom I had known from the Hall Institute, came to my department when he retired in 1958 and stayed, as an NHMRC Senior Research Fellow, for four years. He carried out important work on the neuraminidase of influenza virus.

The number of PhD students graduating also increased substantially, from one in 1956 and 1958 to three in 1960, 1961, 1962, 1963 and 1964, one in 1965, two in 1966 and one in 1967.

Figure 5.6. Academic staff, visiting fellows and students of the Department of Microbiology in 1962, on the roof of Infected Animal House, with Black Mountain in the background

Back row (from left to right): Mary McClain[vf], Rob Webster[st], William Murphy[st], Tom Grace[st], John Roberts[st], Alfred Gottschalk[vf], Betty Ermacora (secretary), Frank Warburton[st], Ric Davern[vf], Stephen Fazekas[ac], Royle Hawhes[st], Allan Logie (head technician).
Front row: Ron Weir[st], Rima Greenland[vf], Brian McAuslan[st], Bill Joklik[ac], Cedric Mims[ac], Gwen Woodroofe[ac], Frank Fenner[ac], Fritz Lehmann-Grube[ac], John Cairns[ac], Ian Marshall[ac], Dennis Lowther[ac].
[t]= technician, [ac]= academic staff member, [st]= PhD student., [vf]= visiting fellow

School Governance

Since 1953, when staff moved to the temporary laboratories, Hugh Ennor had acted as chairman of a committee, comprising the heads of departments, which ran the School. As such, he acted as the primary channel of communication with Florey. In 1957, the University Council formally endorsed this 'School Committee', chaired by Ennor as Dean, as the governing body of the School. With minor concessions to other academic staff, this structure of governance continued until 1967, when a Committee of Council recommended that the John Curtin School should adopt the Faculty/Faculty Board structure that had been adopted as early as 1954 in two other Research Schools. The authors of the history of the first 50 years of the ANU (Foster and Varghese, 1996) comment on the early arrangements: 'the notion of God Professor had extended beyond Olympian heights...when there were rumblings among "other ranks" for regular "academic staff meetings" and the creation of a faculty structure...Ennor, supported by most other members of the School Committee, firmly resisted [these suggestions].'

Other Activities, 1958 to 1967

This decade was the apogee of my career as a bench scientist. Details of my experimental studies during this period are summarized in Chapter 6 and the work of the Department of Microbiology during this decade is well described in a paper by Joklik (1996), which concludes with the comment, referring to the Department of Microbiology and the Laboratory of Cell Biology, US National Institutes of Health: 'The two Departments I have chronicled could hardly be further apart on our globe. But both yielded an extraordinarily rich harvest in discoveries, on the one hand, and, on the other, in scientific alumni; groups that are highly diverse in national origin and culture, but all dedicated to the highest principles of the pursuit of knowledge.'

Travel to India, 2 December 1960 to 17 January 1961

Walter Crocker, who in 1958 had moved back from his position as Australian Ambassador to Indonesia to become, for the second time, Australian High Commissioner to India, arranged for me to go to India under the Colombo Plan. My mission was to visit virology laboratories throughout the country, meeting staff, giving lectures and attending the annual conference of the Indian Virological Society. I had a wonderful trip all around India, by car and plane. When in New Delhi I was the guest of Walter and Claire Crocker and attended several of their delightful dinners. One practice that I noticed was that as soon as the meal was finished, Walter would go to a side room with the various guests with whom he wished to have a serious talk and Claire would look after the other guests, male and female.

My first official visit was to Hyderabad, where over a period of five days I attended the annual meeting of the Indian Council for Medical Research (ICMR), gave a lecture on Genetics of Animal Viruses and talked at length with many medical scientists, including long sessions with Dr C. G. Pandit, Secretary of the ICMR. Then I went to Aurangabad, where I met up with an American couple, a Mexican and an Israeli woman and shared a taxi with them to the Ajanta Caves, 70 miles away. Next day we went to Ellora Caves. Two most enjoyable and interesting days. I then flew down to Trivandrum, in Kerala State, where after looking around the Art Gallery and some of the temples in the morning I went to the Medical College and looked specifically at the level of training of staff and quality of equipment, as matters on which the Colombo Plan might be able to help.

Throughout my travels, I was taken to see local museums, art galleries and temples, a wonderful experience for someone who had never been to India before. My official duties took me in turn to the Pasteur Institute of South India in Coonoor, the Christian Medical College in Vellore, to Bombay, for three days, where I visited the Haffkine Institute, the Grant Medical College and the Indian Cancer Research Centre. I spent three days in Poona, then the major centre of virology in India, giving three different lectures at the Virus Research Centre, the Poona Medical College and the Armed Forces Medical College. In Calcutta I had the very interesting experience of staying with the renowned British scientist J. B. S. Haldane and the statistician P. C. Mahalanobis, as well as visiting and giving lectures at the Infectious Diseases Hospital, the Calcutta School of Tropical Medicine and the Institute of Postgraduate Medical Education and Research. By chance, Mac and Linda Burnet were just coming home from the Nobel Prize ceremonies in Stockholm and we met briefly in Calcutta. Then a week in New Delhi, where I stayed with the Crockers part of the time and lectured at the All India Institute of Medical Sciences, the Maulana Azad and the Lady Harding Medical Colleges and the Patel Chest Institute, at Delhi University. As well as talking with scientists at all of these institutions, I had long interviews with Drs Patel and B. D. Larcia, a member of the Indian University Grants Commission.

During the final week I went to the Institute for Postgraduate Medical Education and Research and Patiala Medical College in Chandigarh, the Central Research Institute in Kasauli, the Agra Medical College and Lucknow University, with side trips to the temples of Patipah Sikri and the sacred sites at Banaras. After I returned home, I wrote a formal report on the trip for the Department of External Affairs (Fenner 1961), the main headings being Medical Education and Research, Weaknesses in Medical Education and Research, Virus Research in India, Postgraduate Education of Indians in Australia and Organization of Visits by Medical Scientists.

While in India, I had arranged to spend a few days in Thailand and then go to Siemrep in Cambodia to look over Angkor Wat (20–23 January). There were very few tourists at Angkor Wat. I wandered around and sometimes used a hired motor tricycle. The whole place, especially the carvings on Angkor Wat and the delicate *apsaras*, were wonderful.

Overseas Fellow, Churchill College, Cambridge, 3 November 1961 to 30 October 1962

I was elected to The Royal Society of London in 1958, and in 1960 I was invited to give the Leeuwenhoek Lecture there. I decided to take a full year's study leave and to take Bobbie and Marilyn with me. It was the only full year's study leave that I ever took. At the time, Sir John Cockcroft, the first Master of Churchill College, was being considered as Chancellor of the ANU, a position in which he served from 1961 to 1965. Presumably because of this association, I applied for a Visiting Fellowship in Churchill College, then in its infancy. I found from the *Churchill Review*, Volume 37, 2000, that although a Visiting Fellow for only one year, I was nevertheless a Founding Fellow of Churchill College.

We put the car on blocks and rented our house to Joe and Sue Johnson (I ascertained that £35 a week was a 'fair rent'). I stayed in University House from 9 September, 1961, and Bobbie and Marilyn went to Adelaide and stayed with my mother. I visited Sydney and Melbourne, saw many old friends and did practise runs of my Lecture in both places, before flying to the United States, arriving in San Francisco on 1 October, 1961. I spent about a month crossing the States, visiting Stanford and the University of California at Berkeley, Seattle for a few days, New York and Princeton, where I spent three days at a Macy Foundation Symposium, Johns Hopkins University, Chapel Hill in North Carolina, Cornell, Harvard and Yale Universities before flying to London on 31 October. Then, on 2 November, I attended a symposium at The Royal Society on Mechanisms of Viral Infection, at which I gave a short paper, and in the afternoon I was admitted as a Fellow and signed the ancient Charter Book. Next day I went to Cambridge and settled into our flat there, one of the very few permanent brick buildings in Churchill College at the time—George Steiner and his family was in the other. My first job was to buy a small car, we had decided on a blue VW Golf.

Bobbie and Marilyn flew to England in mid-November and I collected them from London airport in the new car. With advice from local women, we arranged for Marilyn to go to a small primary school close to Churchill College.

I went to the laboratories of the Department of Pathology but found them much inferior to my own labs in Canberra for work with cultured cells. I therefore decided to use most of my time investigating myxomatosis in Britain and in continental Europe. The story of this is set out in the next chapter, in the description of the production of the book *Myxomatosis* by Fenner and Ratcliffe (1965).

We enjoyed Cambridge, especially the walk from Churchill to the city in spring. We also drove all over England and Scotland. One notable trip was over the Christmas break, when, on a day when there was hardly any other car on the snow-covered road, I drove across to Birmingham, where David Catcheside was Professor of Microbiology. He was being considered for the Chair of Genetics in the John Curtin School. On behalf of the Vice-Chancellor, I discussed the idea of a Research School of Biological Sciences in the ANU with him. Subsequently, he accepted the Chair of Genetics and concurrently acted as Advisor for the Research School of Biological Sciences. In 1967 he became the first Director of this School.

I visited many laboratories all around Britain. One memorable visit was to the Laboratories of the Chester Beatty Research Institute at Pollard Woods, where Sydney graduate Jacques Miller was in the process of discovering the immunological function of the thymus gland, by delicate operations on newborn mice.

Visiting Fellow, Moscow State University, 27 February to 28 March 1964

In May 1963 Ross Hohnen, the ANU registrar, made arrangements with Pro-Rector K. I. Ivanovic of Moscow State University (MSU) for selected ANU senior staff to go to there as Visiting Fellows for a period of six weeks. He told me that the Pro-Rector had issued an invitation for me to go this year. I had already arranged to spend most of February on study leave at conferences in USA, and felt that I could not spend six weeks in Moscow, but agreed to go for a month. I spent a week in London and then flew to Moscow on February 27 and was met by Dr V. Agol, head of animal virology at MSU and Miss Galya Lipskaya, a recent graduate in biochemistry who acted as my guide and interpreter. I realized why they had invited me when I was introduced to Academician Belozersky, Professor of Plant Biochemistry, who was in charge of the development of a new Institute of Molecular Biology at MSU; a new building was almost completed in the grounds of MSU and the Department of Virology was to be part of this Institute.

All expenses were covered by MSU, which also arranged visits to museums and art galleries and evenings at the opera and ballet, in both Moscow and Leningrad. I stayed at the Budapest Hotel in Moscow and the Hotel Europe in Leningrad. Both were centrally situated, comfortable and quiet, and in both of them service in the restaurant was unbelievably slow and the food mediocre. I maintained correspondence with Agol and Lipskaya for several years. My intention of meeting them again at the International Virology Congress in Moscow in 1966 was frustrated, because I had an attack of appendicitis. However, I was able to meet Agol several times on later trips to Moscow.

There were several Research Institutes of the USSR Academy of Medical Sciences on the outskirts of Moscow. I visited several of these and usually gave a lecture in each and two in MSU. I could talk with the younger scientists directly in English, which they spoke remarkably well considering none had ever been outside the USSR. Lectures had to be spoken in paragraphs that were translated, and usually took over two hours, but I found that by speaking without a text and talking slowly and clearly, many members of the audience understood what I was saying and the task of the translator was simplified. Facilities for slide projection were often poor and some of the 'blackboards' and 'chalk' (lumps of stone) impossible.

Galya and I went up to Leningrad by train, in the same sleeping cabin ('equality of the sexes', she said). I remember clearly the seemingly endless birch forests. We went to the ballet the evening after we arrived: Prokoviev, *Romeo and Juliet,* very good. Two excellent art galleries, the Winter Palace with a great collection of classic paintings, Rembrandt, Titian, Rubens and a host of others, and next day to The Hermitage, a marvellous collection of French impressionists up to about 1914. Ballet again that evening: *Spartacus,* music by Katchuchurian. I don't think that it is commonly on display outside Russia. The opera next evening, Rimsky-Korsakov's *Skopsve Maiden,* about the time of Ivan the Terrible, was splendid.

Back in Moscow, I had lunch with the Australian Ambassador and his wife, Mr and Mrs Jamieson. The next day I went to the Institute of Experimental and Clinical Oncology, which has a Laboratory of Virology, with George Svet-Moldavsky in charge. After a splendid lunch with caviar and sturgeon, we had a long conversation on Burnet's character, not so much as a scientist but more generally (he was a local 'god'). Dinner the last evening was with Academician Belozersky and a large number of guests, in a large flat in a large palace of MSU. Much vodka and wine was imbibed, but as I found after I had got back to Canberra, also a dose of *Giardia*. I had an interesting trip back, over the Himalayas from north to south, to arrive in New Delhi.

I produced a ten-page report on my visit, comprising a general overview of the arrangements, an outline of visits to scientific institutions and lectures, comments on the structure of their medical course and the virology institutes operated by the Academy of Medical Sciences, research facilities and availability of modern equipment, their problems in getting chemicals used for tissue culture, and the potential for continuing exchanges of staff, both ways.

Prehistorians' Pilgrimage, 15–30 January 1971

The Pan-Pacific Science Congress was held in Canberra over the 1970–71 Christmas–New Year break. Because of my interest in environmental problems, I attended that congress. Another interested group that held their meeting at the same time was the Far-Eastern Prehistory Association. Because of my early interest in Aboriginal prehistory, I was invited to take part in an archaeological tour of south-eastern Australia. A group of some 40 prehistorians, including all those knowledgeable about Aboriginal prehistory, embarked on a bus tour that included most of the sites of interest in southeastern Australia, including northern Tasmania. On the bus trip down from Canberra to Melbourne, John Mulvaney spoke at length and explained the background, and at each site that we stopped an expert explained the special features of anthropological sites of interest. I found it a most fascinating experience, especially seeing sites in Tasmania, as well as Kow Swamp, Lake Mungo and the rock shelters at Devon Downs, on the River Murray, which I had visited as a university student.

Papua New Guinea, 1962 to 1973

Sir Macfarlane Burnet had a special interest in Papua New Guinea, because his son, Ian, had been a patrol officer there. In 1962, he responded to a request from the Australian Government to act as Chairman of a new committee, the Papua-New Guinea Medical Research Advisory Committee. He asked my close friend, Bob Walsh, to be Secretary, and me to be a member of this committee. In 1966, it was replaced by Council of the Institute of Human Biology, and we all continued to serve on this until an indigenous committee took over in 1973. These meetings involved two annual trips to Papua New Guinea. Sometimes we met in Port Moresby, but usually, when the Institute was set up, in Goroka, where the Institute (later named the Papua New Guinea Institute of Medical Research) was located. Sometimes we travelled to the Highlands. One memorable visit was when Lord De Lisle, Governor-General of Australia, visited Mount Hagen. He travelled around standing in an open car, a tall figure dressed in a white uniform and with a large feather in his hat. Around the show ground danced myriads of New Guineans, short in stature, naked except for skirts, body and face brightly painted and with magnificent bird of paradise plumes on their heads.

Honours and Awards, 1954 to 1967

1954	The Australian Academy of Science was granted its Royal Charter in February 1954; Eccles was a Founding Fellow and Ennor and I were in the first group of elected Fellows.
1957	Harvey Lecturer, Harvey Society of New York.
1958	Fellow of The Royal Society of London.
	Listerian Oration, South Australian Branch of the British Medical Association.
1959	Fellow of the Royal Australasian College of Physicians.
	Walter Burfitt Medal, Royal Society of New South Wales.
1960	Foundation Fellow, Churchill College, Cambridge.
1961	Leeuwenhoek Lecturer, The Royal Society of London.
1964	Mueller Medal, Australian and New Zealand Association for the Advancement of Science.
	Honorary MD, Monash University.
1967	Matthew Flinders Medal and Lecture, Australian Academy of Science.

References

Burnet, M. 1971, *Walter and Eliza Hall Institute 1915–1965*. Melbourne University Press, pp.48–9.

Crocker, W. R. 1971, *Australian Ambassador: International Relations at First Hand*, Melbourne University Press, Carlton.

Fenner, F. 1961, Report on Visit to India by Professor Frank Fenner, FRS, December 1960–January 1961, Basser Library Archives 143/8/5E3.

Fenner, F. 1964, Study Leave Report on visit to Moscow State University, 1964. Basser Library Archives 143/8/5D4.

Fenner, F. 1995, Cecil John Hackett, *Medical Journal of Australia,* vol. 163, p.441.

Fenner, F. and Curtis, D. 2001, *The John Curtin School of Medical Research: The First Fifty Years,* Brolga Press, Gundaroo, 565 pages.

Fenner, F., Sunderland, S. and Robson, H.N. 1956, *Report on Medical Education in Indonesia,* Basser Library Archives 143/8/5E1A.

Foster, S. G. and Varghese, M. M. 1996, *The Making of the Australian National University,* Allen and Unwin, St Leonards.

Miles, M. 2002, Fenner house, 8 Monaro Cescent, Red Hill. (http://www.canberrahouse.com.au/profiles/8monaro.html)

Serle, G. 1995, *Robin Boyd. A Life,* p.133, The Miegunyah Press, Melbourne.

Chapter 6. Professor of Microbiology, John Curtin School of Medical Research, 1949 to 1967: Research

Research in Melbourne, February 1950 to November 1952

As mentioned in the previous chapter, The Australian National University had arranged with the Director of the Walter and Eliza Hall Institute, Sir Macfarlane Burnet, to provide me with two laboratories, on the same floor as his laboratory, for as long as it took to provide laboratories in Canberra. I worked in the room previously occupied by gifted research worker Dora Lush, who had died in 1943 from scrub typhus contracted during her work (Burnet, 1971).

Molecular biology was unknown in the early 1950s, and although I wanted to get back to virology, I thought that I had skimmed the cream from the study of ectromelia virus. Subsequently it was used by several groups in the John Curtin School as their model virus disease and they continue to use it in studies of molecular virology. At first, Burnet suggested that I might like to take over the field that he had been working on, the genetics of influenza virus. However, I did not want the work of my new department to be too closely associated with someone as distinguished as Burnet, so I did not take up his offer.

Studies on *Mycobacterium tuberculosis* and *Mycobacterium ulcerans*

Initially, I carried on working with *Mycobacterium tuberculosis*, the major work being a long review article on the vaccine strain, BCG (Fenner, 1951). With a research assistant, Ronald Leach, I also continued laboratory studies of tubercle bacilli and began serious studies of the 'Bairnsdale bacillus'. The first use of the name *Mycobacterium ulcerans* for this mycobacterium (since then the official name) appears as a footnote in my Inaugural Lecture, given in Canberra on 17 August ,1950, as a new professor in The Australian National University (Fenner, 1950). This organism is characterized by its low ceiling temperature, 34°C (meaning that it will not grow, in culture or *in vivo*, at higher temperatures). It is now known to have a world wide distribution, and is particularly common in tropical Africa, where the disease is known as Buruli ulcer (Asiedu et al., 2000). It is a very interesting organism, not least because the severe skin ulceration is due to a soluble toxin, previously unknown among mycobacteria. Initially, with Leach, I showed that it was antigenically distinct from other mycobacteria. The feature which I investigated in detail, in parallel with my work on myxomatosis over the period 1952 to 1957, was the relation between its ceiling temperature and its pathogenicity (Fenner, 1956). This investigation was made more interesting by comparison with another mycobacterium which was also temperature-sensitive and also produced skin lesions in humans, named *Mycobacterium balnei* by two Swedish workers (Linell and Nordén, 1954). I had a long correspondence with Åke Nordén, before and after I had visited him at the University of Lund in 1953. In mice, both bacteria produced severe skin lesions when inoculated in the manner used in studies on ectromelia, i.e., in the footpad. *M. balnei,* which grew rapidly in culture, produced progressive lesions within 4 days which became very severe within 9 days. On the other hand, *M. ulcerans,* which grew as slowly as tubercle bacilli in culture, did not produce progressive lesions until the fourth week, and they became severe by 7 weeks. However, because of their low ceiling temperature, neither organism produced visceral lesions after intranasal, intraperitoneal or intravenous inoculation, but after a moderate interval in the case of *M. balnei* and long interval in the case of *M. ulcerans,* ulcerating lesions developed on the hairless peripheral parts of the body and on the scrotum. The low ceiling temperature is clearly the reason that in experimental animals, as well as in humans, the lesions are restricted to the skin. *M. balnei* is a saprophyte which is associated with water, sometimes water in swimming pools. *M. ulcerans* also appears to be associated with swamps; cases seem to be more common after disturbances to the water environment. According to his paper (Shepard, 1960), one interesting result of our work was that it led him to successfully exploit footpad inoculation as a way of growing leprosy bacilli in mice. Ron Leach did not come up to Canberra, and I was solely responsible for the later work, described in the 1956 reference.

The History of Myxomatosis

Myxomatosis constituted the major part of my personal research between 1952 and 1967. To put it in perspective, I will begin with a very brief outline of its history, which is covered in detail in Fenner and Fantini (1999). Myxomatosis was first recognized as a virus disease when it killed European rabbits *(Oryctolagus cuniculus)* in Guisseppe Sanarelli's laboratory in Montevideo, Uraguay, in 1896. In 1911, workers in the Oswaldo Cruz Institute in Rio de Janeiro observed the disease in their laboratory rabbits and correctly classified the causative agent as a large virus. Henrique de Beaurepaire Aragão, working at the Oswaldo Cruz Institute, showed that it could be transmitted mechanically by insect bite. In 1942, he showed that the reservoir host in Brazil was the local wild rabbit, *Sylvilagus brasiliensis,* in which the virus produced a localized nodule in the skin (Figure 6.1B). Knowing that the European rabbit was a major pest animal in Australia, and impressed by the lethality of the disease in these rabbits (Figure 6.1A), in 1919 Aragão wrote to the Australian government suggesting that it should be used here for rabbit control, but the quarantine authorities would not permit its importation.

The idea was revived by Jean Macnamara, a Melbourne paediatrician who had worked with Macfarlane Burnet and thus had an interest in virus diseases. In 1934, she went on a world tour to investigate poliomyelitis, which was her main professional interest. In America, she visited the laboratory of Richard Shope, in the Princeton branch of the Rockefeller Institute. He was investigating a tumour in local cottontail rabbits *(Sylvilagus floridanus)*, which he showed was caused by a poxvirus related to myxoma virus. He called it fibroma virus. At the time there was an epizootic of myxomatosis in domestic European rabbits *(O. cuniculus)* in California, which was later found to have a different reservoir host *(Sylvilagus bachmani)*. Shope found that fibroma virus would protect laboratory rabbits against myxomatosis. Learning of this fatal rabbit disease, Macnamara wrote to the Australian High Commissioner in London asking him to help her convince the Government to use the virus for rabbit control.

Francis Noble Ratcliffe

Born in Calcutta in 1904, Ratcliffe studied zoology at Oxford. In 1928, he came to the notice of the London representative of the Council for Scientific and Industrial Research (CSIR), and this led to his invitation to come to Australia as Sir David Rivett's 'biological scout', to study flying foxes and erosion in arid lands, as a result of which he produced a classic book, *Flying Fox and Drifting Sand*. He returned to Britain in 1932 as Lecturer in Zoology in Aberdeen, but was invited back to Australia as a scientific adviser to the CSIR Executive in 1935. In 1937, he was transferred to the Division of Economic Entomology to work on termites. In 1942, he joined the Australian Army and served with distinction as Assistant Director of Entomology. Since I was serving in New Guinea as a malariologist at that time, I saw quite a lot of him then. After demobilization he served briefly as assistant to the Chief of the Division of Entomology, but in 1948 he was appointed Officer-in-Charge of the newly created Wildlife Survey Section of CSIR. Initially he had to work on rabbit control, and after some disappointments succeeded in introducing myxomatosis. Study of this disease preoccupied the Section for several years, but later he was able to broaden studies of the biology of the rabbit and introduce biological studies of native animals as an important part of the work of the Section, which by then had been expanded to the Division of Wildlife and Ecology. He retired from CSIRO in 1969. He played a major role in setting up the Australian Conservation Foundation in 1964, and devoted a great deal of time to its expansion to become Australia's peak environmental non-government organization, until he had to retire for health reasons in 1970 (see Coman, 1998; Mackerras, 1971).

A

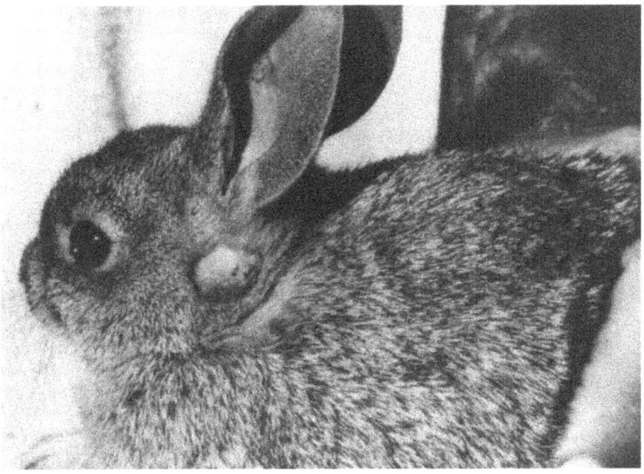

B

Figure 6.1. European and Brazilian Rabbit

Figure 6.1a. European rabbit *(Oryctolagus cuniculus)* 10 days after infection with the Standard Laboratory Strain of myxoma virus.
Figure 6.1b. Brazilian rabbit *(Sylvilagus brasiliensis)* three weeks after infection with the Standard Laboratory Strain of myxoma virus.

The Chief Quarantine Officer was again very reluctant to allow its importation, but allowed scientists in CSIR (which was transformed into the Commonwealth Scientific and Industrial Research Organization, CSIRO, in 1949), to test its species sensitivity against a wide range of domestic and native animals; they found that it infected only European rabbits. Several field trials were carried out, in dry inland areas, but the virus died out. Then came World War II, and in 1943 all investigations were stopped.

With so many country boys in the army, rabbit control, such as it was, had been neglected throughout the period 1939 to 1945, and by 1946 rabbits had increased to unprecedented numbers. Jean Macnamara (now Dame Jean) wrote articles in the rural press highly critical of CSIR/CSIRO for not proceeding immediately to try myxomatosis for biological control of the pest. In 1948, a CSIR/CSIRO scientist, Francis Ratcliffe, was appointed Officer-in-Charge of the newly-established Wildlife Survey Section, but instead of studying the native fauna, Ian Clunies Ross, Chairman of the newly-formed CSIRO, insisted that he should first try out myxomatosis. Several field trials failed, but in the Christmas–New Year period of 1950–51 the disease escaped from one of the four trial sites in the Murray valley and spread all over the Murray-Darling basin, killing millions of rabbits.

Research on Myxomatosis, February 1951 to November 1952

Stimulated by a remark by my friend Professor Hugh Ward, I decided to make myxomatosis the main topic of my research. From Burnet's diaries, I later found that the day I made this decision was 1 February, 1951. I promptly made contact with Ian Clunies Ross, who was delighted to have a virologist working on this topic, for at the time there were no virologists in CSIRO. I also met with Francis Ratcliffe, who was in charge of the new Wildlife Survey Section of CSIRO, and recruited Gwen Woodroofe, MSc, from the Department of Microbiology of the University of Adelaide and Ian Marshall, a new BAgSc graduate of the University of Melbourne, initially as research assistants. Gwen and Ian worked with me on myxomatosis until about 1960. Gwen finished up as a Fellow and Ian as a Senior Fellow, both working on arboviruses.

Gwendolyn Marion Woodroofe

Born in 1918, Gwen graduated BSc at the University of Adelaide in 1940 and gained an MSc degree in bacteriology for work on salmonellosis. In 1951, she joined my department as a research assistant, to work on myxomatosis. She later became a Research Fellow, graduated PhD and was promoted a Fellow until she retired in 1978. She and Ian Marshall were my collaborators in laboratory studies of myxomatosis between 1951 and 1966. She then went to work with Ian on arboviruses. On her retirement, she became very active in work for UNICEF, for which she was awarded a Medal in the Order of Australia in 1997. After my wife Bobbie died in 1995 and I moved into the extension to our house, Gwen became one of several friends who now come around for a drink and a chat on weekend afternoons.

Ian David Marshall

Born in 1922, Ian Marshall served in the Royal Australian Navy in the World War II. After discharge, he graduated BAgSc in the University of Melbourne in 1951 and immediately joined my department to work on myxomatosis. He participated in fieldwork carried out by our CSIRO colleagues and also in laboratory studies. He graduated PhD in 1956, later becoming a Research Fellow, Fellow and Senior Fellow, before formal retirement in 1987. He played a major role in virological investigations of myxomatosis between 1951 and 1959, when he went to work with arboviruses with Bill Reeves at the University of California at Berkeley for two years. While in California he also collaborated with David Regnery of Stanford University in classical studies of the epidemiology of myxomatosis in California, where the host is a different species of *Sylvilagus (S. bachmani* rather than *S. brasiliensis)* and the virus also differs significantly from that found in South America.

When he returned to the John Curtin School, he established an arbovirus laboratory, which became one of the major Australian centres of arbovirus research, on which he continued to work full-time long after his retirement.

The climatic conditions at the time of the outbreak of myxomatosis in the Murray-Darling Basin had been such that there was also an outbreak of encephalitis in that region, similar to X-disease, described by Cleland et al., (1918). As soon as this outbreak was reported, Burnet instructed two of his staff to investigate it, one (Gray Anderson) looking at the epidemiology and the other (Eric French) attempting to isolate the causative virus from the brains of fatal cases. French (1952) was successful in isolating the virus and showed that it was very similar to the one that caused Japanese encephalitis. Burnet made this information widely available to the Government and the press, but the claim that the encephalitis was due to myxoma virus was widely voiced and the Chairman of the Mildura Hospital Board challenged Burnet and R. G. Casey (Minister in charge of CSIRO) to test the harmlessness of myxoma virus on themselves. Burnet consulted me and we decided that it was absolutely safe. I prepared a suspension and injected Burnet subcutaneously with one, 10 and 100 rabbit-infectious doses; Burnet inoculated me. As soon as he heard of this, Clunies Ross said that, as Chairman of CSIRO, he was responsible for the use of the virus, and therefore he should be included in the tests, so I injected him. None of us showed lesions or an antibody response, Casey announced the results of the tests in Parliament and the public was reassured (Burnet, 1968).

Clunies Ross financed the building of a new animal house at the Hall Institute for holding the large numbers of rabbits that I expected to use. When I went to

Canberra, the animal house was initially made available to Geoff Douglas, who was in charge of the myxoma virus inoculation campaigns in Victoria. Later, when Douglas had established the Keith Turnbull Research Institute in Frankston, where continuing research on changes in virus virulence and genetic resistance of rabbits was carried out, the Hall Institute animal house building was converted into laboratories. I also immediately made contact with Ratcliffe's team: Ken Myers, Bernard (Bunny) Fennessy, Alan Dyce, Roman Mykytowycz, Bill Poole and later Bill Sobey, with whom I continued to collaborate over the next 15 years. On one of my frequent trips to Canberra for meetings with the ANU administration, I made contact with Max Day, of the CSIRO Division of Entomology, who was working on insect transmission of plant viruses, and some of my first work, started in 1951, was to collaborate with him on insect transmission of myxomatosis. Other work carried out while I was still in Melbourne involved collaboration with a CSIRO electron microscopist on a comparison of the morphology of myxoma and vaccinia viruses and comparison of the pathogenesis of myxomatosis with that of mousepox.

Research on Myxomatosis, 1953 to 1967

Gwen Woodroofe came up to Canberra a few weeks before I did, and Ian Marshall a little later, and they had the laboratory equipment organized within the new temporary laboratories, in which I had arranged to have a large animal house for infected rabbits. I resumed work on myxomatosis as soon as Bobbie and I had settled into the house that I had rented from the ANU. Our studies had two components: Gwen and I worked on various aspects of the disease that could be studied in the laboratory; and Ian worked in the field with the CSIRO scientists and in the lab with me, our basic interest being the evolution of virulence of the virus and the genetic resistance of the rabbit. Each of them also did some independent work related to their PhD requirements.

In the laboratory, we studied the active immunity conferred by previous infection (in the rare rabbit that recovered) and following vaccination with fibroma virus, and passive immunity in kittens borne by immune does. Gwen collaborated in the later work done with Max Day on insect transmission and Ian and I wrote two long papers on the topics of major importance in considering the evolution of virus and host (Fenner and Marshall, 1957; Marshall and Fenner, 1960). There were also some important papers that involved close collaboration with us in the laboratory and the CSIRO scientists in the field (Myers et al., 1954, Fenner et al., 1957).

Another aspect of the work on myxomatosis followed my first study leave in 1953, when I met Harry Thompson and learnt more about myxomatosis in England. The virus that spread through Europe (the Lausanne strain) was more virulent that the one used in Australia, and in 1957 it was introduced in Australia.

In 1961–62, during study leave as an Overseas Fellow of Churchill College, Cambridge, I spent most of my time investigating myxomatosis in Europe.

Harry V. Thompson

Born in 1918, Harry Thompson graduated in zoology at the University of London in 1940 and then worked at the Bureau of Animal Population at the University of Oxford, where he came under the influence of the famous British ecologist, C. S. Elton, with whom Francis Ratcliffe also worked when on leave in England in 1948. In 1946, Thompson joined the Ministry of Agriculture, Fisheries and Food, where he became head of the department dealing with wild animals and birds affecting agriculture. Inevitably he became interested in rabbits, and he was at the forefront of work on myxomatosis in Britain. In 1959, he set up the Ministry's Worplesdon Laboratory at Guildford, Surrey, and remained its Director until retiring in 1982 to become a private consultant. Besides serving on most committees dealing with rabbits and myxomatosis and numerous national and international bodies concerned with wildlife and conservation, Thompson published numerous scientific papers and two important books on the European rabbit, in 1956 and 1994. Over the period 1952–65, I always tried to see him when I went to England, and after he retired I maintained a steady correspondence with him.

Deliberate spread of myxomatosis was made illegal in Britain in 1954. Nevertheless, the disease spread all over the country. A meeting with Paul Chapple, an English virologist who was working on myxomatosis, led to a study of the evolutionary changes in the Lausanne strain of the virus between 1954 and 1962 (Fenner and Chapple, 1965). I also visited scientists involved with myxomatosis in France, where the Lausanne strain was initially introduced by Dr P. F. Armand Delille, by inoculating two wild rabbits on his estate at Maillebois on 14 June 1952, whence it spread all over Europe.

It is impossible to cover the work on myxomatosis in this autobiography. As well as two substantial books, Fenner and Ratcliffe (1965) and Fenner and Fantini (1999), I wrote several review articles and book chapters on it and I used it as the topic for a Harvey Lecture in New York in 1957 and the Florey Lecture to The Royal Society in 1983.

The production of my second book, *Myxomatosis*

In 1949, at Burnet's request, I had been a co-author of the second edition of his book, *The Production of Antibodies,* but my contribution to this was minor, mainly looking up references on transplantation immunology. In 1957, I thought that the time was ripe to begin a book on myxomatosis. Since most of my papers on the subject had been published in one of the Cambridge University Press (CUP) journals, the *Journal of Hygiene*, I wrote to CUP in July 1957 to suggest that they might publish such a book and followed this up when I was in Cambridge in September 1957. The response of the Syndics was positive, but for various reasons I did not start it until 1960, when I knew that I would be going to Churchill College, in Cambridge, for a year in 1961–62. In a letter dated 23 June, 1960, I set out a rough outline of the book and told them that Francis Ratcliffe had agreed to be a co-author and that this would ensure that there would be adequate coverage of the ecological aspects. We signed a Memorandum of Agreement on 22 September, 1961; my first guess was that it would be about 100 pages long and that it would be finished in March 1963.

However, I had not allowed for the large amount of additional material I was able to get about myxomatosis in Europe during my year at Churchill College through travel in the United Kingdom and the Continent, as well as the extensive correspondence initiated with scientists there who had information on the disease. It was a pleasure to work with Francis on this book. He lived in Mugga Way, just up the street from my home in Monaro Crescent, and we met at his place after work to plan it and discuss its progress. The manuscript and figures were sent to CUP early in 1964 and it was finally published in October 1965. CUP provided us with copies of the reviews. There were four in Australian journals, one in an Austrian journal, four in French journals, seven in German journals, two in Italian journals, two in Romanian journals, two in South African journals, and eight in journals in the United States. There were also three in Australian newspapers and 16 in newspapers in the United Kingdom. I would like to quote from two of journal reviews. *The Lancet* said:

> It is a splendid book. Not a word and not a picture are wasted, and it is a pleasure to read. The authors have drawn on every available source of information, as much by personal contact as from the printed word. The result is a complete story which ranges from the introduction of wild rabbits into Australia (domestic strains failed to take root) to the changes in virulence of the virus and susceptibility of the host which are still taking place.

Science said:

> Without doubt man's own evolution has been greatly affected by racial experience with plagues of various types, ranging from malaria, typhus and smallpox to tuberculosis and other similar diseases; great die-offs in population create conditions favorable for evolutionary change. Nearly all virulent diseases, newly introduced, have become attenuated with time by mutual adaptations of host and parasites. The Australian investigators are to be congratulated on providing such a lucid and well-documented account of how such modifications actually take place.

Genetic Studies of Poxviruses

By 1957, it was clear that genetic changes in the virulence of myxoma virus were the major factor in the changing epidemiology of the disease and made possible changes in the genetic resistance of rabbits. However, it was also clear to me that although myxomatosis was a superb 'natural experiment' in evolution, myxoma virus was not a good virus with which to study viral genetics. I therefore initiated work on this with a survey of various marker properties of several orthopoxviruses, mostly different strains of vaccinia and cowpox viruses, which were ideal agents for laboratory investigations (Fenner, 1958). Following selection of two with contrasting characters, I demonstrated, for the first time, intramolecular genetic recombination between animal viruses (Fenner, 1959). Travelling across USA on study leave in 1957 and discussing this work, as mentioned earlier, I was quickly convinced by Salvador Luria, then at Urbana, Illinois, that it would be impossible to delve deeply into mechanisms of recombination if I used two different wild type viruses. As with bacterial viruses, which he had studied, it was essential to use a suite of viruses derived from a single parent. Fortunately, I had such material on hand, the white pock mutants of rabbitpox virus (Gemmell and Fenner, 1960), and initiated work on these which later extended to the use of host-cell dependent and temperature-sensitive conditional lethal mutants (reviews, Fenner and Sambrook, 1964; Fenner, 1970).

The Reactivation of Animal Viruses

What Burnet (1960) described as 'the first example of what may be called genetic interaction between animal viruses' was the reactivation of heat-inactivated myxoma virus by infection of the same cells with live rabbit fibroma virus (Berry and Dedrick, 1936). This was later found by Hanafusa et al., (1959) and workers in our laboratory (Fenner et al., 1959) to be a non-genetic reactivation and was a general phenomenon among the poxviruses. It is now thought to be due to the fact that promoter sequences of an early gene are conserved among poxviruses; they are destroyed by heat inactivation, but may be supplied by another poxvirus infecting the same cell (review: Fenner, 1962).

My Work Pattern at the Bench

From childhood, I have been an early riser, going to bed about 10 pm and getting up when I woke at about 5 am. During most of the time covered in this chapter, I would come in to work shortly after a breakfast of fruit and cereal. Since the distance between my home and the John Curtin School building was only 6 kilometres and there was very little traffic at that time of the day, I usually arrived at the School between 6 and 7 am.

Figure 6.2. Frank Fenner at the bench, inoculating chick-developing embryos with a virus suspension

Throughout the period that I did bench-work (1946–67) biological experimentation was much simpler than it became after the expansion of molecular virology in the 1960s. Most of my papers had only one or two authors, there were no such things as animal ethics committees and most of my research involving experimental animals were carried out with the outbred Walter and Eliza Hall strain of mice and with captured wild rabbits or domestic rabbits bred in the ANU Animal Breeding facilities under the supervision, in Canberra, of a veterinarian, Wes Whitten. In those days we all wore laboratory gowns, but did not use gloves, so I would put on my gown and then look at the experimental animals and do autopsies when they were needed and take down eggs that had been inoculated on the chorioallantoic membrane and count the pocks, or else look at the tissue culture plates that had been inoculated a couple of days earlier. All results were entered in exercise books, in which the relevant experiments had been recorded. I would then consult Ian Marshall and Gwen Woodroofe and hear what they had to say, and also my PhD students. I rarely had more than two students at any one time and after a few months in my lab to learn the basic techniques and decide on the topic on which they would work they would proceed on their own, but consult me whenever they wanted advice. At about 10.30 am we usually had morning coffee, on the lawn just outside our seminar room, and talk with my colleagues. I would then usually go up to the Library to look over all new periodicals dealing with viruses or infectious diseases.

I would often write drafts of papers as soon as I had an idea of what I wished to report, since this would give me a good idea of what additional experiments were needed. I read over all draft PhD theses and papers coming from members of the Department, usually in the evenings, and discussed them with the authors a few days later. I followed Burnet's practice of never putting my name on a paper unless I had carried out some of the bench work. Even from my earliest days in the laboratory, I produced review papers whenever I thought it appropriate, usually as sole author; on ectromelia in 1949, BCG in 1951, myxomatosis in 1954, 1959 and 1964 and the genetics of animal viruses in 1964 and 1970.

Of course, as head of a Department, on some days I would have to spend a good deal of time at meetings of the School Committee or the Board of the Institute of Advanced Studies, but I always had a few hours early in the morning to keep up with the lab work.

Book: *The Biology of Animal Viruses*

In December 1963, I received a letter from Kurt Jacoby, the Vice-President of Academic Press, telling me that Burnet had suggested to him that I should revise the second edition of his book, *Principles of Animal Virology*, which had been published in 1960. After discussions with Burnet, who said that he had switched to immunology and did not want to be involved, and considerable thought, I told Jacoby that I would not undertake a revision of Burnet's book, but would write a new book with much the same coverage, with the title, *The Biology of Animal Viruses*. After correspondence with many overseas virologists, it took me about two years to write, but eventually it was published in 1968 as a two volume book of 845 pages (excluding the subject and name indexes). It received excellent reviews and sold well enough for Academic Press to ask me to prepare a second edition. During the two years that I was writing this book, I did much less bench work than usual, and this influenced my decision to apply for the position of Director of the JCSMR when Hugh Ennor resigned in February 1967.

References

Asiedu, K., Scherpbier, R. and Raviglione, M. (eds) 2000, *Buruli ulcer;* Mycobacterium ulcerans *infection*, Global Buruli Ulcer Initiative, World Health Organization, Geneva.

Berry, G .P. and Dedrick, H. M. 1936, Method for changing the virus of rabbit fibroma (Shope) into that of infectious myxomatosis, *Journal of Bacteriology,* vol. 31, pp. 50–1.

Burnet, F.M. 1960, *Principles of Animal Virology,* Second Edition, Academic Press, New York.

Burnet, M. 1968, *Changing Patterns, an Atypical Autobiography,* William Heinemann, Melbourne, pp. 107–12.

Burnet, M. 1971, *Walter and Eliza Hall Institute 1915–1965,* Melbourne University Press, pp. 48–9.

Cleland, J. B., Campbell, A. W. and Bradley B. 1918, The Australian epidemic of acute polioencephalomyelitis (X disease), *Report of the Director-General of Public Health,* 1917, Sydney. pp. 150–280.

Coman, B. 1998, Francis Ratcliffe, pioneer conservationist, *Quadrant,* vol. 42, pp. 20–6.

Fenner, F. 1950, The significance of the incubation period in infectious diseases, *Medical Journal of Australia,* vol. 2, pp. 813–8.

Fenner, F. 1951, Bacteriological and immunological aspects of BCG vaccination, *Advances in Tuberculosis Research,* vol. 4, pp. 112–86.

Fenner, F. 1956, The pathogenic behavior of *Mycobacterium ulcerans* and *Mycobacterium balnei* in the mouse and the developing chick embryo, *The American Review of Tuberculosis and Pulmonary Diseases,* vol. 73, pp. 650–73.

Fenner, F. 1958, The biological characters of several strains of vaccinia, cowpox and rabbitpox viruses, *Virology,* vol. 5, pp. 502–29.

Fenner, F. 1959, Genetic studies with mammalian poxviruses. II. Recombination between two species of vaccinia virus in single Hela cells, *Virology,* vol. 8, pp. 499–507.

Fenner, F. 1962, The reactivation of animal viruses, *British Medical Journal,* vol. 2, pp. 135–42.

Fenner, F. 1968, *The Biology of Animal Viruses,* Vol. I, *Molecular and Cellular Biology,* pp. 1–474, Academic Press, New York.

Fenner, F. 1968, *The Biology of Animal Viruses,* Vol. II, *The Pathogenesis and Ecology of Viral Infections.* pp. 475–845, Academic Press, New York.

Fenner, F. 1970, The genetics of animal viruses, *Annual Review of Microbiology,* vol. 24, pp. 297–334.

Fenner, F. and Chapple, P. L. 1965, Evolutionary changes in myxoma virus in Britain. An examination of 222 naturally occurring strains obtained from

80 counties during the period October–November 1962, *Journal of Hygiene*, vol. 63, pp. 175–85.

Fenner, F. and Fantini, B. 1999, *Biological Control of Vertebrate Pests. The History of Myxomatosis–an Experiment in Evolution*, CABI Publishing, Wallingford, 339 pages.

Fenner, F. and Marshall, I. D. 1957, A comparison of the virulence for European rabbit *(Oryctolagus cuniculi)* of strains of myxoma virus recovered in the field in Australia, Europe and America, *Journal of Hygiene*, vol. 55, 149–99.

Fenner, F. and Ratcliffe, F. N. 1965, *Myxomatosis*, Cambridge University Press, London, 379 pages.

Fenner, F. and Sambrook, J. F. 1964, The genetics of animal viruses, *Annual Review of Microbiology*, vol. 18, pp. 47–94.

Fenner, F., Holmes. I. H., Joklik, W. K. and Woodroofe, G. M. 1959, Reactivation of heat-inactivated poxviruses; a general phenomenon which includes the fibroma-myxoma virus transformation of Berry and Dedrick, *Nature*, vol. 183, pp. 1340–1.

Fenner, F., Poole, W. E., Marshall, I. D. and Dyce, A. L. 1957, Studies in the epidemiology of myxomatosis. VI. The experimental introduction of the European strain of myxoma virus into Australian wild rabbit populations, *Journal of Hygiene*, vol. 55, 192–206.

French, E. L. 1952, Murray Valley encephalitis: Isolation and characterization of the causative agent, *Medical Journal of Australia*, vol. 1, pp. 100–5.

Gemmell, A. and Fenner, F. 1960, Genetic studies with mammalian poxviruses. III. White pock (u) mutants of rabbitpox virus, *Virology*, vol. 11, pp. 219–35.

Hanafusa, H., Hanafusa, T. and Kamahora, J. 1959, Transformation phenomena in the pox group viruses, II. Transformation between several members of the pox group. *Biken Journal*, vol. 2, pp. 85–91.

Linell, F. and Nordén, Å. 1954, *Mycobacterium balnei:* a new acid-fast bacillus occurring in swimming pools and capable of producing skin lesions in humans, *Acta Tuberculosa Scandinavica, Supplement 33*.

Mackerras, I.M. (1971). Francis Ratcliffe (1904-1970). *Search*, 2(3), 74–5.

Marshall, I. D. and Fenner, F. 1960, Studies in the epidemiology of myxomatosis. V. Changes in the innate resistance of Australian wild rabbits exposed to myxomatosis, *Journal of Hygiene*, vol. 56, pp. 288–302.

Myers, K., Marshall, I. D. and Fenner, F. 1954, Studies in the epidemiology of myxomatosis. III. Observations on two succeeding epizootics in

Australian wild rabbits on the riverine plain of south-eastern Australia 1951–1953, *Journal of Hygiene*, vol. 52, pp. 337–60.

Shepard, C.C. 1960, Acidfast bacilli in nasal excretions in leprosy, and the results of inoculation of mice, *American Journal of Hygiene*, vol. 71, 147–57.

Chapter 7. Director of the John Curtin School of Medical Research, 1967 to 1973

My Appointment as Director

Hugh Ennor, who had been knighted in 1965, was Dean of the John Curtin School in 1966 (he preferred the title of Dean, rather than Director). He had been appointed Deputy Vice-Chancellor in 1964 and hoped to be appointed Vice-Chancellor. However, Sir John Crawford was the obvious choice for that post and, in February 1967, Ennor accepted an invitation from Senator John Gorton to become the first Secretary of the Commonwealth Department of Education and Science.

Professor Colin Courtice was appointed Acting Dean, and the position of Director of the John Curtin School was advertised in Australia and overseas. I had spent a large part of my time in 1966 and 1967 working on the book, *The Biology of Animal Viruses*, and felt out of touch with bench work, so I decided to apply for the position, and was appointed, with the title of Director, for a period of seven years. At the time I thought that I would want to spend some time at the bench and arranged that I should also have a chair and access to a laboratory in the Department of Microbiology. I also arranged for Colin Courtice to be my deputy, if I were to go abroad.

Governance

An interesting comment on the governance of different Research Schools in the ANU was published in the *Sydney Morning Herald* at the time of Ennor's farewell dinner:

> If Senator Gorton was looking for a strong man to head his new Commonwealth Department of Education and Science, he has certainly found such a man in Sir Hugh Ennor. Sir Hugh is a big, hearty biochemist whose administrative talents have come into full flower at the Australian National University. Since 1953 Sir Hugh has been Dean of the John Curtin School of Medical Research, and since 1964 he has been Deputy Vice-Chancellor. Unlike the humanities schools at the Australian National University, the John Curtin School has no faculty structure. It is run entirely by the Dean. The School of Social Sciences, under Professor P. H. Partridge, is referred to jocularly as an Athenian democracy; the School of Pacific Studies, headed by Sir John Crawford until he recently became Vice-Chancellor, as a guided democracy and the John Curtin School of Medical Research as an Oriental despotism.

One of my first jobs on becoming Director was to introduce the Faculty/Faculty Board structure into the School. Fortunately, I had been a member of the Committee appointed by Vice-Chancellor Huxley and chaired by economist David Bensusan-Butt, as had Frank Gibson, Professor of Biochemistry and three senior non-professorial staff—Desmond Brown, from Medical Chemistry, Hugh Mackenzie, from Physical Biochemistry, and Bede Morris, from Experimental Pathology—along with Sir Robert Madgwick, former Vice-Chancellor of the University of New England and Geoffrey Sawer, Professor of Law in the Research School of Social Sciences. In May 1967, the Butt Committee produced a report recommending that there should be a Faculty consisting of all academic staff of and above the rank of Research Fellow, and a Faculty Board, composed of Heads of Departments and four members of the non-professorial academic staff elected by Faculty, with the Director as Chairman. For my term of office, as recommended by the Butt Committee, I acted as Chairman of both Faculty Board and Faculty. After my resignation in May 1973, Faculty Board decided that the Chairman of Faculty should be elected by Faculty, and Ian Marshall, my first PhD student and by then Senior Fellow, was elected to that position.

Changes in Existing Departments and Units

Department of Genetics

Early in my term as Director, decisions had to be made on two departments and one unit, after their heads had retired or resigned. David Catcheside had been appointed Professor of Genetics in the John Curtin School in 1964, on the understanding that he would act as Adviser to the University on the development of biological research, as distinct from medical research, in the ANU. The Research School of Biological Sciences was established in October 1967, with Catcheside as Director, and he and most members of his staff were transferred to its payroll. However, they continued to occupy their existing laboratories in the John Curtin School until the building of the new School was completed early in 1973. Robert Kirk, a human geneticist who had taken up duty as a Senior Fellow in June 1967, elected to remain in the John Curtin School as the sole academic staff member of the Human Genetics Group, which was attached for administration to the Department of Clinical Science until the Department of Human Biology was established in 1970 (see below).

Department of Medical Chemistry

My most difficult decision concerned the future of the Department of Medical Chemistry after Adrien Albert retired, at age 65, in December 1972. The Department occupied all four storeys of the rear west wing, a situation justified in the early days by the fact that it was the only department of chemistry in the University and there was a need for a two-storey section to accommodate production equipment. However, a Research School of Chemistry was established in the Institute of Advanced Studies in 1964 and its building was completed in 1967. By the late 1960s, there was growing pressure for additional laboratory space in the John Curtin School, especially for the developing work in pharmacology, and proposals for the School's submission to the Australian Universities Commission for the 1972–74 triennium were urgently required. In 1969, after extensive discussion by Faculty and Faculty Board, including a subcommittee comprising the Director, Professor Geoffrey Badger (University of Adelaide), Professor R. D. Wright (University of Melbourne and long-time member of the ANU Council), Professors Frank Gibson and Alexander Ogston (Biochemistry and Physical Biochemistry in JCSMR), it was decided that when Albert retired, at the end of 1972, no replacement would be sought and the Department would be contracted to a Group. I fully expected that some members of this Group would transfer to the Research School of Chemistry, but that did not occur and the Group was not disbanded until 1985.

Electron Microscope Unit

In July 1967, Edgar Mercer, Professorial Fellow and Head of the Electron Microscope Unit, decided to move to the United States and make his living as a sculptor. The Unit was reorganized and in 1971 absorbed into the Department of Experimental Pathology.

Establishment of New Departments

My term as Director was a time of expansion in the ANU. In the JCSMR, this was represented by the establishment of four new Departments: Clinical Science, Human Biology, Immunology and Pharmacology.

Department of Clinical Science

The Head of the Department of Clinical Science, Professor Malcolm Whyte, had been appointed in 1966. However, because of delays in the completion of new laboratories for the Department in the Canberra Community Hospital, he worked in the School building and in the field in Papua New Guinea before moving into the Hospital in September 1967. In the Canberra Community Hospital, besides providing a clinical service, by referral, for both ambulatory patients and in-patients, the Department conducted a program of clinical and laboratory research oriented towards problems associated with coronary heart disease. Initially there was a good deal of suspicion among doctors in the hospital that the presence of the Department of Clinical Science was the thin edge of the wedge, the introduction of government-paid medicos into a purely private practice hospital. Malcolm and his staff negotiated these problems effectively.

Department of Human Biology

A Department of Human Biology was established in 1970 by bringing together Robert Kirk, a human geneticist who had elected to remain within the JCSMR when Catcheside transferred to the Research School of Biological Sciences, and the Urban Biology Group, headed by Stephen Boyden, which had previously been located in the Department of Microbiology (see Chapter 5). Research in the new department was carried out in these two fields, each concerned with human population biology. In 1972, the Urban Biology Group, in collaboration with the University of Hong Kong, initiated a study of the human ecology of Hong Kong, then the most densely populated city on earth. This was a groundbreaking study of the 'metabolism' of a city, which was completed in 1977 and which led to Boyden's involvement in UNESCO's Man in the Biosphere program.

Kirk's interest had long been in human genetics, and in Canberra he undertook population genetic studies of Australian Aborigines, natives of Papua New Guinea and, with local collaborators, inhabitants of India and South Africa. In 1973, he was appointed Head of the Department of Human Biology.

Department of Immunology

In 1970, Council approved the establishment of a Department of Immunology, and Bede Morris, who had come to Canberra with Colin Courtice in 1958, was appointed Professor and moved from the Department of Experimental Pathology to the new department, with three other academic staff and four PhD students. Initially dispersed in several parts of the building, early in 1973 they came together in the space previously occupied by Catcheside and his staff. Their studies were focused on self/non-self discrimination, with a particular concentration on transplantation biology, using a unique and particularly useful system of pregnant sheep and their foetuses. This made it possible to cannulate the lymphatic system of both mother and foetus and study the development of circulating lymphoid cells in both animals.

Department of Pharmacology

In 1966, David Curtis, Eccles' first PhD student, had been appointed a Professor within the Department of Physiology. In 1973, when space became available in the laboratories occupied by the former Department of Medical Chemistry, he was appointed Professor of Pharmacology, and with his staff moved into some of the vacated laboratories. Research in the new department was concentrated on *in vivo* studies of the effects of known chemicals administered close to single neurones in various portions of the nervous system of the cat and the relationship of the effects observed to synaptic inhibition or excitation.

Developments in Older Departments

The principal developments in the established departments were the appointment in 1966 of Frank Gibson as Professor of Biochemistry, replacing Hugh Ennor; of Gordon Ada as Professor of Microbiology in 1968, replacing me; of Laurie Nichol in 1971, replacing Sandy Ogston; and of Peter Bishop in 1967, replacing Jack Eccles.

Building Activities

For the first time since the occupation of the new building in 1957, substantial new building activities, other than the fitting out of vacant wings for Physical Biochemistry and Genetics, were carried out in 1967–73.

Specific-Pathogen-Free Animal House

A special building for breeding specific-pathogen-free rats and mice, located to the west of the main Animal Breeding House, was commenced in 1971 and completed in 1972, but because of difficulties with the mechanical services it was not commissioned until early 1973.

Wing F Animal House

In 1968, a Radioisotope Suite, with chemical laboratories and rooms for handling with safety large and small animals that had been inoculated with highly radioactive materials, was built on the eastern end of what had been the lawn-covered roof of Wing F, the Experimental Pathology/Immunology animal house. Also, on the ground floor of the same building, a well-equipped Animal Hospital was constructed, with operating theatres for large and small animals.

The Library and the Space Beneath It

In 1957, the Library, on the top floor, occupied the area east of the passageway connecting the front and rear wings and there was a small open balcony extending towards the west. By 1970, the balcony had been converted to library space extending 6 metres to the west, and the Common Room, in the corresponding area on the ground floor, was also extended. The additional space on the floor level below this was used for storage space, a School Computer Centre and a seminar room.

Accommodation for the Department of Human Biology

Just as laboratories had been constructed on the roof on Wing F in 1968, in 1972, laboratories to accommodate the Department of Human Biology were built on the lawn-covered roof of Wing E, the Animal House for Infected Animals. The layout of the laboratories was designed by Kirk with advice from Boyden. In addition to laboratories and studies, a Seminar Room and adjacent space which served as a coffee room were provided.

Overseas Trips, 1967 to 1973

I travelled overseas once during each year of my period as Director, usually in response to an overseas invitation which covered my expenses. The timing and reasons for the trips are set out below.

18 November to 10 December 1967

The primary purpose of this trip was to make enquiries in England and the United States about possible candidates for the Chair of Microbiology: none were found. I also attended a small working party on Smallpox, sponsored by the USA-Japan Cooperative Medical Science Program and held in Honolulu.

6–23 November 1968

Financed by the US National Institutes of Health, this trip was to attend two conferences, one in Princeton, New Jersey, on 'Influenza Virus Genetics and Vaccines' and the other, in London, on 'Microbiological Standardization of Rubella Vaccines'. At their request, I made a four-page summary of the latter conference for the Commonwealth Serum Laboratories.

23 March to 30 April 1969

In retrospect, this was one of the most important trips that I made, for it was my introduction to the Intensified Smallpox Eradication Program of the World Health Organization. What was called 'The WHO Informal Conference on Monkeypox' was held in Moscow and brought together for the first time a number of poxvirus experts. Their role at this meeting was to decide whether human monkeypox, recently discovered in West Africa, might indicate that there was an animal reservoir of smallpox virus, which would have made eradication impossible. I was rapporteur for the conference, which concluded that monkeypox and variola viruses were different species of orthopoxviruses.

I then went to England, where I made enquiries in London, Oxford and Leeds connected with vacant chairs and new developments in the John Curtin School and, at the request of David Catcheside, the Research School of Biological Sciences. In the United States, I gave a lecture on Conditional Lethal Mutants of Animal Viruses at the National Institutes of Health, and then went to the University of California at Davis for two weeks, as Life Sciences Lecturer. Besides visiting laboratories and giving seminars to graduate students, I gave two public lectures: 'Evolutionary Changes in an Infectious Disease' and 'Civilization and Infectious Diseases: the Effect of Social Organization on Human Infections'.

3 June to 3 July 1970

The main purpose of this trip was to give the Lilly Lecture of the Royal College of Physicians, on Genetic Aspects of Virus Diseases. It was first given as the concluding item of a conference on Virology for General Physicians held at the College headquarters in London and repeated a week later in Sheffield.

3 September to 7 November 1970

On 9, 10 and 11 September, 1970, I gave the CIBA Lectures at Rutgers University and three lectures on different aspects of the genetics of animal viruses. I then went on to spend about a month as Scholar-in-Residence at the State University of New York, spending four or five days at each of its campuses: Stony Brook, Albany, Syracuse, Buffalo, Binghamton and Downstate.

12 June to 6 July 1971

The main purpose of this trip was to attend the Second International Congress for Virology in Budapest. Much of my time was taken up with meetings of the International Committee on the Nomenclature of Viruses, of which I was then President. I also visited and gave seminars at the All-India Institute of Medical Sciences in New Delhi, the Max Planck Institut für Virusforschung in Tübingen, in Germany, the Institute for Microbiology and Epidemiology in Prague and the Institute of Virology in Bratislava.

14 December 1971 to 24 March 1972

This was the first stage of a 12-month long Fogarty Fellowship of the US National Institutes of Health (NIH). This was a prestigious award, with all expenses paid, an allowance of $US30,000 per annum and a medal. I was not able to take it for the full 12 months, but arranged with the authorities to take in three sessions. For this first session I lived in Stone House, the original residence on the property that became the campus of the NIH. Other Scholars in residence at the time included an old friend from my Rockefeller Institute days, Rollin Hotchkiss, and Nobel Prize winner Ragnar Granit, of Sweden. I spent almost all of my time there working on the second edition of *The Biology of Animal Viruses*, but gave the luncheon address at the Eighth Gustav Stern Symposium in New York and seminars at Johns Hopkins University, Cold Spring Harbor and the National Institutes of Health. I also gave the opening address at a symposium in Israel and three seminars at the Hadassah Medical School, and then attended the fifth meeting of the Scientific Committee on Problems of the Environment (SCOPE) in London.

Lectures in Australia

One important role of the Director of the JCSMR at that time was to give lectures widely in Australia on matters of interest to the general public as well as others to explain the activities of the School. I gave a number of such lectures, few of which were published, as well as many on various aspects of virology. Several deal with environmental problems, because at the time I was a Vice-President of the Australian Conservation Foundation and Chairman of the National Committee on Problems of the Environment of the Australian Academy of Science.

They were:

1968	'The Structure and Activities of the John Curtin School', at the Melbourne, Monash, Sydney and Queensland Universities.
1970	'Man and his Environment in Australia', to the First International Congress on Domiciliary Nursing.
1970	'James Cook and the Prevention of Scurvy', to the Royal Society of Queensland, on the Cook Bicentenary.
1970	'Infectious Disease and Social Change', to the Canberra Postgraduate Committee on Medicine.
1971	'How Many Australians? Immigration and Growth', to the Australian Institute of Political Science Summer School (Fenner, 1971).
1971	'Population and Resources, Local and Global', to Canberra Hospital Medical Seminars.
1971	'Abortion, Medical Questions', to the ANU Centre for Continuing Education.
1971	'Migration and Australia', to the Australia Party.
1972	'Is There an Environmental Crisis?' an ANU Public Lecture.
	'The World Situation in Resources and Some Implications for Australia', for *The BHP Review*, Summer 1972.
1972	'The Environmental Crisis—Population, Resources, Pollution', at Cessnock, New South Wales.

Writing Textbooks

Before I became Director, I had thought that I would like to continue with some laboratory research and had made arrangements to have space in the Department of Microbiology. However, I found that if I were to be available whenever wanted for committee meetings and meetings with staff, I could not have the long breaks I needed for bench research. Further, I knew that I could not carry out laboratory work through research assistants or PhD students, without 'getting my hands dirty'. The one PhD student that I had at the time I accepted the Directorship was Bob Blanden, who was completely independent and produced some excellent papers from his thesis material. However, at that time, the directorship was not a full-time job. I had already assisted Burnet with a book and written two others, *Myxomatosis* and *The Biology of Animal Viruses,* so I thought that I would write some more books.

A Textbook on Medical Virology

I had always felt somewhat guilty being called a Professor, having never given a course of lectures (I was never invited to lecture to students in the Faculties), and now I was never likely to. So I thought that I might occupy the spare time I had, always early in the morning before anyone else arrived and at irregular times on most days, to write a textbook on virology for medical students. I persuaded David White, an early PhD graduate (1960) from the Department of Microbiology, who was by then Professor of Microbiology at the University of Melbourne, who I knew to be a first-class teacher of virology, to be a co-author. The first edition, entitled *Medical Virology* and 390 pages long, was published by Academic Press in 1970. It received excellent reviews and sold very well. M. A. Epstein, writing in *Nature*, said, amongst other things:

> Despite the risk of writing a rave notice, I am still filled with wonder on reflecting on the seeming ease with which extremely complicated topics [have] been reduced to an orderly survey of the basic facts involved, together with suitable explanations of their significance. This book can be used, therefore, both as an introductory textbook and as a very reliable and solid reference book.

We also received many complimentary letters and expressions of interest in using the book as a textbook. A Spanish edition was published in 1973.

The Biology of Animal Viruses, Second Edition.

The first edition of *The Biology of Animal Viruses*, published in 1968, had sold very well, but by the early 1970s clearly needed updating. The rapid advances were primarily in molecular virology, in which I lacked expertise, so I enlisted the help of three former students (Brian McAuslan, Joe Sambrook and David White) and one former staff member (Cedric Mims) as co-authors. We agreed on who should take primary responsibility for each chapter, then all authors should read each chapter, and, finally, I edited them so that there was a uniform style and a minimum of overlap. The Memorandum of Agreement with Academic Press was signed in March 1971 and the second edition published in hardback in 1974. The reviews were good; Kenneth Berns in *American Scientist* said:

> In spite of the multiple authorship, the text is uniform and cohesive and may come close to being the universal source in animal virology. Its general sections serve admirably as an introduction for anyone who has had a basic biology course, while other sections contain well-organized compendia of the known facts concerning specific viruses and other aspects of virology...The book is a must...for all individuals interested in animal virology.

Since many of those interested in the book complained that the price was too high, after several thousand copies of the hardback had been sold the publishers produced a soft-cover 'Students' Edition' for about half the price. Total sales exceeded 11,000 copies. A Russian edition, translated by my friend Vladimir Agol, who I had met in Moscow in 1964, was published in 1976. Between 1976 and 1981, I received 36 requests from other authors wishing to use diagrams or tables contained in that book.

Other Activities, 1967 to 1973

Chairman, Committee to Examine the Possibility of Establishing An Undergraduate Medical School in the ANU

An Advisory Committee on Undergraduate Medical Education in the Australian Capital Territory (ACT) was first established in 1965 and a major conference on Medical Practice and the Community held in the JCSMR on 26–30 August, 1968. In 1968, a new Committee on Medical Education in the ACT was set up and the Vice-Chancellor asked me to act as Chairman, but the detailed plan was carried out by a subcommittee chaired by Malcolm Whyte. Its report was debated by an expanded Committee in November 1969, circulated to the Boards of the Institute of Advanced Studies and of The Faculties in 1970. After much further discussion, the Vice-Chancellor told Council that the proposal was academically sound, that questions of shielding the rest of the University from any adverse effects had been 'fully answered in policy terms by the Australian Universities Commission (AUC) and by the Government' and that talks which Professors Fenner and Whyte had had with the health authorities and medical fraternity pointed to the probability of a satisfactory agreement on the integration of medical education and the local health services'. The proposal was submitted to the AUC in April 1971, but a decision was deferred until completion of the report of a Committee on Medical Schools set up by the Minister for Education and Science in June 1972. In July 1973, this report recommended deferment of the ANU proposal for a further three years, although new medical schools were approved at James Cook University and at the University of Newcastle. A Medical School was finally set up in the ANU in 1998 and accepted its first students in 2004. Its administration is housed in a building that was named the Frank Fenner Building in May 2003 (it should have been the Malcolm Whyte Building!).

Committee to Examine the Possibility of Establishing A Centre for Natural Resources (see Fenner, 1973a, 1979)

The notion that the ANU might embark on some activity in an area called 'resources' arose at meetings of the ANU academic boards in 1965, when the Deputy Vice-Chancellor reported on the conclusions of the committee that had looked into the question of medical education in the ACT (see above). Some members of the Board suggested that other faculties such as agriculture, veterinary science or rural science should also be considered. The committee that was set up dismissed these suggestions, but the Universities Commission supported the idea that something should be done about a body to examine renewable natural resources. In November 1967 it was decided that a committee should be established to examine the feasibility of establishing a 'Centre for

Natural Resources' in the ANU. Following receipt of a positive recommendation from this committee, the Universities Commission approved the Statement of Intent in its 1969 Report. The Vice-Chancellor, Sir John Crawford, then set up a new committee, which he asked me to chair, to consider the establishment of a Centre and Research School for Natural Resource Studies. Recast in a more modest form, a proposal for a Centre for Natural Resources (CNR), dated 1970, was part of the ANU submission to the AUC for the 1973–1975 triennium. After the report was submitted the committee decided that the word 'environment' should be included in the title, and the need for a convenient acronym gave rise to the name Centre for Resource and Environmental Studies (CRES). In its Fifth Report, published in August 1972, the AUC explicitly endorsed two proposals in environmental studies, one in ANU (CNR/CRES) and one in the University of Melbourne.

CSIRO and Medical Research

Among my extra-curricular responsibilities as Director was membership of the Advisory Council of CSIRO, and at its meeting in May 1972 the Council asked me to report to it on CSIRO and Medical Research. I submitted a draft report at the Advisory Council meeting in May 1973, which was approved in a general sense by the Council. During the next few months I had extended correspondence and personal discussions with Dr D. N. Everingham, the Minister for Health and Dr J. R. Price, Chairman of CSIRO. The final report (Fenner, 1973b) was submitted to the Advisory Council in October 1973, and contained three recommendations to CSIRO, which were accepted by the Executive, and two to the Australian government concerning medical research, which were not accepted. The recommendations to CSIRO were:

1. to establish a CSIRO Committee on Medical Research,
2. to review its official attitude on medical research in CSIRO,
3. to establish a Division of Human Nutrition.

The last and most important recommendation was implemented in 1975, when the former Division of Biochemistry and Nutrition, located in Adelaide, was converted into the Division of Human Nutrition.

International Committee on the Nomenclature (Taxonomy) of Viruses

I had long been interested in viral taxonomy (Fenner, 1953) and in 1966 I was elected a foundation member of the International Committee on the Nomenclature of Viruses, which was established at the International Congress for Virology in Moscow in September 1966. The driving forces were C. H. Andrewes and André Lwoff. Although I was booked to go to that Congress, I could not attend, since a few weeks earlier I had contracted appendicitis while in the United States, then flown to Glasgow where I was confined for two weeks in the Glasgow Infirmary before returning immediately to Australia. The International Committee operated through a number of Study Groups, each consisting of specialists in a particular group of viruses. I was a member of the Poxvirus Study Group from 1966 until 1990, when I resigned.

It was arranged that apart from correspondence, the Committee would meet every five years, at each International Congress of Microbiology. Because of commitments at ANU, I was unable to get to the next International Congress in Mexico City in 1970, but in my absence I was elected President of the Committee until the next Congress, which was in Madrid in 1975. As Chairman, I attended meetings of the Executive Committee in London in 1968 and 1973. At the latter meeting it was agreed that a number of changes should be recommended to the next meeting of the Committee, in 1975, including that its name should be changed to the International Committee for the Taxonomy of Viruses (ICTV). Two meetings of the Executive Committee were held in Madrid, at the first of which a plant virologist, R. E. F. Matthews, was elected President (because of concern that the plant virologists would decide not to operate through ICTV, I pushed hard for Matthews' election). A meeting of the full International Committee, spanning two days, passed a large numbers of amendments to the Rules of ICTV. As outgoing President, I was elected a life member of ICTV.

After each meeting, the outcomes in terms of rules and names and classification of vertebrate, invertebrate, plant, fungal and bacterial viruses were published in *Archives of Virology*. In addition, the outgoing president produced a book setting out descriptions of the currently agreed genera and species (Wildy, 1971; Fenner, 1976; Matthews, 1979). As an illustration of the enormous increase in knowledge of viruses in the 20 years since 1980, the seventh report (van Regenmortel et al., 2000), is 1,162 pages long.

Honours and Awards

1967	Britannica Australia Award for Medicine.
1969	Life Sciences Lecturer at the University of California at Davis.
1970	Lilly Lecture, Royal College of Physicians, London.
	CIBA Lecturer in Microbial Biochemistry, Rutgers University, New Jersey.
1971	Victor Coppleson Lecturer, Australian Post-Graduate Medical Foundation.
	Fogarty Scholar, US National Institutes of Health.
1973	David Memorial Lecture, Australian and New Zealand Association for the Advancement of Science.

Overview: the Role of the Director

My period as Director coincided with the Vice-Chancellorship of Sir John Crawford, whom I had met 10 years before as a colleague in a Saturday tennis group. My wife, Bobbie, became a friend of his wife, Jess, and the four of us used to meet on Saturday nights, alternately at each other's homes, for bridge. Crawford was a man of great ability and industry, a superb chairman of committees and a creative Vice-Chancellor. In that capacity, he called on senior administrators such as the directors of research schools to act as Chairmen of university-wide committees to examine new proposals. He asked me to be Chairman of Committees to examine proposals to establish an Undergraduate Medical School and a Centre for Natural Resources (later the Centre for Resource and Environmental Studies) in ANU, as described earlier.

On 1 May, 1973, I resigned from my position as Director of the John Curtin School to become Director of the Centre for Resource and Environmental Studies. Since there was a meeting of the ANU Council that day and I was scheduled to present my annual 'Director's Report to Council', I decided to give an overview of my six years as Director. After outlining the changes described earlier in this chapter, I set out my views on the role of the Director in a research school like the John Curtin, as follows. What follows must be viewed in the context of the times, the 1960s and 1970s:

> Finally, may I record my personal views on a matter that is widely debated in the Institute, the role of the Director. It is clear that the requirements and opportunities differ in the different research schools, and at different periods during the development of each School. In the early stages there is no doubt of the necessity of having a senior academic as the first Head of School. In JCSMR, Florey fulfilled this role in its formative first decade of development. In a developed school (if such ever exists!) I am not so sure that one need have a person of academic eminence if the job is looked at purely from the point of view of running the School. Looking back on my five and a half years of office, there have been few occasions on which I have been conscious of providing academic leadership. In JCSMR interdisciplinary work, quoted in the paper setting out the duties of a Head of School as one of his responsibilities, is regarded with considerable suspicion by most members of the School. It is a School of highly diverse and self-contained departments, and it is the departmental heads who must provide the real scientific leadership. As I have already indicated, there have been several new academic developments in the School over the last six years. These have all been the result of discussion and debate in Faculty Board and Faculty; as Director I have merely acted as Chairman of these bodies. The only positive suggestion that I made, for a Department of Human

Genetics to replace the Department of Genetics when that department transferred to the Research School of Biological Sciences was transformed into the rather different subject of Human [Population] Biology.

Much of the time of the Director is occupied in chairing meetings of committees within the School, in representing the School on University Committees, and with making decisions on what are essentially minor details of administration, such as the ranking of laboratory craftsmen, technicians or secretaries. Any experienced academic with administrative competence could do these jobs. However, it is likely that only a distinguished medical scientist would command the support of heads of departments within the School, and the respect of the Vice-Chancellor and his colleagues on the Heads of Schools and Budget Advisers' Committees to the extent that is necessary to represent the School's viewpoint on these bodies.

In addition to his responsibility to Council for running the research school, I believe that the Director of a developed School like JCSMR has real opportunities to make constructive contributions to the University as a whole; and on a wider scene, if he has the mind to do it, to Australian public life. I believe that this outside work will continue to be perhaps the major attraction of the position. The position of Director of a Research School in the ANU is highly respected in Australia and overseas, and the Director is free from the pressing need that a departmental head has of providing continuing scientific leadership in a narrow discipline. In my own term of office, I believe that apart from writing two major books on virology, my major satisfaction has come from the contributions I have been able to make to the University outside JCSMR, in four areas of innovation. In none of these would I claim to have been the prime mover, but I believe that I have been able to exert a useful influence on each. These are: (a) the development on an experimental scale of the Human Sciences Program, initiated this year in the School of General Studies [Faculties]; (b) support for the planning, primarily by Professor H. M. Whyte, and for the promotion through University Committees and the Australian Universities Commission of the proposal that the ANU should develop a Faculty of Medicine in the near future; (c) the development of the idea, published as an appendix of the Fourth Report of the Australian Universities Commission, that ANU should initiate activity in relation to renewable natural resources to its realization as a more broadly based Centre for Resource and Environmental Studies; and (d) the establishment, following initiatives of Professors Bishop and Whyte, of the Postgraduate Committee in Medicine of the ANU.

I believe that a similar range of opportunities exists for the next Director of the JCSMR. If the government agrees to proceed with the development of a Faculty of Medicine, it is of the utmost importance to the JCSMR in particular and to the University as a whole that this development should proceed with full harmony and understanding between the John Curtin School and the Faculty of Medicine. There is an opportunity for major interaction of mutual value to occur between these two large and important components of the University, and the next Director could play a crucial role in realizing the full potential of a situation that is unique in Australia, perhaps in the world.

I believe that Headships of Departments are the most important positions in the John Curtin School. They require sustained and full-time effort if the departmental head is to supply the leadership within his department and in Australian science as a whole that his position calls for, for these are the premier positions in their respective fields in Australia. I therefore wholeheartedly support Council's view that a Head of School should not be at the same time the Head of a Department. As far as the term of appointment of a Director is concerned, five or at most seven years should be long enough for a man to make any contributions that he is able to make; after that time a new point of view is needed. It is also long enough, if I may say so, for someone to go on attending to the rather boring and trivial administrative details that inevitably fall to the lot of a Head of School. However, except in exceptional circumstances I believe that the term of office should not be less than five years, if a Head of School is able to follow through any of the developments initiated during his term of office. Looking back over the last six years, I find that all the new academic developments that have occurred in JCSMR over that time were set out explicitly in the Australian Universities Commission Submission for the 1970–72 triennium, which was produced during my first six months in office...The worst decision of all, in a large and diverse School like JCSMR, would be to accept the notion of rotating 'deanships'. By deflecting the energies of successive departmental heads from their primary role of scientific leadership of their departments, such a procedure would effectively weaken a series of departments in turn. It would also be likely to provide a guarantee that the *status quo* would be maintained indefinitely, in a world in which the demands made on the University and research schools are constantly changing.

In conclusion, I would like to pay tribute to the generous support that has always been provided to the activities of JCSMR by Council and by the administrative officers of the University; especially, since our terms of office largely coincided, to our recent Vice-Chancellor, Sir John Crawford.

References

Fenner, F. 1973, The Centre for Resource and Environmental Studies. *ANU News,* July 1973, pp 1–3.

Fenner, F. 1979, *The Centre for Resource and Environmental Studies: 1973–79.* The Australian National University.

Fenner, F. and White, D. O. (1970). *Medical Virology*. Academic Press, New York.

Fenner, F., McAuslan, B. R., Mims, C. A., Sambrook, J. and White, D. O. 1974, *The Biology of Animal Viruses,* Second Edition. Academic Press, New York.

Fenner, F. 1976, Classification and Nomenclature of Viruses. Second Report of the International Committee on Nomenclature of Viruses. *Intervirology,* vol. 7, 1–2.

Matthews, R. E. F. 1976, Classification and Nomenclature of Viruses. Third Report of the International Committee on Taxonomy of Viruses. *Intervirology,* vol. 12, 3–5.

Van Regenmortel, M. H. V., Fauquet, C. M. and Bishop, D. H. L. 2000, *Virus Taxonomy. Seventh Report of the International Committee on Taxonomy of Viruses.* Academic Press, San Diego.

Wildy, P. 1971, Classification and Nomenclature of Viruses. First Report of the International Committee on Nomenclature of Viruses. *Monographs in Virology*, Vol. 5, Karger, Basel.

Chapter 8. Activities Associated with the Australian Academy of Science

Introduction

The two most important organizations with which I have had an association since July 1949 have been The Australian National University, especially the John Curtin School of Medical Research and the Centre for Resource and Environmental Studies, and the Australian Academy of Science, and it is appropriate that I should include a chapter on the Academy in my autobiography. I was in the first group of scientists, apart from the Foundation Fellows, to become a Fellow (by election) in 1954, the year in which the Academy received its Royal Charter. From 1958 to 1961, I served as Secretary (Biological Sciences) and initiatives established then led to my appointment as Chairman of the Flora and Fauna Committee from 1967–74 and a member of the Fauna Committee from 1974–81. Even before my appointment as Director of the Centre for Resource and Environmental Studies in 1973, I had become involved in some of the Academy committees on conservation: member of the Standing Committee on National Parks and Conservation from 1970–82 (Chairman, 1970–72), Chairman of the National Committee for SCOPE, the Scientific Committee on Problems of the Environment (an international committee), 1971–78, and member of the Standing Committee on the Environment, 1970–78.

I also served as a member of several Academy committees that warrant no more than a mention here: member of the Public Lectures Committee, 1977–83, the National Committee on the Environment, 1979–82, the National Committee on the History and Philosophy of Science, 1980–82, Chairman of the National Committee for Medical Sciences, 1980–84, the Fellowship Establishment Committee, Australian Foundation for Science, 1990–91 and the Library Committee, 1995–2005. I was a Director of the Australian Foundation for Science from 1992–2001.

In 1970, the royalties that I received from the sale of the textbook *Medical Virology* were such that I started to make annual donations to the Academy to set up an Environment Fund, which by 1983 totalled $19,050. From 1984–97, Bobbie and I made more substantial donations ($230 000) to fund annual Fenner Conferences on the Environment, which were started in 1988. In 1992, I became involved in a video history project of the Royal Australasian College of Physicians and subsequently initiated the Video Histories of Australian Scientists (now called Interviews with Australian Scientists) project within the Australian Academy of Science.

Secretary, Biological Sciences, 1958 to 1961

In 1957, I was elected to Council and the next year to the position of Secretary, Biological Sciences. In those early days, just after completion of the Dome, the officers' jobs were not as demanding as they are now, and my appointment greatly broadened my views of biological science. As early as 1955, the first Secretary (Biological Sciences), entomologist Dr A. J. Nicholson, had brought the need for a proper study of Australia's fauna to the attention of Council. Apart from general support by the Directors of the Australian Museum in Sydney (Dr John Evans) and the Western Australian Museum (Dr David Ride), little was done until 1959–60, when two initiatives from Fellows resident in Adelaide, Professors J. G. Wood and W. P. Rogers, resurrected the idea of a national biological survey in a wider context and asked the Academy to look at problems of the flora as well as the fauna.

It took the success of a proposal by physical scientists for Commonwealth support for a large optical telescope in Australia to prompt a letter to me, as Secretary (Biological Sciences) from several Fellows who were biologists, pointing out the discrepancies between government support for major projects in the physical sciences and the biological sciences. In March 1960, I therefore set up a Fauna and Flora Committee, of which I was initially a member and, from 1967–74, Chairman, 'To prepare a detailed scheme for the implementation of major research projects in the biological sciences in Australia, with special emphasis on the fauna and flora'. Several initiatives arose from the work of this committee (Fenner, 2005a). In July 1961, the President of the Academy (Professor Tom Cherry) wrote to the Vice-Chancellor of The Australian National University, Leonard Huxley, pointing out the desirability of establishing a Research School of Biological Sciences in the Institute of Advanced Studies. Instead of posting this letter to the Vice-Chancellor, who was also Secretary (Physical Sciences) at the time, I handed it to him. There was no reply for several months, and it transpired that Huxley had put my letter into a drawer until he had persuaded the Prime Minister to support the establishment of a Research School of Chemistry. Both were eventually established, Chemistry in 1964 and Biological Sciences in 1967.

In January 1962, Cherry wrote to the Prime Minister with two more proposals from the Fauna and Flora Committee: the establishment of a Research Museum of Australian Biology in Canberra, to carry out a National Biological Survey of Australia; and the appointment of an editor and staff to compile a comprehensive Flora of Australia. After long delays the Prime Minister finally replied, in April 1965, saying that 'Government finance...cannot be provided at present'. In 1967, I wrote to Bob Walsh, then Secretary (Biological Sciences), saying that I had written an informal letter to my old colleague Hugh Ennor, then Secretary of the Department of Education and Science, bringing to his attention the urgency of action on the Fauna and Flora Committee's Report. Walsh reactivated the

Committee and established botanical and zoological subcommittees. I was a member of the Committee and the zoological subcommittee. The latter then visited all State museums to ascertain the position of their zoological collections. I found this a very interesting exercise. The final Report of the Fauna and Flora Committee was published in 1968 (Flora and Fauna Committee, 1968). In March 1969, a deputation from the Academy met the Minister for Science and Education (Malcolm Fraser) and Macfarlane Burnet, then nearing the end of his term as President, told the minister that he thought the proposal for an Australian Biological Resources Study was the most important project to be proposed by the Academy. No decision was reached at this time, but, in August 1972, I gave evidence on behalf of the Academy's Fauna and Flora Committee to the House of Representatives Select Committee on Wildlife Conservation, which subsequently recommended 'that a biological survey be established by the Commonwealth Government to undertake on a continuing basis surveys of birds, mammals and reptiles and their ecology and to establish a national collection of wildlife species'. In August 1973, the new Labor Minister for Science (W. Morrison) announced the establishment of the Australian Biological Resources Study, which now operates within the Parks Australia Division of the Department of the Environment and Heritage, and provides grants to researchers on taxonomy, maintains substantial databases and produces a wide range of publications.

Committees on Environment and Conservation

Standing Committee on the Environment

In October 1969, I wrote to Council suggesting that the Academy might be able to act as a catalyst to promote the preparation of reports on the Australian environmental situation, similar to some produced at that time by the US National Academy of Sciences. In response, Council set up a Standing Committee on the Environment and a National Committee on the Environment. I was a member of both Committees (1970–78 and 1979–82 respectively). For several years, the Standing Committee was very active in establishing working groups and *ad hoc* committees that prepared reports on many environmental problems. Later, it was difficult to maintain the impetus, because the more obvious problems (at that time) had been covered by either the Committee's activities or governmental action.

The Botany Bay Project

Following its successful working symposia on the Murray River (Frith and Sawer, 1974), the Consultative Committee of the three learned academies (Humanities, Science and Social Sciences; the Academy of Technological Sciences and Engineering was not established until 1976) sought another project that would require and promote collaboration between them and the various disciplines that they encompass. The idea of looking at problems of environmental change in some long-occupied part of Australia won support and it was decided to focus on Botany Bay because of its historic associations. After discussion by the Councils of the three Academies, with the active participation of Noel Butlin, Professor of Economic History in the ANU, a Botany Bay Project Committee, of which I was elected Chairman, was established with four members of each Academy.

Discussions between the President (Bob Robertson) and Biological Secretary (Bob Walsh) of the Academy of Science and the Minister for Environment Control of the New South Wales government (Mr J. G. Beale) secured State government support and, in 1972, discussions between ministers of the newly elected Whitlam government and representatives of the three Academies secured a promise of $1 million over five years. A Ministerial Policy Committee was set up and attended by Professors Butlin, Hancock and me, to oversee financial matters and the appointment of a Director. Butlin was persuaded to apply, was appointed to the position in January 1974, and moved to an office in the University of New South Wales in April.

An early activity of the Project Committee was to seek proposals relevant to the Project from staff of the ANU and universities located in Sydney. 18 projects, each involving unpaid services and contracts, were approved, and over time produced a great deal of useful information. Between April 1974 and January 1975, five academics—whose fields of expertise covered chemical engineering, economics, demography, geography, microbiology, political history and water engineering—and three research assistants were appointed.

However, by October 1974, Butlin resigned because of difficulties in dealings with the New South Wales government and, to a lesser extent, with the Commonwealth government. After much discussion, the project was terminated, with only $380,027 of the $1 million Commonwealth grant spent. Butlin continued working on the project within the Research School of Social Sciences until 1978 and a substantial number of books and papers were produced (Botany Bay Project, 1972–78).

Histories of the Australian Academy of Science

In 1978, A. L. G. (Lloyd) Rees, a physical chemist who had been elected in 1954 and served as Secretary (Physical Sciences) from 1964–68, and I (who had a similar experience on the biological side) suggested to the Academy Council that it would be useful to take advantage of the fact that so many of the founders of the Academy were still alive and well, to put together a short history of its first 25 years, in preparation for the Silver Jubilee of the Academy in 1979. To do this, we enlisted the help of many Fellows and the secretarial staff of the

Academy. A hard-cover book of 282 pages (Fenner and Rees, 1980) was published and distributed to all Fellows and to other selected organizations.

In 1993, I suggested to Peter Vallee, the Executive Secretary of the Academy, that I should update the extensive Appendices of that book, in preparation of publication of a history celebrating the Academy's jubilee in 2004. However, noting that *The First Twenty-five Years* was then out of print, he suggested that I should update and revise the whole book. The co-editor of *The First Twenty-five Years,* Lloyd Rees, had died in August 1989, and I agreed to do this. With the guidance of a small Advisory Committee (Fellows R. W. Crompton, L. T. Evans, N. H. Fletcher and the Editor of *Historical Records of Australian Science*, R. W. Home), the book was published in 1995. In the same format as the previous book, it grew from 282 to 503 pages, largely because of the inevitable expansion of the appendices. In preparing the first two versions of the history, I spent a great deal of time working in the basement of the Dome, searching through the many files of archives stored there. Once again, I depended on the assistance of a number of Fellows and members of the secretariat; it was essentially an edited book. By the time that the second version, *The First Forty Years*, was prepared, even more assistance was provided by Fellows and members of the Academy's secretariat.

In 2002, in anticipation of the Jubilee celebrations in 2004, I began the update of *The First Forty Years* to *The First Fifty Years*. With the expert help of the publications Manager, Maureen Swanage, we greatly improved the format, largely on the basis of the arrangement of the fourth edition of *Medical Virology*, which David White and I had published in 1994. By judicious editing, the length of the text was reduced from 309 to 305 pages, but inevitably the length of the appendices and indexes increased, from 194 to 245 pages. On this occasion, however, instead of spending a great deal of time reading through relevant archives in the basement of the Shine Dome, I was able to consult the material on the Academy's website. As with the earlier versions, the task would have been impossible without the help of other Fellows, members of National Committees and the Secretariat.

Fenner Conferences on the Environment

As mentioned in the introduction to this chapter, by 1988 the money available in the Fenner Environment Fund was sufficient, given adequate external sponsorship (shown in brackets in the below list), to initiate these conferences, which were nearly always held, over a period of three days, in the Shine Dome. They have been very successful in providing an opportunity for people from academia, government, industry and non-governmental organizations with a wide range of interests and responsibilities in environmental problems to meet in the Shine Dome of the Academy. The following list gives an idea of the wide range of topics covered. The majority were published; a list of publications can be found in Fenner (2005b).

1. Australian Alps National Parks—World Heritage area? (Australian Alps National Parks Liaison Committee) 13–15 September, 1988.
2. Chemicals in agriculture (Australian Institute of Agricultural Sciences) 16 January, 1989.
3. Conservation in management of the River Murray system—making conservation count. (South Australian Department of Environment and Planning) 5–7 September, 1989.
4. Ultraviolet-B radiation impacts (Australian Society of Plant Physiologists and the Australian Marine Sciences Association) 19–21 September, 1990.
5. The constitution and the environment (Centre for Natural Resources and The Centre for Comparative Constitutional Studies of the University of Melbourne) held at the University of Melbourne, 29–30 November, 1990.
6. Protection of marine and estuarine areas—A challenge for Australians (Australian Committee for the International Union for the Conservation of Nature and Natural Resources and the Australian National Parks and Wildlife Service) 9–11 October, 1991.
7. Biological diversity—its future conservation in Australia (Department of the Arts, Sport, the Environment and Territories and the Ecological Society of Australia) 11–13 March, 1992.
8. A conservation strategy for the Australian Antarctic Territory (Australian Antarctic Foundation and the Centre for Resource and Environmental Studies, Australian National University) 8–9 February, 1993.
9. International trade, investment and the environment (Faculty of Engineering and Applied Science, Gold Coast University College) 27–29 July, 1993.
10. Sustainability: principles to practice (Department of the Environment, Sport and Tourism) 13–16 November, 1994.
11. Environmentally responsible defence (Australian Defence Studies Centre) 8–10 November, 1995.
12. Risk and uncertainty in environmental management (Environment Protection Agency) 13–17 November, 1995.
13. Linking environment and economy through indicators and accounting systems (Institute of Environmental Studies, University of New South Wales) 30 September–3 October, 1996.
14. Developing strategies for sustainable habitation in the rangelands (CSIRO Division of Wildlife and Ecology) 29–30 October, 1996.
15. Ethics of manipulative research and management practices in world heritage and environmentally sensitive areas; policy and practice (Great Barrier Reef Marine Park Authority) 26–28 November, 1997.
16. Future Australian landscapes—visions of harmonious environment (Bureau of Resource Sciences) 2–5 May, 1999.

17. Biodiversity conservation in freshwaters (Cooperative Research Centre for Freshwater Ecology) 5–8 July, 2001.
18. Nature tourism and the Australian environment (Cooperative Research Centre for Tourism and Griffith University) 3–6 September, 2001.
19. Redesigning agriculture for the Australian environment (Johnstone Centre, Charles Sturt University) 31 July–1 August 2002.
20. Understanding the population-environment debate: Bridging disciplinary divides (Australian Academy of Science) 24–25 May, 2004.
21. Wildlife Health Workshop (Wildlife Disease Association, Australasian Section) 11 July, 2005.
22. Urbanism, Environment and Health (National Centre for Epidemiology and Population Health, ANU) 25–26 May, 2006.

Video Histories of Australian Scientists

In 1992, the Royal Australasian College of Physicians became involved in a project initiated several years earlier by the Royal College of Physicians of London, by which video histories (audiovisual records) were prepared depicting selected Fellows of the College. In March 1992, Dr Brian Gandevia, the Honorary Librarian of the Royal Australasian College of Physicians, contacted me to say that Dr Max Blythe, of Oxford Brookes University, who had carried out the interviews for the London College, was coming to Sydney with his cameraman, and he wanted me (as a Fellow of both Colleges) to come down to Sydney for an interview. Another visit to Sydney was planned for September 1993.

Having seen the results of these interviews, in which those interviewed talked about their early lives, the development of their interest in science, their research work and other aspects of their careers, I thought that it would be a good idea if the Academy of Science should become involved and a small committee, of which I was Chairman, was set up by the Council. I suggested that the Academy should arrange for Blythe to come to Canberra to interview Fellows whenever he came out to Australia. In February 1994, Blythe and his cameraman interviewed several Fellows of the College in Sydney, of whom three were also Fellows of the Academy. They then came up to Canberra and interviewed nine more Fellows, all biological scientists, since this was Blythe's background. This occurred on several occasions. Subsequently, a number of Fellows acted as interviewers and local firms undertook the recordings. In March 2001, Council of the Academy resolved, over time, to make such recordings of as many Fellows as possible. In addition, with outside funding, 15 female scientists who were not Fellows and another 20 early-career scientists, including nine women, were interviewed in 2001 and 2002. Edited transcripts of the interviews, together with accompanying teachers' notes, are progressively added to the Academy's website. The total number of visitors to these entries in 2003–04 was 167,921, an increase on the previous year's total of 121,429.

Fenner Medal for Plant and Animal Sciences

I was executor of the will of Alfred Gottschalk, FAA, who died in 1973, and at his request I arranged for $35,000 to be donated to Australian Academy of Science as an endowment to support the award of a medal to a young scientist, who was not a Fellow, for distinguished research in medical or biological science. The first award was made in 1979. 20 years later, looking over the list of recipients, I realised that biomedical scientists had received by far the most awards, and I saw the need for another similar award for non-medical biology. Given Gottschalk's background in biomedical research, and my broad interests in the conservation of biological diversity as well as virology and preventive medicine, I decided, after consultation with the Academy Council, to set up a fund for a medal in biological science, the Gottschalk Medal being awarded for biomedical research. Having donated enough to cease further support for the Environment Fund, I started in 1997 to make donations for a medal in biological sciences, which Council, following precedent, named the Fenner Medal for Plant and Animal Sciences, to be awarded annually for research in biology, excluding the biomedical sciences, carried out by a scientist under the age of 40 years who is not a Fellow of the Academy (Fenner, 2005c). The first award was made in 2000 and the endowment reached the required total of $100,000 in 2001.

References

Botany Bay Project, (1972–78). MS 125, Basser Library, Australian Academy of Science.

Fenner, F. (ed) 2005a, *The Australian Academy of Science. The First Fifty Years.* The Australian Academy of Science, Canberra, pp. 133–8.

Fenner, F. (ed) 2005b, *The Australian Academy of Science. The First Fifty Years,* The Australian Academy of Science, Canberra, pp. 447–8.

Fenner, F. (ed.) 2005c, *The Australian Academy of Science. The First Fifty Years,* The Australian Academy of Science, Canberra, pp. 37, 384.

Fenner, F. and Rees, A. L. G. (eds) 1980, *The Australian Academy of Science: The First Twenty-Five Years,* The Australian Academy of Science, Canberra.

Flora and Fauna Committee 1968, Proposal to Establish a Biological Survey of Australia. Australian Academy of Science.

Frith, H. J. and Sawer, G. (eds) 1974, *The Murray Waters,* Angus & Robertson, Sydney.

Chapter 9. Director of the Centre for Resource and Environmental Studies, 1973 to 1979

Establishment of CRES

In August 1969, Vice-Chancellor Crawford asked me to chair a committee to prepare a submission to the Australian Universities Commission (AUC) for the 1973–75 triennium, advocating the establishment in the ANU of a Centre for Natural Resources (later and henceforth called the Centre for Resource and Environmental Studies (CRES)). The AUC approved the proposal, envisaging that the Centre would grow from 2 senior academic staff at the end of 1973, to 5 senior, 7 short term and 13 support staff by 1975, and when fully developed in 1977, a core of 8 senior academic staff with tenure appointments, 15 non-tenure appointments at levels from post-doctoral fellow to senior research fellow, and several short-term senior visiting fellowships, supported by a non-academic staff of 19 (Fenner, 1979a). As well as PhD students, there was to be an MSc program by course-work.

My Appointment as Director of CRES

I had been appointed Director of JCSMR in September 1967, with a term of seven years. This appointment would therefore terminate in September 1974, and well before that I had to decide whether I wanted a further term as Director of JCSMR, or to resume bench work in the Department of Microbiology, or do something else. CRES had received approval and funding as from January 1973, and the Directorship of the new institution was advertised late in 1972. Initially, because of my involvement in the planning, I was named as a member of the selection committee, but before it met, having decided that I would apply for that position, rather than remain in the JCSMR, I withdrew from the Committee. It is of some interest to insert here a copy of my letter to the Vice-Chancellor, dated 11 September, 1972, which sets out my reasons for this application.

> Dear Vice-Chancellor,
>
> I am writing to tell you that I am interested in the position of Director, Centre for Resource and Environmental Studies. I therefore wish to withdraw from the Electoral Committee for that post. It may be useful to you if I outline how a person whose scientific career has been in animal virology is now interested in such a post. In the first place, my family background gave me a broad interest in natural sciences, and the medical course itself provided a broad if superficial training in human biology. During my medical course I took Botany 1 and Geology 1 as additional

science subjects, and spent much of my spare time working with anthropologists and others in the South Australian Museum. After graduating I spent about half of my army career (which extended from June 1940 to February 1946) as a malariologist, a post which involved the supervision of units for malaria control and entomological research as well as the diagnosis of malaria and scrub typhus. Subsequently my first scientific contacts were with F. M. Burnet and R. J. Dubos, men with whom I have maintained close associations and who have become elder statesmen in the environment-resource area. Likewise, although I have been interested in the cellular and molecular biology of viruses as well as the pathogenesis of viral diseases, my principal experimental work for a number of years (1951–64) involved collaborative studies with animal ecologists and entomologists in CSIRO, on the ecology of myxomatosis.

My appointment in 1958 as Secretary, Biological Sciences, of the Australian Academy of Sciences further broadened my responsibilities and interests. In that position I was in part responsible for producing reports on the desirability of establishing the Research School of Biological Sciences in the ANU, and on setting up a Museum of Australian Biology and Biological Survey of Australia. In 1967–68 I resumed contact with the Biological Survey as Chairman of the Flora and Fauna Committee of the Academy, that produced a second report, published in 1969, that is now under active consideration by Government.

I moved into the more deliberate consideration of environmental problems during the preparation of a paper delivered at a symposium in September 1969, given in honour of Macfarlane Burnet's 70th birthday. In the same year I became Chairman of the ANU Committee on Natural Resources (now the Centre for Resource and Environmental Studies) that prepared the report that was subsequently accepted by Council and the AUC. Since then I have been Chairman of the Working Group of the Centre for Natural Resources and its representative on the Users Committee for the Life Sciences Library Building. During the last three years I have also become involved in several other committees concerned with environmental and resource problems. As I indicate, these vary in the intensity of their activities, but they have brought me into close contact with problems of resource management and with people involved with environmental questions.

The Australian Academy of Science

- Chairman, Flora and Fauna Committee since November 1967 (intermittently active)

- Chairman, Committee on National Parks and Conservation, since 1970 (relatively inactive)
- Chairman, National Committee for SCOPE, since 1971, (relatively inactive)
- Member, Standing Committee on the Environment, since 1970 (active)

National Activities

- Vice President, Australian Conservation Foundation, since October 1971 (active)
- Chairman, Study Group on International Aspects of the Human Environment (Australian Institute of International Affairs), since September 1972 (just being activated)
- Chairman, Three Academies Project Committee on Botany Bay, since August 1972 (active)
- Member, Advisory Council of CSIRO, since 1970 (intermittently active)
- Member, National Committee on Man and the Biosphere (UNESCO), since 1971 (relatively active)
- Member, Executive Committee of the Fact Finding Study on the Alligator Rivers Area, Northern Territory, since August 1972 (active)

International Activities

- Member, Scientific Committee on Problems of the Environment (SCOPE) since 1971 (intermittently active)

I have also given a few addresses on environmental topics that have been published.

- 1969 Brahma, Shiva and Vishnu, three faces of science. *Australasian Annals of Medicine* , vol. 18, 351–60.
- 1970 The effects of changing social organization on the infectious diseases of man. In *The Impact of Civilization on the Biology of Man* (S. V. Boyden, ed.), pp. 48–68, ANU Press.
- An overview of man and his environment in Australia. *Proceedings of the First International Congress on Domiciliary Nursing* , pp. 1–8.
- 1971 The environment. In *How Many Australians? Immigration and Growth*. Proceedings of the 37th Summer School of the Australian Institute of Political Science, pp. 37–60. Angus & Robertson, Sydney.

I attach a curriculum vitae and a list of publications.

Yours sincerely,
Frank Fenner

I was appointed and took office in May 1973, with a term that extended to my retirement on 31 December, 1979. Initially, I took up office in the old Nurses' Home just across the road from the JCSMR, my secretary from the JCSMR, Margaret Mahony, coming with me. In 1976, CRES moved to the two top floors of the recently completed Hancock Library building.

Activities as Director

My first priority was to attract a number of well-qualified senior staff members, and this was successful, Stuart Harris commencing work as Professor of Resource Economics in February 1975, and Peter Young as Professorial Fellow in Applied Systems Analysis in January 1975. David Ingle (Dingle) Smith was appointed as a Senior Fellow in January 1976, to be responsible for the course-work degree of Master of Environmental Studies as well as research in the Hydrology/Water Quality Program. However, in 1976, University funds had ceased to grow, for the first time in its history, and the prospect of appointing enough research fellows to support the work of these senior staff members looked bleak. Two events offered some relief. The first was an agreement with the Director of the JCSMR to transfer to CRES, with funds, the Urban Ecology Group of the Department of Human Biology in that School (leader, Professorial Fellow Stephen Boyden). The second was the recognition by the Heads of Schools Committee, on the initiative of the Director of the Research School of Biological Sciences, Professor R. N. Robertson, of the plight in which the freeze on funds had caught CRES. An additional continuing grant of $50,000 per annum allowed three additional research fellows to be appointed.

In mid-1976 two distinguished Visiting Fellows, Dr H. C. Coombs, founding father of the ANU, and Dr A. B. Costin, pioneer of ecological studies in the Kosciusko National Park, joined CRES as long-term Visiting Fellows, supported by external funds.

Because of the need to use research assistant and research fellow positions in CRES to support the major research activities of other senior staff members, I worked without such support and thus spent a large part of my time on *ad hoc* problems rather than a major research project. On appointment to CRES, I resigned from membership of the Epidemiological (Standing) Committee, National Health and Medical Research Council, but during the next few years I continued or accepted membership of several other national and international committees concerned with environmental problems, as follows:

- Member, Advisory Council of CSIRO, 1970–75.
- Member National Committee on Man and Biosphere (UNESCO), 1971–78.
- Vice-President, Australian Conservation Foundation.
- Chairman of the Three Academies Project Committee on Botany Bay, 1972–74, and its replacement, the Botany Bay Project Management Committee, 1974–77.

- Member, Scientific Committee on Problems of the Environment (SCOPE), 1971–76 and Editor-in-Chief, SCOPE Publications, 1976–80.
- Member, Senior Scientific Advisory Board, UN Environment Program Project, 'The State of the Environment: Ten Years after Stockholm, 1978–82.

My lecturing and writing activities became increasingly diverse, often through committees set up by the Australian Academy of Science (see Chapter 8). However, I continued to take an interest in the work of the WHO Intensified Smallpox Eradication Program and attended many meetings of its expert committee on poxviruses (see Chapter 10). I also continued my book-writing, producing, with co-author David White, a second, completely revised, edition of *Medical Virology*, published in 1976; later that year the *Second Report of International Committee on the Taxonomy of Viruses*. In 1977, I teamed up with a former colleague, Adrian Gibbs, in producing a book of essays by 15 experts, *Portraits of Viruses: A History of Virology*. In 1978 Lloyd Rees and I co-edited *The First Twenty-five Years*, a history of the Australian Academy of Science.

The Scientific Committee on Problems of the Environment (SCOPE)

SCOPE was set up in 1970, as the International Council of Scientific Unions (ICSU) committee devoted to environmental problems, and I was elected a member of SCOPE at a meeting in Canberra in August 1971. I attended meetings of the Scientific Committee in Paris in January 1973 and March 1975, and of the General Assembly and Scientific Committee in Kiel, Germany in October 1973 and in Paris in May 1976.

One of the principal functions of SCOPE was to produce reports by small groups of internationally chosen natural and social scientists about important environmental problems of a scientific nature. Six of these had been produced between 1971 and 1975, but there were difficulties in distributing them. After the May 1976 meeting the Executive Committee decided to secure a commercial publisher, and in December 1976, concluded a contract with John Wiley and Sons Ltd. (UK), initially for four years. The contract was renewed in 1980. As part of the agreement SCOPE undertook to appoint an Editor-in-Chief to be responsible for the editorial content of the Reports. I was appointed to this post by the Executive Committee in October 1976.

Apart from attending all meetings of the Executive Committee and visiting the offices of John Wiley and Sons in Chichester, England, after these meetings, I developed an extensive correspondence with Gilbert White, the President of SCOPE, Ronald Keay, Executive Secretary of The Royal Society, who was Treasurer of SCOPE, Vassily Smirnyagin, the Executive Secretary of SCOPE and especially with Dr Howard Jones, John Wiley's Molecular and Earth Sciences Editor (the Jones' file is more than one centimetre thick in my Basser Library files, MS 143/12/1 to MS143/12/5). The changed arrangements led to much better production and sales (in addition to copies distributed by SCOPE), the figures rising from about 300, before the new arrangements, to between 750 and 1,400.

Because I had taken on the major task of writing an account of the global eradication of smallpox earlier that year, I handed over my responsibilities as Editor-in-Chief to Dr R. E. Munn on 31 December, 1980.

Overseas Travel 1973 to 1979

Between 1974 and 1979, I undertook many short trips related to meetings of either the Executive Committee of SCOPE or in connection with the WHO Smallpox Eradication Program. In the main, these were funded by the organizations concerned. Those involving CRES or SCOPE business are listed here; those solely or primarily concerned with smallpox are listed in Chapter 10.

27 September to 15 October 1973

The main purpose of this trip was a meeting of SCOPE in Paris. On the way, I stopped off in Kathmandu, Nepal, for three days (as a tourist, very interesting) and in London, to interview applicants for positions in CRES and the Botany Bay Project. Peter Young came to CRES as Professorial Fellow in Applied Systems Analysis as a result of this visit. Then I went to Kiel for the General Assembly and Scientific Committee of SCOPE, from 5–10 October. After the meeting Dr Otto Fenner (a second cousin) and his wife Elizabeth, who live in Hamburg, picked me up for dinner.

19 November 1973 to 15 May 1974

This was the second stage of my Fogarty Fellowship, this time with Bobbie, leaving my son-in-law Arthur Marshall and daughter Marilyn to look after the house at 8 Monaro Crescent. We made it a holiday trip over and back, stopping off at Fiji, Tahiti, Peru, with a trip to Cuzco, and Guatemala, for several days; a total of 45 days of most enjoyable and interesting sight-seeing. In Washington, we were met by Dr Haggerty, the man in charge of the Fogarty Scholarships, and driven to Stone House. Other scholars in residence at the time were Michael Sela, an immunologist from Israel, Dr and Mrs Darling, who were involved in investigations of the Hiroshima bombing, a Finnish histochemist and Margaret Mead.

We had a wonderful time there, travelling extensively around the States, to see sights like Williamsburg and visiting academics involved with environmental studies in universities in many of the eastern states. We also spent a lot of time visiting the many galleries and museums in Washington, an experience which led me to become a Foundation donor (of $1,000) to the National Gallery of Australia when I returned home. We also saw a lot of Bob and Beth Chanock, who became our best friends in the United States. I had stimulating talks with René Dubos when I went to New York. Bobbie and I drove up the Adirondacks to Saranac Lake to see George Mackaness (former Head of the the Department of Experimental Pathology in JCSMR) and Alan Logie (former Head Technician of the Department of Microbiology), now Director and Manager of the Trudeau Institute. I persuaded Bob Blanden, a South Australian who had worked with Mackaness in Adelaide and went with him to the Trudeau Institute, to come to Canberra as a PhD student.

On 6 January, I flew to England for 12 days, primarily to attend the Ditchley Conference, which dealt with social aspects of the environment, and interview further prospective staff for CRES and the Botany Bay project. I went to Victoria Harbor in Canada for a week in February for the SCOPE 5 Conference on Environmental Impact Assessment, which resulted in a report of some 240 pages (Munn, 1975), a second edition of which was published in 1979. Looking over the diary of the visit, I wrote many letters and saw a great many people interested in environmental problems.

We left Washington for London on 11 April, where we stayed with Cecil and Beattie Hackett, and then went to Amsterdam, where we spent a lot of time in the Van Gogh Museum. After spending couple of days in Vienna we went to Athens, where we had booked for a five-day trip around the Adriatic on *MS Aquarias*. Marvellous. Then we went to Cairo, where we were met by Harry Hoogstraal. He showed us around Cairo, then we flew up to Luxor and spent two days there before returning to Cairo and then back to Canberra, with a day in Hong Kong, mainly to see Stephen Boyden's students, Keith Newcombe and Sheila Millar, at work on the 'Metabolism of Hong Kong'.

14–20 October 1974

This was a brief trip to Pattaya, in Thailand, to participate in the Expert Group Meeting on Environmental Studies and Development, organized by the UN Asian Institute for Economic Development and Planning. In my diary I note that I 'felt that my contributions were useful and used'.

15–29 November 1974

SCOPE meeting in Moscow. I went there via Tokyo. As well as the five day meeting, I went to the Kremlin Theatre twice, to *Don Carlos* and *Rigoletto*. Another day I was collected by my virologist friend of 1964, Vladimir Agol, and taken to his flat in a newly developed part of Moscow for the evening. On 23 November, I returned via New Delhi, where by chance I met Jack Crawford and dined with him, the Australian High Commissioner and Hedley Bull, of ANU, who was on study leave. Spent time at the Jawaharal Nehru University, discussing among other things, the proposed Australia-India Science Agreement.

6–22 March 1975

This trip was made to attend a meeting at the International Institute of Applied Systems Analysis (IIASA) in Vienna and a SCOPE meeting in Paris. Again, opportunities to go the opera, *Karmen* and *The Barber of Seville*. I went home via London and interviewed D. I. (Dingle) Smith, who was then appointed a Senior Fellow in CRES, with responsibility for the MSc (Environmental Studies) program.

7 August to 11 October 1975

This was a long and complicated trip, which included five main items: a climate conference in Norwich, the 13th Pacific Science Congress in Vancouver, a virology conference in Madrid, a visit to the UN Environment Program Headquarters in Nairobi, a Climate Conference, and a visit to Kruger National Park in South Africa. I went via Bangkok to Kathmandu, mainly to replace a topaz earring that I had purchased there on an earlier trip. This took a few days, so I worked for most of the day on a climate change report and went to some Nepalese dance shows in the evenings. Then to Tehran via New Delhi. I went on tour to Isfahan, with its superb Blue Mosque. Then to Vienna, where I left my luggage at the airport and took a bus to Bratislava to see Zlata Wallnerova, who had worked in the Department of Microbiology for two years. She and her husband later came to Canberra as immigrants. Back to Vienna airport and on to London and by train to Norwich and the University of East Anglia, where I attended the week-long World Meteorological Organization–International Association of Meteorology and Atmospheric Physics, on Long-Term Climatic Fluctuation. There were several Australian experts there; I was very much a learner. On 23 August, I took a Great Circle flight to Vancouver, for the 13th Pacific Science Congress, where I gave a lecture in a Symposium on Mankind's Future in the Pacific (Fenner, 1976a). Looking at the text almost 30 years later, I am surprised how much in tune it was with present-day views.

Then to the Virology Congress in Madrid, where I spent most of my time with the International Committee on the Taxonomy of Viruses (ICTV), of which I was then President. Besides agreeing on the substance of the Second Report (Fenner, 1976b), I was successful in getting a plant virologist, New Zealander R. E. F. Matthews, elected President. This was important because at the time there was a serious possibility that the plant virologists would break away from the International Committee. After the Congress I travelled around Spain, visiting many wonderful places including Toledo and the Alhambra palace in Granada.

Then by plane to Nairobi, to visit the headquarters of the United Nations Environment Program (UNEP). This was useful visit; we discussed interactions between SCOPE, as the principal environmental non-governmental organization (NGO) and UNEP. Then a minibus trip to a National Park and then to 'Treetops' for an overnight stay, to see all the African animals coming to drink or lick salt, under brilliant floodlights. Then to Pretoria and met by George Bornemissza (dung beetle expert); later met with South African members of SCOPE and visited the local Department of Planning and the Environment. Then to the Conference on Climate Change at the Transvaal Museum. After talking with staff of the University of Witswatersrand I went on a long drive through Kruger National Park. This was most interesting. Then three very interesting days in Capetown before embarking for the long flight back to Sydney.

27 November to 6 December 1975

Short trip to Tokyo via Manila to attend a meeting of the Council of the United Nations University in Tokyo, where I gave a talk on Resources and Reserves.

15–22 May 1976

This trip was to Paris to attend a meeting of the General Assembly of SCOPE, at the UNESCO Building in Paris. It had been called to ratify the new Constitution. I was made Chairman of the Resolutions Committee, which discussed future programs, so I was kept quite busy.

1–24 October 1976

The prime purpose of this trip was to attend a meeting of the Executive Committee of SCOPE in Washington. As usual, besides attending the meeting and seeing friends in Washington, especially Bob and Beth Chanock, I stopped off in Hawai'i and Phoenix, Arizona, on the way over and went to Toronto to see the arrangements for environmental studies at the University of Toronto and York University. On the way over I stopped in Hawai'i with my old friend Edgar Mercer, who lived on the Big Island. He drove me all around the island, and to the crater of the active volcano, Kilauea, a wonderful site for someone interested in geology. Then, on the way across America, I went to Flagstaff, Arizona, and from there got a bus to see the Grand Canyon. I spent three days there, walking around the rim and going on a plane flight. A marvellous sight. Then to Washington, where I stopped to attend a meeting of ICSU and the Executive Committee of SCOPE, both held at the National Academy of Sciences. I stayed at the Cosmos Club, in central Washington. It was there that I had dinner with Gilbert White, the new President of SCOPE and Ronald Keay, Treasurer of SCOPE, and Gilbert suggested that I should take on the job of Editor of SCOPE Publications; I agreed. I also visited the museums and National Gallery, went to a concert and then to a reception at the National Academy of Sciences, where I met several other SCOPE friends and also Michael Stoker, the new Foreign Secretary of The Royal Society, and Carleton Gajdusek.

From Washington I went to Trentin, New Jersey, where I was met by George Mackaness, then President of the Squibb Institute of Medical Research. He was living in a converted barn, three storeys high, built in 1790. Then a brief stop in New York to see friends at the Rockefeller University and on to Toronto, where I met the President of the University of Toronto and gave a lecture on CRES. The next day, I met with eight Directors of Institutes in the University of Toronto (like our Centres) who saw themselves as the cutting edge of new interdisciplinary developments in the university. I spoke about the position in ANU. After a day at the Faculty of Environmental Studies at York University I returned to Australia.

1 April to 6 May 1977

The principal reason for this trip was to participate in the International Commission for the Certification of Smallpox in India (see Chapter 10). After spending three weeks there I had a week to fill in before attending a meeting of the SCOPE Executive Committee in Paris. So I flew to Dubrovnik, a wonderful old city on the Adriatic and spent two days there and on a bus trip along the shores of the Adriatic to the south of Dubrovnik. Then on to England, where I went down to Chichester to see Dr Howard Jones, of John Wiley and Sons, in my new job as Editor-in-Chief of SCOPE publications. After a day in London, during which Bobbie rang me while I was with Cecil and Beattie Hackett to say that I had been elected a Foreign Associate of the US National Academy of Sciences, I flew to Paris for a meeting of the Executive Committee of SCOPE.

26 September 1977 to 3 November 1977

The initial purpose of this trip was a meeting of the Executive Committee of SCOPE in London. It was a busy meeting that went on for four days, including a full day at Wileys in Chichester. Then I flew to Geneva for a major meeting on the Certification of the Eradication of Smallpox.

3–20 May 1978

After a few days in England, during which I went to Chichester for a day to see Dr Jones about SCOPE publications and interviewed Professor Maynard Smith as possible Director of the Research School of Biological Sciences (on behalf of Vice-Chancellor Low), I went to Warsaw for a week before going to a meeting of the SCOPE Executive Committee in Moscow. As well as seeing a good deal of the fascinating history of Warsaw and Cracow, I gave a lecture on smallpox eradication at the Institute of Medical Microbiology in Cracow, another on the activities of SCOPE at the Silesian Centre for Environmental Studies and a third on environmental studies in Australia at the Polish Academy of Science in Warsaw. Then flew to Moscow, where we had a four-day meeting of the SCOPE Executive Committee. Meals at the Moscow hotels were poor, but each evening members of the Committee were taken to dine at very good restaurants. One evening Gilbert White, President of SCOPE, spoke of his days as a Quaker travelling around Europe during the World War II.

6 November to 14 December 1978

Once again, a trip encompassing smallpox and SCOPE. I went first to Geneva for meetings of the Global Commission and the Monkeypox Consultative Group. After three weeks in Geneva, I went to England and down to Chichester to see Wileys and then flew to Nairobi, where the SCOPE Executive Committee met with the officials of the United Nations Environment Program. After four days there I flew back to Geneva for the first meeting of the Global Commission for the Certification of Smallpox Eradication, of which I had been elected Chairman.

7 June to 1 August 1979

This rather long trip began with a meeting of the SCOPE Fourth General Assembly, followed by a meeting of the Executive Committee, in Stockholm, from 9–15 June. Then I went to London, where I stayed with Cecil and Beattie Hackett, who were celebrating their 40th wedding anniversary. I went to Cambridge to see Peter Young (just recruited to CRES as a systems analyst) and Joseph Needham, the great authority on Chinese science, to discuss early Chinese experience of smallpox. Then to Geneva for a few days to discuss my forthcoming trip to China concerning smallpox eradication there and to work on the agenda for a meeting of the poxvirus expert committee to be held in Atlanta. I visited Lloyd Thomson, who had worked with Bobbie in the blood bank at 2/2 Australian General Hospital in the Middle East and Queensland, he was then Australian Ambassador in Switzerland. Then to Atlanta, where I worked on the final report of the Commission for the Certification of the Global Eradication of Smallpox. I went with Walter Dowdle to his home in the woods and then to Stone Mountain, where gigantic carvings of the heads of some US presidents have been carved on the side. Then New York, where I stayed at the Rockefeller University and saw old friends, René Dubos, Merrill Chase and Jim Hirsch. Talked with Barsky (Academic Press) about *Medical Virology*, a possible *Veterinary Virology* and a possible third edition of *The Biology of Animal Viruses*; the latter two to be post-retirement options. Then to Washington, I went with Chanocks to the National Gallery and the next day to the National Institutes of Health to see scientists and enquire about a third and six-month spell of my Fogarty Fellowship, probably in 1982. On 11 July, Joel Breman and I flew to Japan, en route to China, to check their smallpox eradication program.

18 November to 14 December 1979

This trip was initially to Geneva, to work on smallpox final report and also, in another section of WHO, to discuss the United Nations Environmental Program 1982 Report. Then to Paris, for three days, for another meeting of the SCOPE

Executive Committee. Then Geneva again for the final meeting of the Commission for the Certification of the Global Eradication of Smallpox.

Lectures in Australia (excluding Lectures on Smallpox)

1973

'Environmental Problems: Approaches to Understanding and Solution at Local, National and International Levels'. Public lecture to the National Parks Association of the ACT, 22 March, 1973.

'Population, Pollution, Poverty and Prosperity', ANZAAS, Adelaide, 23 June, 1973.

'Comment on Treasury Economic Paper No. 2.' Lecture to the Economics Society of Australia (ACT Branch), 19 June 1973.

'The Importance of Port Phillip Bay.' Lecture to a symposium on the Environmental Study of Port Phillip Bay, 8 September, 1973.

1974

'A Lateral Arabesque: from Virology to Environmental Studies.' Address at A Tribute to Sir Macfarlane Burnet on the Occasion of his 75th Birthday, 3 September, 1974.

'Strategies for Energy Conversion: Resources and Constraints.' School Seminar on Growth, Research School of Social Sciences, October, 1974.

1975

'Environmental Impact Assessment: History, Rationale and Methodologies.' Zoology Section, ANZAAS, January 1975.

'Environmental Implications of Economic Development in the ESCAP Region.' Address to MADE, June 1975.

'The Last Five Years and the Future: Environmental Studies,' Octagon Lecture, University of Western Australia. July 1975.

1976

'Siglitis: the Activities of ICSU and SCOPE,' CRES Workshop 14 December, 1976.

1977

'The Role of Science in Environmental Policy,' Lecture to Australian Water and Waste Water Society, Canberra. July 1977.

Closing Lecture. International Symposium on Microbial Ecology, Dunedin, New Zealand, August 1977.

'The Australian Environment,' Lecture to National Environmental Education Conference, Lorne, Victoria. 14 November 1977.

1978

'Aggression and the Environment,' Lecture to Second Austral-Asian Pacific Forensic Sciences Conference, Sydney, July 1978.

Introductory address to Symposium: 'Fire and the Australian Biota' on 'The SCOPE Project: Ecological Effects of Fire'. 9 October, 1978.

1979

'Some Ecological Effects of Viral Infections', Lecture for Australian Society for Microbiology (ASM), Adelaide, 14 May 1979.

'Control of Infectious Diseases.' Lecture for Australian Society for Microbiology (ASM), Adelaide, 15 May 1979.

Honours and Awards

Both of the awards I received during my period with CRES recognized work done while I was in the JCSMR.

In 1976 I was appointed a Companion of the Order of St Michael and St George (CMG), for services to Medical Research.

In 1977 I was elected a Foreign Associate of the US National Academy of Sciences.

Overview, the Activities of CRES, 1973 to 1979

I was scheduled to present my annual Director's Report to Council at the meeting of the ANU Council on 9 November, 1979, and since I was to retire at the end of the year, I decided to give an overview of my six years as Director, as set out here (Fenner, 1979b):

> CRES was established to undertake high quality policy-oriented applied research in the fields of natural resources and the environment, and to provide a master's degree program in these fields. After my appointment as Director on 14 May, 1973, I set out to recruit four senior colleagues to fill tenured positions: in resource economics, urban ecology, applied systems analysis and as a coordinator for the projected master's degree course. After the inevitable delays that occur in the early stages of a new activity, four excellent appointments were made: Professor Stuart Harris (February 1975), Dr Stephen Boyden (January 1976), Dr Peter Young (January 1975) and Mr D. Ingle Smith (January 1976). I delayed making appointments of research fellows until these men had settled in and defined their requirements; unfortunately by this time (mid-1975) the financial situation had deteriorated greatly and funds were available for only four such posts. Fortunately, action by the Heads of Schools

Committee enabled a further three research fellowships to be filled by early 1977. Since then the position has remained static, except for posts funded from outside grants. The latter source has been important in providing support for some senior long-term visiting fellows (Dr H. C. Coombs and Dr A. B. Costin) and similar appointments have been made at the research fellow level (Dr M. Brandl) and one is in the course of advertisement.

The academic staff is supported by a total of 16 research assistantships (seven on part time appointment), seven of which are supported on outside funds, an excellent secretarial staff, a programmer and a librarian. In my view the tenured staff of CRES is adequate, for the time being, but the support staff (both research fellows and research assistants) should be increased if the incoming Director is to be able to make the contribution to university, national and international affairs that CRES should make over the next decade.

Philosophy of Research.

The opinion was expressed by the Review Committee that most of the staff of CRES should be employed for most of their time in a succession of 'major projects', each lasting two or three (never more than five) years. I disagree with this view. I believe that the prime task of CRES is to carry out good research in the resources/environment field, directed mainly to matters of public policy. There are three components in this task; to do good multidisciplinary work on relevant topics, to develop new ideas and concepts, and to build up a strong base of knowledge and information in the fields in which CRES works. Nevertheless, I believe that most of the staff of CRES are now ready to devote part of their time to work on a common program, and the incoming Director should be provided with at least two research fellowships to assist him in launching such a program

The Performance of CRES

In the research schools one can judge performance, to a large extent, by the quality and volume of output of published articles and books as judged by the appropriate international peer group. CRES research has to satisfy this criterion, but has an additional need to devote a large part of its effort to problems of national importance, including advice to government on policy matters.

Considering that project work did not start seriously until 1976 and was not fully developed (with the available staff) until 1977, I believe that CRES has performed very creditably. Over the period of three and a half years during which CRES has been operating at anything like full strength, 163 article in journals or chapters in books have been produced,

99 internal CRES Reports and 19 books. A further nine books are in preparation.

	Articles and chapters	Books	CRES reports
Resources	49	7	37
Applied systems Hydrology	25	2	33
Land use	28	2	4
Environmental management	13	5	0
Human ecology	48	8	20

Teaching

Besides PhD students, CRES provides a Master's degree by coursework. Dingle Smith has done splendid work as the Coordinator of this course, which is now on its second run. CRES has provided a focus for Australian interest in such courses and initiated a meeting of Australian Course Coordinators in 1976, followed up by a UNESCO-sponsored course (jointly with the Centre for Environmental Studies of the University of Melbourne) in September 1979.

Accommodation

Present accommodation is adequate for the present staff and could accommodate a few more, but if an incoming Director is as successful in attracting visitors and students as I would expect, additional accommodation will be required before his term expires. From the point of view of providing more suitable accommodation than that presently available for the Hancock Library, and to accommodate CRES, a strong case can be made to extend the first two floors of the Library to the South, for the use of the Library, and allow CRES to occupy the whole of Floor Level Three.

Affiliation

It is highly desirable that CRES should be included in the information-exchange system of the university; the best method to achieve this would be to invite the Director to all meetings of the Heads of Schools Committee.

Outside Funding

If it were necessary to do so, an enterprising Director could raise large sums of money for consultant work by CRES. I have eschewed such action, nevertheless the funds that have been obtained for the support of work that is academically desirable and are within the research programs of CRES have been very important in enabling project work to proceed.

References

Fenner, F. 1976a, Options for man's future: a biologist's view. in *Mankind's Future in the Pacific*, 13th Pacific Science Congress (R.F. Scagel, ed.) pp. 140–60, University of British Columbia Press, Vancouver.

Fenner, F. 1976b, Classification and nomenclature of viruses. Second Report of the International Committee on Taxonomy of Viruses, *Intervirology*, vol. 7, pp. 1–115.

Fenner, F. 1979a, *The Centre for Resource and Environmental Studies: 1973–79,* The Australian National University, 76 pages.

Fenner, F. 1979b, The activities of CRES: 1973 to 1979, Report to Council of ANU, 9 November, 1979, Basser Library 143/13/11/6A.

Chapter 10. Smallpox and its Eradication, 1969 to 1980

Introduction

Why have a whole chapter on smallpox? The reader will know when he reads this chapter and portion of the next. For almost the whole of my career at the laboratory bench—excepting my time as a pathologist during my army service—I worked on poxviruses. Initially, with Macfarlane Burnet at the Walter and Eliza Hall Institute, I worked on infectious ectromelia, which we were able to rename 'mousepox'. From 1950 until 1965, I spent most of my time working and writing about myxomatosis, and have written and lectured about it, off and on, ever since. In 1957, I began work on the genetics of vaccinia virus as a simpler laboratory model, with the ultimate goal of studying the genetics of virulence in myxoma virus. This gave some very interesting results, but the problem was much too complex to be solved by the techniques then available. In 1967, I was appointed Director of the John Curtin School of Medical Research and gave up research at the bench and the supervision of research students.

The World Health Organization (WHO) Intensified Smallpox Eradication Program was initiated in 1967, and in 1969 WHO set up a small committee of scientists expert in the laboratory study of poxviruses, called the 'Informal Group on Monkeypox and Related Viruses'. I was appointed a member of that committee and served as rapporteur at its first meeting in Moscow in 1969 and Chairman for the meetings in 1976, 1978 and 1979. I missed the meetings in 1971 and 1973.

Clearly, as countries appeared to have succeeded in stopping transmission of smallpox within their jurisdictions, there needed to be an independent assessment of these claims, and International Commissions for the Certification of Smallpox Eradication were established. In April 1977, I served on the International Commission for the Certification of Smallpox Eradication in India, where I was appointed rapporteur for the final meeting. In March 1978 I was a member of the International Commission for Certification in several African countries, including Malawi. There were 23 such International Commissions and a few less formal investigations that needed to be consolidated; in October 1977 a large 'Consultation on the Worldwide Certification of Smallpox Eradication' was established, and I was appointed Chairman of the Consultation and later, in December 1978 and December 1979, of its successor, the Global Commission for the Certification of the Smallpox Eradication. At its second and final meeting, the Global Commission affirmed that smallpox had indeed been eradicated and produced a substantial report which contained 19 recommendations for WHO responsibilities post-eradication. As Chairman of the Commission, I presented the report and its recommendations to the 33rd World Health Assembly in May

1980, where it was approved unanimously. To put this work into context, I set out below a brief account of the disease and efforts to control it before describing my own involvement in this program.

Smallpox: the Disease

Some thousands of years ago, for humankind as a whole, smallpox was an emerging disease, and it certainly followed this pattern during the European-American exchanges and African-American exchanges that followed the discovery of the Americas by Europeans.

Clinical Features

There were several forms of the disease, with different case-mortality rates. Here I will describe only the commonest form (Figure 10.1). Infection occurred via the respiratory tract but there were no signs for 10 to 12 days after a person was infected and, during this incubation period, he or she was quite well. Then there was a sudden onset of fever, malaise, headache and backache (1 to 3 days),

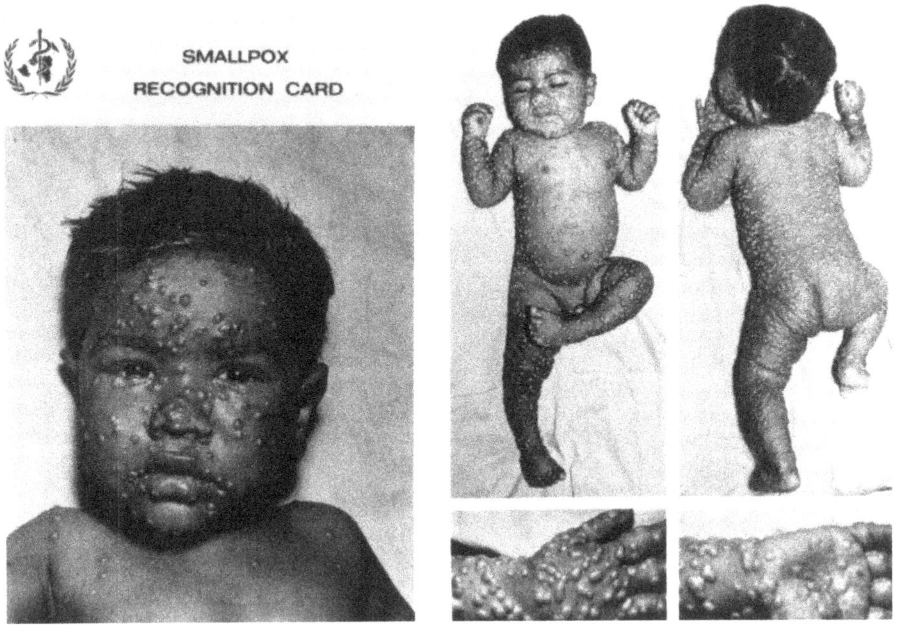

Figure 10.1. The Smallpox Recognition Card, showing a child with smallpox at the pustular stage

followed by the development of a rash through stages of macules (2 to 4 days), progressing through vesicular, pustular and crusted stages over the following two weeks. All the lesions were at the same stage of development and there was a greater concentration on the face and limbs with a predominantly centrifugal distribution; each lesion was full of virus particles.

Epidemiology

Smallpox was a specifically human disease, and the vast majority of infections resulted in disease with obvious clinical manifestations. Patients became infectious at the onset of fever and infectivity increased when the rash appeared, because lesions on the buccal mucous membranes broke down and virus was excreted in the saliva and pharyngeal secretions. Infectivity declined rapidly as these healed and the rash scabbed. Although there was a large amount of virus in the scabs, they were not an important source of infection. Recovery was followed by lifelong immunity and recurrences never occurred. These features meant that smallpox could be maintained as an endemic disease only in rather large populations, of the order of 200,000, so that small isolated populations—for example, of Amazonian Indians or the inhabitants of oceanic islands—never constituted a continuing focus of infection.

Infection occurred by close face-to-face contact, such as is common between members of the same household. Rarely, smallpox could be spread by fomites and, even more rarely, if the index case had a severe cough early in the course of the disease, infection could be airborne over moderate distances, within a bus or, rarely, a hospital, for example.

The History of Smallpox

We know a good deal more about the history of smallpox than about most diseases, because all infections resulted in a disease with a characteristic rash, which left visible pockmarks in many of those who recovered. Smallpox did not occur in the Americas, or most of Africa south of the Sahara, until these places were invaded by Europeans or people from the Middle East. It was probably endemic in the Ganges and Indus basins from as early as 500 BC, and is thought to have entered China from the west in 48 AD, moving to the Korean peninsula in 583 AD and occurring for the first time to Japan two years later. To the west of the Eurasian landmass, the major spread of smallpox occurred during the great Islamic expansion across North Africa and into the Iberian Peninsula in the 8th and 9th centuries.

By 1000 AD, smallpox was probably endemic in the more densely populated parts of the Eurasian land mass from Japan to Spain and including the African countries on the southern rim of the Mediterranean. Over the next few centuries, with the Crusades, smallpox became well established in Europe. In Africa, caravans crossing the Sahara to the densely populated kingdoms of West Africa carried smallpox as well as Islam with them, and the disease was repeatedly introduced into the port cities of eastern Africa by Arab traders.

By the 16th century, smallpox was becoming steadily more serious in European countries and statistics of smallpox deaths began to be collected at about this time, in Geneva, London and Sweden. The stage was set for the next explosive spread of the disease, with the development in Europe of ocean-going ships and the movement of European explorers and colonists to the newly discovered continents.

Smallpox played a crucial role in the Spanish conquest of Mexico and Peru and in the successful settlement of North America by the English and the French. In Africa, smallpox was introduced into Angola by the Portuguese and into Cape Town by contaminated bed-linen from a ship returning from India, with disastrous results for the aboriginal inhabitants, the Hottentots.

Immunization

As well as having an ancient history as a disease, smallpox is unique among infectious diseases in that it has a very long history of effective means of immunization. The first method was by the inoculation of pus from smallpox pustules, a process called variolation, which appears to have arisen independently in India and China in the 11th and 12th centuries. In 1721 this procedure was introduced from the Ottoman Empire into Bohemia and England, by Dr Johann Adam Riemann and Lady Mary Wortley Montagu respectively. Although it produced fearsome skin lesions at the inoculation site and had a case fatality rate of 1-2 per cent, it was deemed much better than natural smallpox, which by this time was an almost universal disease in Europe with a case-fatality rate of over 40 per cent in babies and about 25 per cent overall.

As a boy, Edward Jenner had been variolated and he was himself a variolator. But his astute observations in the early 1790s that cowpox appeared to protect milk-maids against smallpox, and then his experiments in 1796–98 showing that this was indeed the case, forever altered the history of infectious diseases. Vaccination was introduced and where it was assiduously practiced had a dramatic effect on the incidence of smallpox.

Early Attempts to Eradicate Smallpox

In 1793, just before the introduction of vaccination but at a time when variolation was widely practised in Britain, John Haygarth published a remarkable pamphlet in which he proposed to 'exterminate' smallpox from that country by a scheme that included systematic variolation throughout the country, isolation of patients, decontamination of potentially contaminated objects, supervised inspectors responsible for specific duties, rewards for the observance by poor persons of rules for isolation, fines for transgression of those rules, inspection of vessels at ports and prayers every Sunday. A few years later, shortly after publication of his *Inquiry*, Jenner had a similar vision: 'it now becomes too manifest to admit of controversy, that the annihilation of the Small Pox, the most dreadful scourge of the human species, must be the final result of this practice.'

However, in spite of the practice of vaccination in many countries for almost a century, in 1900 smallpox was endemic in almost every country in the world. Australia, New Zealand and small islands of the Pacific Ocean and elsewhere were kept free of the disease because of their isolation and efficient quarantine services. So were the Scandinavian countries, thanks to well-conducted vaccination campaigns and containment of outbreaks.

The prevalence of the disease was greatly reduced in Europe until the disruption and mass movements of people associated with World War I exacerbated the disease in Russia, from whence it spread to Germany, Austria and Sweden—in 1919, there were some 300,000 cases in Europe. After 1923, statistics on whether smallpox was endemic in various countries was collected by the Health Commission of the League of Nations and from 1948 by the World Health Organization (WHO).

After World War II, there was a rapid fall in the numbers of endemic countries in Europe and North America, from which continents the disease was eliminated by 1953. A few of the smaller countries in Asia were also freed of smallpox, but it remained a common disease in most countries of Africa and Asia. Heartened by these results, in 1950, Fred Soper, a great enthusiast for the eradication of vector insects (*A. gambiae* and *Ae. aegypti*) and infectious diseases, persuaded the Pan American Sanitary Organization to undertake the eradication of smallpox from the Americas. Then, in 1953, the first Director-General of the WHO, Canadian Brock Chisholm, tried unsuccessfully to persuade the World Health Assembly to agree to a program for the global eradication of smallpox. The Assembly dismissed his proposal as unrealistic. Yet two years later, the same body, mesmerized by the new approach of using DDT for the control of *Anopheles* mosquitoes, agreed to a vastly more difficult program which called for the global eradication of malaria.

It was not until 1958 that global eradication of smallpox was again considered by the World Health Assembly. By that time, the malaria program, which had been strongly supported by the United States, had run into serious difficulties. The Soviet Union, which had been excluded from the Assembly for several years, returned in 1958 and, wishing to make its presence felt, put forward a carefully planned proposal for the global eradication of smallpox, which on this occasion was endorsed by the Assembly. The suggestion was that this could be achieved in 4-5 years, essentially by vaccinating and re-vaccinating up to 80 per cent of the population of each endemic country and thus producing a level of herd immunity sufficient to break the chains of transmission. The World Health Assembly resolution was accompanied by arrangements for the provision of freeze-dried vaccine to endemic countries, many of which were encouraged to set up national smallpox eradication programs. However, no WHO funds were set aside to provide for coordination of the program.

During the next few years country-wide elimination was achieved in several more countries in Asia, Africa and the Americas, but by 1966 it was clear that progress along these lines would never achieve global eradication. The countries in which smallpox remained endemic were the hard core of the problem; the Indian subcontinent, Ethiopia, sub-Saharan Africa, Indonesia and Brazil.

The Intensified Smallpox Eradication Program, 1967 to 1977

In 1966, the World Health Assembly adopted, by a narrow margin, a resolution which included acceptance of the need for coordination of the programs of individual countries and for WHO finance from its regular budget. This resolution was put into effect in 1967, by the establishment of the Intensified Smallpox Eradication Program, to be coordinated by a Smallpox Eradication Unit at WHO Headquarters in Geneva and with the goal of global eradication within 10 years. The men who led this effort were Donald A. Henderson and Isao Arita. Success was achieved by October 1977, just a few months after the target date. What problems did the Intensified Program encounter, what were the strategies used to solve them and what were the features of smallpox that made this feat possible?

The Extent of the Problem

In 1967, the disease was still very common. Although surveillance and reporting were initially very poor, data gathered during the Program showed that smallpox was then endemic in 31 countries and imported cases had been reported in another 11 countries, and there were about 20 million cases of smallpox and two million deaths annually.

The Vaccine

Although potent and stable freeze-dried vaccine had been available commercially since 1955, in 1967 many of the endemic countries still used liquid vaccine whose stability under tropical conditions was such that only 15–20 per cent of the vaccine used in the field was of acceptable potency. To solve this, the Smallpox Eradication Unit undertook a major campaign to upgrade the quality and increase

Donald A. Henderson

Born in Cleveland, Ohio in 1928, D. A. Henderson graduated from Oberlin College and the University of Rochester School of Medicine, later receiving a Master's Degree in Public Health from the Johns Hopkins School of Hygiene and Public Health. After three years of residency training he joined the Communicable Disease Center of the US Public Health Service in 1955, serving successively as Chief of the Epidemic Intelligence Service, Chief of the Surveillance Section and, finally, as Chief of the Centers for Disease Control (CDC) Smallpox Eradication Program. In 1966, he was appointed Chief Medical Officer for the WHO Smallpox Eradication Program, a position he held until 1977, when he resigned from WHO to become Dean and Professor of Epidemiology of the Johns Hopkins School of Hygiene and Public Health. In 1990, he was appointed by the first President Bush as Associate Director for Life Sciences in the White House Office of Science and Technology. From 1993-95, in the Clinton administration, he served as Deputy Assistant Secretary for Health and Science Adviser to the Secretary of the Department of Health and Human Services. On return to Johns Hopkins in 1995, he founded a Center for Civilian Biodefense Strategies and played a central role in highlighting the need for medical and public health expertise to assume major responsibilities for the development of public health emergency preparedness. In November 2001, the Secretary of Health and Human Services asked him to direct the new office of Public Health Emergency Preparedness in the Office of the Secretary. With a budget of $3 billion, major initiatives were taken to strengthen all components of the health infrastructure and to do the needed research in development of diagnostics, vaccines and anti-microbial agents. He returned to Baltimore in 2004, where he is now Resident Scholar at the Center for Biosecurity of the University of Pittsburgh Medical Center (the former Hopkins Center for Civilian Biodefense Strategies) and Professor of Medicine at the University of Pittsburgh. As well as sharing the Japan Prize in 1988, he is a recipient of the Presidential Medal of Freedom (2002), the National Medal of Science (1986), and other awards and symbols of recognition from a number of governments and institutions. We have been very good friends for many years, and I always stayed with him and Nana when I visited Baltimore.

Isao Arita

Born in Kumamoto, Japan, in 1926, Isao Arita graduated in medicine at Kumamoto Medical School in 1950. Between 1951 and 1961, he was employed as Medical Officer in Infectious Disease Control in the Ministry of Health and Welfare in Japan. In 1962, he joined the World Health Organization and worked in Africa. In 1966, he became a member of the newly established Smallpox Eradication Unit, becoming Chief of the Unit when D. A. Henderson resigned in 1977 and continuing in that position until 1985, when he retired and returned to Kumamoto. He was active in the work of the Unit at Headquarters and in the field, being responsible, among other things, for the supply and quality control of smallpox vaccine used in the eradication program, related poxvirus research, and development of the epidemiological surveillance and containment system, notably in the last endemic country in the world, Somalia. After the declaration of the global eradication of smallpox in 1980, he was responsible for overseeing the operation of the 19 recommendations made by the Global Commission, and for the production of the book *Smallpox and its Eradication*. In 1988, he shared the Japan Prize (with Henderson and Fenner) for this work.

From 1985 to 1992, he was the Director of the Kumamoto National Hospital, and on retirement in 1993 he established the Agency for Cooperation in International Health (ACIH), which for many years organized meetings, in which I often participated, designed to promote the use of vaccines against communicable diseases, especially in Third World countries. I have the happiest memories of working with him at WHO Headquarters subsequently and at meetings of ACIH.

the quantity of vaccine. WHO Reference Centers for Smallpox Vaccine were established at the Connaught Laboratory in Toronto and the National Institute of Public Health in Bilthoven, The Netherlands. An international meeting of producers was held, training courses established and regular quality control testing of all vaccine was introduced. In a way that was unprecedented—and was opposed by some WHO administrators on the grounds that it breached national sovereignty—quality control testing by the WHO Reference Centres was applied not only to vaccine donated through the WHO, but to that donated under bilateral aid programs and to vaccine produced in the endemic countries themselves. By 1970, the testing program had ensured that most of the vaccine used in both developed and developing countries had reached WHO standards for potency, heat stability and bacterial content.

Priorities

With a fixed sum of only $2.4 million a year guaranteed from the WHO Regular Budget and no assistance from bodies like UNICEF, which had been disillusioned by the failure of the malaria eradication program, the Smallpox Eradication Unit had to decide upon regional priorities. The Pan American Health Organization program had been successful in reducing the endemic countries in the Americas to one, Brazil. Assistance here was a top priority. In Africa, the US Agency of International Development had already initiated a 'measles control and smallpox eradication program' in a group of 18 countries in West and Central Africa, for which WHO was able to provide assistance of a kind not possible under a United States bilateral arrangement. Smallpox had been eliminated from the Dutch East Indies (since 1949, Indonesia) in 1937, only to re-enter and spread after the War. Assistance here held out hopes for early success. However, the major attention of the program had to be concentrated on the ancient strongholds of the disease in the Indian subcontinent, with the still-endemic countries of Africa as another continuing problem. When smallpox had been eliminated from the whole of Asia and most countries of Africa by early 1975, the full resources of the Program were concentrated on the Horn of Africa, where the disease remained endemic in Ethiopia and had been exported to Somalia.

Strategies for Eradication

There were a number of factors in the strategy developed by the Smallpox Eradication Unit that led to the ultimate success of the program. The major change from that enunciated by the Soviet Union delegation a decade earlier, which relied on mass vaccination, was the elevation of surveillance and containment to a pre-eminent place. This required that in every endemic country programs for surveillance and notification had to be set up or greatly strengthened. Discovery of a case was to be followed by containment by vaccination of all contacts, in ever-increasing distances from the affected household, and the discovery and follow-up of the sources of infection. In India in particular, nation-wide 'house-hold searches' were carried out periodically, which revealed vastly more cases of smallpox than had previously been suspected. Then, as the incidence of smallpox fell to a low level, a system of rewards for reporting cases, offered to both the general public and the public health workers, ensured that the great majority of cases of smallpox were reported.

Certification of Eradication

By systematically applying these strategies, and by dint of a great deal of hard work and some good luck, smallpox was progressively eliminated from each of the countries in which it had been endemic in 1967.

Concurrently with the elimination of smallpox from countries, groups of countries or continents, a system of certification of eradication by teams of independent international experts was developed, culminating in 1977 with the establishment of the Global Commission for the Certification of Smallpox Eradication, of which I was elected Chairman. On the afternoon of Sunday December 9, 1979, after four days of intensive discussion, all members of the Commission accepted its final report, which stated that the world had been freed of smallpox and made 19 recommendations concerning the post-smallpox world, on vaccination, vaccine stocks, monkeypox, publications and the like. Five months later, on 8 May, 1980, I had the honour of presenting this report to the World Health Assembly, where it was accepted without change.

Informal Consultations on Monkeypox and Related Viruses

First Meeting, Moscow, 26–31 March 1969

10 outbreaks of a poxvirus disease had occurred in laboratory colonies of primates in Europe and North America between 1958 and 1966, which investigations showed to be caused by an Orthopoxvirus and the disease had been called 'monkeypox'. To exclude the possibility that it might represent an animal reservoir of smallpox virus, WHO organized a meeting of poxvirus experts from Australia, Japan, the Soviet Union, the Netherlands and the United States, called an 'Informal Group on Monkeypox and Related Viruses', which met first in Moscow in April 1969 and biennially thereafter. The first meeting, for which I acted as rapporteur, agreed that monkeypox virus could be readily distinguished from all other orthopoxviruses by its biological properties. A year later, sporadic cases of a human disease clinically very similar to smallpox were discovered in Côte d'Ivoire, Liberia, Nigeria and Zaire, all in areas where there had not been a case of smallpox for over a year. All of these were shown to be caused by monkeypox virus.

Fourth Meeting, Geneva, 10–13 February 1976

I had missed the second and third meetings of the Informal Group on Monkeypox and Related Infections in 1971 and 1973, but was asked to be chairman of the meeting in Geneva in 1976. The group had been expanded from nine to 17 experts in poxvirus virology and epidemiology. During the interval, there had been 21 cases of human monkeypox in six countries in West and central Africa and these had all been carefully investigated. Two were possibly transmitted from one person to another. Blood had been collected from monkeys and other wild animals, notably squirrels and rodents in the vicinity, for serological tests and attempts to isolate virus; antibodies had been detected in several species of monkeys and in rodents.

A more puzzling observation was that four isolates of a poxvirus that was indistinguishable from variola virus (called 'whitepox' virus) had been made from the kidneys of a chimpanzee, a monkey and two rodents in an area of Zaire near where seven cases of human monkeypox had occurred. The isolates had been made by virologists, headed by Dr S. Marennikova, in the Moscow Research Institute for Viral Preparations, which was one of the two WHO Collaborating Centres on Smallpox and Other Related Infections which had been established by Henderson in 1967.

Fifth Meeting, Geneva, 9–10 November 1978

The main objective of this meeting, of which I was again Chairman, was to discuss the findings of Dr Marennikova's group that viruses having some of the characteristics of whitepox virus arose as variants of monkeypox virus. Comparisons were drawn between the known white pock mutants of both cowpox and rabbitpox viruses, on which I had carried out a lot of work. It was pointed out that in those studies every mutant appeared to be different, whereas all the reported monkeypox 'white pock mutants' were identical, and, like the original 'whitepox' isolates, indistinguishable from variola virus. Arrangements were made for a virologist from the Moscow laboratory to work on the problem in the other WHO Collaborating Centre on Smallpox and Other Related Infections, in Atlanta, Georgia.

Study Group on Orthopoxviruses

First Meeting, Atlanta, USA, 26–28 June 1979.

This group was appointed by WHO on the recommendation of the meeting of the Global Commission for the Certification of Smallpox Eradication in December 1978 (see below), to continue investigations into possible animal reservoirs of variola virus and determine the importance to humans of monkeypox and whitepox viruses. It was reported that application of the relatively new technique of restriction endonuclease analysis had shown that all the 'whitepox' viruses and some of the white variants of monkeypox virus isolated in the Moscow laboratory were indistinguishable from variola virus. However, none of a large number of white variants isolated from monkeypox virus by Dr K. R. Dumbell in London had the restriction endonuclease maps or phenotypic characteristics of 'whitepox' virus. Plans were made for continued research on these problems.

Certification of Smallpox Eradication

With the imminent eradication of smallpox from Brazil, the last endemic country in the Americas, in 1971, it became necessary for the Smallpox Eradication Unit to develop procedures for assessing claims that smallpox had indeed been eradicated. Because of the widespread use of variola virus in laboratories in many parts of countries where smallpox had been endemic, it was decided that eradication should not mean ascertainment that no virus remained but rather that transmission had been interrupted. It was decided that this should, wherever possible, be undertaken by groups of independent public health experts, epidemiologists and virologists, with assistance from WHO and from the health authorities of the country involved. To allow the country's health authorities to prepare detailed documentation of freedom from smallpox, it was decided that such independent assessments should be made not less than two years after the last known case in the country concerned. 22 International Commissions for the Certification of Smallpox Eradication operated between August 1970 and October 1979. I served on two of these Commissions, and also, with help from one other person in each case, assessed the situation in South Africa and in the People's Republic of China.

India, 4–23 April 1977

It was generally agreed among those involved in the eradication campaign that, if smallpox eradication could be certified in India, the prospects for global eradication were very good. The largest of all of the 22 formal International Commissions, with 16 members from as many different countries, assembled in New Delhi to plan field visits. During these, each member was accompanied by national and WHO personnel who had either been members of the Indian National Certification Commission or were WHO epidemiologists who had worked in India during the pre-certification activities. Each small group then visited different areas in one or two states or union territories over a period of some two weeks before reassembling in New Delhi. I went to Himachal Pradesh and Rajasthan. It was a wonderful experience, travelling through countryside I had never been able to visit in previous trips to India, with temples and palaces and the planned city of Chandigarh. We stopped wherever there were local health authorities, at state, district, or municipal primary health care levels, who had assembled data on the gradual elimination of smallpox from their area of responsibility. I was also able to go to any place I chose; in Rajasthan I picked the most isolated and primitive place that could be imagined.

All members of the team then reassembled in New Delhi and the brief reports produced by various Commission members or teams were discussed. I had been appointed rapporteur for this session, so I was kept busy all Thursday and that evening preparing a draft report. In the main, this was accepted next day, and a redraft late that morning brought it to a state that was acceptable to all members, and we had a celebratory dinner that evening. The next day, the Commission assembled in the Health Minister's office and presented him with the declaration that smallpox had been eradicated from India.

Malawi, Mozambique, Tanzania and Zambia, 6–29 March 1978

I met with other members of the International Commission in Maputo, the capital of Mozambique, on 6–7 March to discuss our program; I acted as rapporteur. We split into four groups and arranged to meet again for final discussions in Zambia on 27–29 March. Over the intervening two weeks a WHO official, Dr Abou-Gareeb, and I, together with two European medical scientists who had been assisting the local health authorities prepare for our visit, travelled around Malawi. There was particular concern about a suspected case of smallpox in 1972 in a very remote area of the country. After a long drive over 'non-roads', in heavy rain, we arrived at the village and found Mary Joseph, the suspected case. She had facial pockmarks from an attack of smallpox in 1972, which had never been reported. We found three other children in a nearby village who also had facial pockmarks, probably also dating from 1972, but there was no evidence of any more recent cases. During the next few days we drove all over Malawi, calling in at villages and hospitals, and by 19 March I was able to begin my report. We then drove to Lusaka, the capital of Zambia, and met the other members of the Commission. After a meeting lasting two and a half days (as usual I acted as rapporteur) we agreed that on the evidence that we were able to obtain, smallpox transmission had been interrupted in all four countries.

South Africa, 23 January to 18 February 1978

The most populous country in southern Africa, since the end of World War II only variola minor had occurred there and it was regarded as a trivial disease. Because of their apartheid program, at the time South Africa was not represented on the World Health Assembly. Henderson went there in 1972 to persuade them to cooperate with the WHO program, but they did not establish a national smallpox eradication program. In October 1977, the Director-General of WHO wrote to the South African Minister of Health suggesting that I should visit them early in 1978, a proposal they accepted. The purpose of the visit was threefold: (1) to advise South African health officials on how to prepare a country report for submission to the Global Commission; (2) to emphasize the importance of chickenpox surveillance and make arrangements for the collection of specimens for laboratory examination by WHO Collaborating Centres; and (3) to obtain an impression of the recent history of smallpox and rural health services in South Africa (including in some of the 'black homelands').

I arrived in Johannesburg on 23 January and was met by Dr James Gear. The next day I visited the National Institute of Virology. I discussed the problem of destroying stocks of variola virus with Gear and the Institute Director, Professor O. W. Prozesky. On 24 January, I talked at length to members of the Health Department to work out an itinerary for my visit to South Africa and Namibia (Figure 10.2).

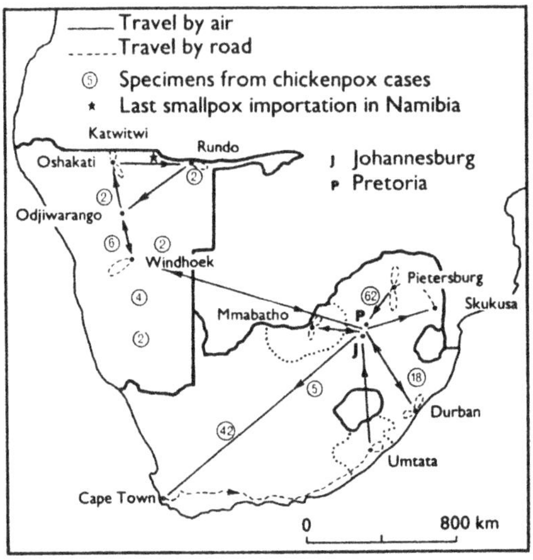

Figure 10.2. Map showing my travel by air and road around South Africa and Namibia in 1978

As I reported to WHO when I went to Geneva on 20 February:

> Everywhere that I went I was received with great cordiality, and there were frequent expressions of satisfaction that WHO had arranged the visit. No effort was spared to enable me to see what I regarded as vulnerable areas, such as the more densely populated parts of Namibia, near the Angolan border and the northern parts of Transvaal (Lebowa) and Bophuthatswana, near Botswana. I was driven around the countryside and to clinics and hospitals by senior health officials, health inspectors or nursing sisters over a distance of more than 3,000 kilometres, and a chartered plane was made available for three days. Such travel provided opportunities for extensive informal discussion of the health conditions and the complex policies of these parts of Africa, as well as observation of the countryside and living conditions in the rural areas, and the rural hospitals and clinics.

Although the United Nations had terminated the South African mandate over Namibia in 1966, that country was still controlled and administered by South Africa, under the name 'South West Africa'. I spent five days there (see Figure 10.2), and met health officers and staff in hospitals and laboratories, obtained information on health services, especially the surveillance and control of communicable diseases, visited schools and conducted a limited vaccination scar survey.

South Africa, Second Visit, 11–12 October 1978

WHO had suggested to the South African authorities that they should submit their national report on smallpox by 1 June, 1978. Since nothing had arrived by September, Dr Arita asked me to revisit South Africa briefly in mid-October. I flew there across the Indian Ocean and was met, as usual, by my good friend Dr James Gear. As well as stimulating Dr de Beer, the Secretary for Health, about getting their report to Geneva soon, I discussed at some length the disposal of their stocks of smallpox virus. I then went to Geneva, where I talked with Arita and Dr Nicole Grasset about the latter's proposed visit to South Africa in November to collect their national report.

People's Republic of China, 14–31 July 1979

In the 1970s, the People's Republic of China was not recognized by WHO, and it was not possible to arrange a visit by an International Commission. Dr Joel Breman was an American who served as a member of the Smallpox Eradication Unit from 1977–80. It was decided that he and I should visit China, make extensive enquiries and arrange for a national report on the situation to be sent to WHO in Geneva as soon as practicable. Guided by Dr Jiang Yu Tu, who had trained at the Johns Hopkins School of Public Health, we spent two and a half weeks there, travelled extensively around the country, and consulted with health authorities in Beijing. Detailed records (in Chinese) were sent to WHO, translated and examined by staff of the Smallpox Eradication Unit.

Eradication from Africa, 16 October to 6 November 1979

I had been selected to be a member of the last International Commission, that devoted to the certification of smallpox eradication in Somalia, where the last field case in the world had occurred in October 1977. This took place in October 1979. However, a preliminary medical clearance was required, and my physician had reported that I had an inguinal hernia, and because of the rough terrain and distance from medical facilities, WHO refused to allow me to take part, to my great disappointment. However, it had been decided that the certification of eradication from Somalia, and therefore Africa, and the world, was an event not to be missed, and special celebrations were planned to take place in Nairobi. Although the celebratory function was to be held on 26 October, I arrived there on 19 October, because there were also a number of other important meetings, among them a meeting of the four International Commissions that had worked in Africa in October 1979: Djibouti, Ethiopia, Kenya and Somalia. I attended those meetings part of the time, but also met and talked with Henderson, Arita and Gordon Meiklejohn, who had a contract with WHO to prepare a draft Final Report of the Global Commission. I spent most of my spare time reading through Meiklejohn's report, but also met with Svetlana Marennikova and told her that on the grounds of genome mapping, I thought that her 'whitepox' viruses and most 'white mutants' of monkeypox virus that had been described by her laboratory were due to laboratory contamination with variola virus.

The celebratory function was held in the main hall of the Kenyatta Centre, a very impressive room, on Friday 26 October. There were speeches by the Director-General of WHO, Halfdan Mahler, and Jan Kostrzewski, who had chaired the meeting of the four International Commissions, followed by a press conference. Later, an ABC reporter came to my hotel room and taped a discussion for use in Australia. Then there were speeches by the Ministers of Health of Somalia and Kenya, some impressive Kenyan dances and a fine meal. I flew to Geneva on the Sunday for a meeting with the Director-General of WHO, in preparation for the Global Commission meeting in December and the World Health Assembly meeting in May 1980.

Consultation on the Worldwide Certification of Smallpox Eradication, 11–13 October 1977

Coincidentally with the later certification visits, just described, I was involved with the larger committees set up by the Director-General of WHO, Dr Mahler, to consolidate the picture and ultimately to produce a report which could be presented to the Governing Body of WHO, the World Health Assembly, hopefully at their meeting in May 1980.

Following the usual WHO pattern, the Director-General initially called together a group of experts who might participate in the large committee as a 'Consultation' on the Worldwide Certification of Smallpox Eradication, to be held in October 1977. I spent the day before the meeting discussing the situation with Dr Henderson and Dr Arita, and they outlined the position and suggested that I should chair both the Consultation and the subsequent Global Commission.

The next day I was 'elected' Chairman of the Consultation; the participants included 17 experts on epidemiology, virology and public health from 15 countries: three from Africa, three from the Americas (including Henderson, who had resigned from WHO earlier in 1977), four from Asia, one from Australia and six from Europe, plus eight WHO advisers, from Geneva and Regional offices, and five other members of the Smallpox Eradication Unit, of which Isao Arita had succeeded Henderson as Chief. Unlike the meetings of the smaller WHO committees with which I had been involved, this meeting, and subsequent meetings of the Global Commission, adopted the United Nations convention of having all discussion translated into each of the six UN official languages: Arabic, Chinese, English, French, Russian and Spanish, although all papers were in English and most speakers used English in their presentations.

The aim was to reach, as quickly as possible, a situation at which it could be certified that smallpox had been eradicated globally. After four days of intensive discussions it was decided to divide the countries of the world into four categories:

Category 1—those countries requiring formal certification by international commissions of experts, who would visit the countries concerned and assess their smallpox-free status by examining records and making field visits to determine whether surveillance activities would have been adequate to detect a case of smallpox if one had occurred during the previous two years.

Category 2—those countries requiring the visit of selected experts to verify and document the smallpox incidence since 1960, the last known outbreak and control measures employed, and procedures for handling suspected cases.

Category 3—certification through submission of detailed country reports.

Category 4—official statements by countries declaring their smallpox-free status during the previous two years and signed by government health authorities.

It was planned to have a meeting of the successor to this Consultation, the Global Commission for the Certification of Smallpox Eradication, in December 1978. The next two years were largely devoted to collecting the material outlined above in time for what was hoped would be the final meeting of the Global Commission in December 1979.

First Meeting of the Global Commission for the Certification of Smallpox Eradication, 4–7 December 1978.

I had been in Geneva early in November 1978 at the fifth meeting of the Consultation on Monkeypox and Related Viruses, and took the opportunity to discuss the arrangements for the First Meeting of the Global Commission with Isao Arita and Joel Breman. Then, after a week in England and a visit to Nairobi for meetings of SCOPE (see Chapter 9), I went back to Geneva for the First Meeting of the Global Commission. As for the Consultation, this was a large committee, with 20 members, eight WHO advisers, eight WHO Regional Office staff and 13 WHO Headquarters staff, including nine from the Smallpox Eradication Unit. I acted as Chairman, except when I reported on the situation in South Africa and Namibia, Jan Kowscrzewski of Poland was Vice-Chairman, and Arita served as secretary.

We met for four days and considered the numerous reports that had been received from International Commissions and from and about individual countries which had been listed in categories two, three and four (above). In the majority of cases, eradication was certified; reports by International Commissions were expected in 1979 for 10 countries, mostly in Africa, and additional information was sought from five countries: China, Democratic Kampuchea (Cambodia), Iraq, Madagascar and South Africa. In addition, all countries which had not been asked to provide detailed information—for example, all countries in Europe and the Americas—were asked to provide formal statements detailing when the last known case of smallpox had occurred in their country.

On human monkeypox, the Global Commission concluded that, although four cases of possible person-to-person infection had occurred between 1970 and 1979, this disease 'did not constitute a threat to the permanence of smallpox eradication', nor did the Commission regard the presence of human monkeypox in a country to be a justification for continuation of vaccination.

Noting that the number of laboratories holding variola virus had been reduced from 76 (in 1976) to 10, it recommended that these holdings should be destroyed or transferred to one of the two WHO Collaborating Centres and asked for a report from a group of scientific experts at the 1979 meeting. The Commission also discussed and made recommendations about the cessation of vaccination, vaccine reserves, surveillance after global certification, and documentation of the Smallpox Eradication program.

Second and Final Meeting of the Global Commission for the Certification of Smallpox Eradication, 6–9 December 1979.

As was usual with my overseas trips, this meeting came at the end of two other commitments, discussion of a UN Environment Program (UNEP) report and a meeting of the SCOPE Executive Committee in Paris. The meeting with UNEP was in Geneva, and I took advantage of this to spend several days discussing the forthcoming Global Commission meeting with Arita and Breman.

Early in 1979, Arita had arranged for Gordon Meiklejohn, Professor of Medicine at the University of Colorado, who had been a WHO consultant on smallpox almost every year from the early 1960s, had served on several International Certification Commissions and had attended the first meeting of the Global Commission, to produce a draft Final Report of the Global Commission. This was a substantial book of 65 A4 pages of text and 19 appendices, and was made available to all members before the meeting.

The participants in the meeting and the procedure were almost the same as in 1978, but the draft report was the focus of discussion. The current certification status was reviewed by Arita. Of the 79 countries considered by the Commission in 1978, 64 had been certified prior to 1979 and, of the other 15, 10 had been certified during 1979 and the action recommended by the Commission in 1978 had been completed in three of the other five. The situation in China was discussed at some length, note being taken of the fact that Fenner and Breman had spent three weeks in China in July 1979 and a substantial report had subsequently been sent to WHO. The Commission noted that the last cases in China had occurred in 1960, in the border provinces of Yunnan and Tibet, and that pockmark surveys there confirmed this. Routine primary vaccinations had been practiced since 1950 and mass re-vaccinations at six year intervals until 1978. On the basis of the documentation provided, the adequacy of the health services at the local level, satisfactory disease reporting and the long period since the last cases, the Commission certified China free of smallpox. The other country needing consideration was Democratic Kampuchea (Cambodia) where smallpox was last reported in 1959. Although there had been some problems in obtaining formal endorsement of the country report that had been prepared by WHO, on the basis of the information supplied Democratic Kampuchea was verified as free from smallpox.

The other matters considered at some length were human monkeypox and the 'whitepox' virus. It was noted that of the 45 cases of human monkeypox reported since 1950, more than 50 per cent were among children less that five years of age, the age-group least adequately vaccinated. Of 66 unvaccinated individuals with face-to-face contact with prior cases, only four possible secondary cases occurred. WHO had just begun a six-year ecological study of monkeypox in Zaire. Preliminary results showed that 32 per cent of 234 monkey sera, from several species of monkeys, contained orthopoxvirus antibodies. It was decided that the existence of human monkeypox did not present a problem for smallpox eradication and did not justify maintenance of vaccination in African countries. The Commission noted that the other virological problem, the nature of the 'whitepox' viruses isolated from African animals, was almost certainly due to laboratory contamination with variola virus; confirmation was expected to be obtained by molecular biological experiments currently in progress.

These discussions occupied the first day of the meeting. The next three days were spent in detailed consideration of the Final Report and agreement was reached on all aspects. Commission members then signed a document indicating their agreement with the final conclusion of the report, namely that smallpox had been eradicated throughout the world.

Smallpox Eradication Program, Meeting of the World Health Assembly, May 8, 1980

I arrived in Geneva on Sunday 4 May, and spent the next few days talking with Arita, with Henderson and others visiting his new house, out in the country, looking at the conference room at the Palais des Nations where the Assembly would meet, and revising my speech. Since there had never been a ceremony like this at a World Health Assembly meeting, a number of those involved in the action had a rehearsal at 2 pm. Then, at 4 pm, we gathered for the real thing. There were hosts of photographers, television crews and the like. I sat on the rostrum with the Director-General and Deputy Director-General, the President of the Assembly and the President of the Executive Board. The President of the Assembly spoke first, to introduce me, I then spoke as Chairman of the Global Commission (Figure 10.3) and then the President of the Executive Board presented

Resolution One, accepting the global eradication of smallpox. The President of the Assembly then asked: 'Are there any objections?' After waiting for 30 seconds, during which there was no response, he banged his hammer and signed the new Assembly parchment, followed by the Director-General. Representatives of the six regions of WHO then spoke and finally the Director-General gave another short speech. Arrangements were made outside the conference room for all delegates to sign the parchment for eventual binding, and the Global Commission parchment and Resolution were placed on display in a glass case. The Director-General then hosted a reception after which I went back to my hotel and had a good sleep.

Figure 10.3. Frank Fenner addressing the World Health Assembly on 8 May 1980

Chapter 11. Visiting Fellow, John Curtin School of Medical Research, From 1980

Introduction

This is the longest chapter in the autobiography, because it extends for a longer time—26 years—than I spent in any other position. In 1979, the rule in Australian universities was that staff had to retire (from paid positions) at the end of the year during which they reached the age of 65. For me, that meant 31 December, 1979, because I was born on 21 December, 1914. Fortunately, the ANU was willing to provide office space and access to all facilities except the laboratory (unless the retiree had an independent grant to cover laboratory expenses) if he/she wanted to write up work. Space was at a premium in the CRES building and, in any case, I did not want to get in the way of my successor, so I arranged to retire in the John Curtin School. This had the advantage that I would be able to use the Eccles Library for much of the work that I planned to carry out. The School was very generous, providing me with professorial offices that happened to be unoccupied at the time, and I am still here, in such an office, in 2006. Further, University Fellowships for retirees, granting $10,000 annually for three years, had just been introduced and I was awarded one of these.

Clearly, the work of the Smallpox Eradication Unit and its need of support from personnel outside WHO did not stop with the declaration of eradication at the World Health Assembly in May 1980. Over the next few years, a number of committees were set up and I served on most of these. As mentioned on the last page of the previous chapter, one of the recommendations of the Global Commission for the Certification of Smallpox Eradication was that an appropriate book should be produced describing the campaign. Fulfilling this requirement kept me fairly busy for the first eight years of my retirement; *Smallpox and its Eradication* was published in 1988. It was followed by two specialist books dealing with topics which could not be adequately covered in that book, namely a book on the orthopoxviruses generally and one on monkeypox virus in particular.

Another book I worked on during the 1980s was a companion book to *Medical Virology* for veterinary students; with the help of veterinary co-authors the first edition of *Veterinary Virology* was published in 1987 and a second edition in 1993. David White and I also published a fourth edition of *Medical Virology* in 1994.

In 1987, David White persuaded me to produce, with contributions from 320 other microbiologists, a *History of Microbiology in Australia;* it was published, a book of 610 pages, in 1990. This whetted my appetite for the history of science and I edited two books on the history of the Australian Academy of Science, published in 1994 and 2005, with Bernardino Fantini as co-author, *The History of Myxomatosis*, in 1999, and with David Curtis, *The John Curtin School of Medical Research, the First Fifty Years*, published in 2001.

I have been involved in many other activities during the last 26 years, some arising from the threat of smallpox as a bioterrorism weapon, contributing chapters in books on virology, lectures and addresses to conferences, many trips overseas, often at a more leisurely pace than before 1980, and of course my home life and overseas trips with my wife Bobbie and the experience of her death, in 1995. Also, in 1984, Bobbie and I decided that we could spare money for donations of various kinds, and initiated these by establishing annual conferences on medical research in the John Curtin School and on environmental problems in the Academy of Science.

The Committee on Orthopoxvirus Infections

Concurrently with the declaration of the global eradication of smallpox in May 1980, the Global Commission was dissolved and, in 1981, a new six-member committee—the Committee on Orthopoxvirus Infections—was established by the Director-General of WHO. Its membership was drawn from the Global Commission, and I was appointed chairman of the Committee. Except in 1985, this Committee met annually in Geneva from 1981 to 1986, critically reviewed the progress made in the post-smallpox-eradication program, and advised WHO on the implementation of the Global Committee's recommendations. The Committee was serviced by a reduced Smallpox Eradication Unit, with Arita as Chief, two and later one other medical officer, an operations officer and support staff, all of whom had had extensive experience in the eradication program.

First Meeting, 3–5 March 1982

The meeting reviewed the implementation of WHO's post-smallpox eradication policy under seven groups of topics: vaccination policy, reserve stocks of vaccine, investigation of suspected cases of smallpox, laboratories retaining variola virus, human monkeypox, laboratory investigations of orthopoxviruses, and archives and publications. The results of the meeting were published (Arita and Gromyko, 1982), some of the more important of which are outlined below. Although vaccination had been discontinued in 149 of the 158 Member States, it was still being used for military personnel in some countries. The Committee advised that this practice should be discontinued. Since eradication, there had been 63 rumours of new smallpox cases in 1979, 31 rumours in 1980, and 30 in 1981. All were carefully investigated; none had been smallpox. The Committee endorsed a plan for continued monkeypox surveillance in four countries in West and Central Africa, and it noted with satisfaction that the records of the eradication program had been established in archival mode and that there was a comprehensive inventory and guide indicating what was in each of the files. It also endorsed the establishment of a five-person editorial group to prepare for the proposed book on the eradication program (see below).

Second Meeting, 15–17 March 1983

The Committee highly commended WHO for diligently implementing the recommendations of the 33rd World Health Assembly regarding activities to be undertaken in the post-smallpox eradication era. Matters for note were: (1) that WHO had established two refrigerated depots for vaccine storage, in Geneva and New Delhi, with a reserve sufficient for vaccinating 200 million persons; (2) that there were now only three laboratories retaining stocks of variola virus and that it was possible that the stocks held in South Africa would be destroyed before the next meeting; (3) that continuance of surveillance of monkeypox should be continued in Zaire beyond 1985; and (4) that DNA fragments from three strains of variola virus had been cloned into recombinant plasmids and that these materials were available through WHO for more detailed analysis of the variola virus genome, which could be safely conducted outside maximum containment laboratories.

Third Meeting, 28–30 March 1984

The Committee again commended WHO for effectively implementing the recommendations of the 33rd World Health Assembly regarding post-smallpox activities. It noted that 160 of the 162 Member States had now discontinued routine vaccination. However, because problems had arisen with storage facilities in New Delhi, the stocks of vaccine stored there had been transferred to Lausanne. I had visited South Africa in September 1983, in the hope of persuading the authorities there to destroy their stocks of variola virus while I was there. The scientists were willing, but the Minister for Health would not bow to outside pressure. However, they did destroy their stocks in the presence of Keith Dumbell, a member of the Committee, in December 1983. The only remaining stocks of variola virus were those in the WHO Collaborating Centres in Moscow and the Center for Disease Control, Atlanta, Georgia. A WHO team had visited the high containment laboratories in each of these centres and found that they fulfilled WHO requirements for such laboratories. Surveillance for human monkeypox in Zaire had been greatly intensified and more cases had been discovered. The Committee recommended that surveillance there should be intensified and continued until 1989. A summary of the current situation was published by the Committee in 1984 (WHO, 1984). In addition, it saw the investigation of smallpox rumours and the maintenance of smallpox vaccine reserves as major activities that would need to be continued beyond 1985.

Fourth Meeting, 24–26 March 1986

This was an important meeting, because the Committee had been established with the intention of operating for no more than five years after its establishment in 1981. The principal problem it tackled was whether all stocks of variola virus should be destroyed. In preparation for this, Keith Dumbell, on behalf of the Committee, had consulted some 60 virologists in 21 countries, only five of whom thought that stocks should be maintained indefinitely. Based on this, the availability from WHO of plasmids contained cloned fragments of the variola genome, and their own analysis, the Committee concluded that there was now no need to retain stocks of variola virus. A new discovery from the monkeypox surveillance project in Zaire was that squirrels were an important animal reservoir of monkeypox virus (WHO, 1986).

Surveying the situation in 1986, the Committee considered that there was no further need for its existence, but suggested that WHO might need to set up *ad hoc* committees at various times to consider specific problems that might arise. Such occasions arose in 1990 and 1994.

Ad Hoc Committee on Orthopoxvirus Infections

First Meeting, 11–13 December 1990

A somewhat enlarged group, of which I was elected Chairman, met in December 1990 for the express purpose of deciding whether all existing stocks of variola virus (thought to be those in the WHO Collaborating Centres in Moscow and Atlanta) should be destroyed. The *ad hoc* committee recommended that all stocks of variola virus and materials containing variola virus should be destroyed by 31 December, 1993. It endorsed proposals from the two collaborating centres that they should determine the nucleotide sequences of two strains of both variola major and variola minor before that was done. It also suggested that an expert technical committee should be established to oversee these sequencing efforts.

Second Meeting, 9 September 1994

The same group that had met in December 1990 met again in 1994. The publication of the previous meeting, which had recommended that all stocks of variola virus should be destroyed by December 1993, had engendered a debate between those supporting and those opposing this proposal, and an open international scientific forum was organized by the Ninth International Congress of Virology in Glasgow in August 1993. The unanimous recommendation of the Committee was that the last remaining stocks of variola virus should be destroyed by 30 June, 1995, i.e., just after the meeting of the World Health Assembly in May 1995. A press release outlining the reasons for this recommendation was given wide publicity and there were articles about it in many newspapers, all around the world.

Destruction of Variola Virus, the Final Outcome

The WHO Executive Board, meeting in January 1995 in preparation for the World Health Assembly meeting in May, withdrew the item about variola virus stocks from the agenda. After consulting all members of the *ad hoc* committee that made the recommendation in 1994, I sent a letter to the Director-General of WHO and members of the Executive Board in March 1996 that was signed by seven of the 10 members of that Committee, suggesting that 'since 1996 is "the Year of the Vaccine" and 14 May, 1996, the bicentenary of Jenner's first vaccination against smallpox, destruction of all virus stocks in June 1996 would send a message of reassurance to all nations that smallpox will never occur again and that possession of smallpox virus would be a crime against humankind.'

In the event, the item was placed on the agenda of the World Health Assembly, which agreed but deferred the date to 1999. In 1999, the Assembly again agreed with the proposal but deferred the date to June 2002. In addition, it established a WHO Advisory Committee on Variola Virus Research, with a large membership, most of whom were poxvirus experts, to oversee the proposed research.

First Meeting of the WHO Advisory Committee on Variola Virus Research, 6–9 December 1999

This Committee consisted of 16 members, most of them virologists, 10 expert advisers, 5 observers and the secretariat. I was a member; Dr André Plantinga, from the Netherlands, took the chair. The discussion focussed on experimentation with variola virus, which was to be carried out in approved high security laboratories. It was decided that the main topics to be investigated were: more DNA sequences, monoclonal antibodies for variola virus diagnostic tests, antiviral drugs, novel smallpox vaccines, and the need for a non-human primate model to evaluate antiviral drugs and novel vaccines. Since I had suffered from pulmonary embolism on the previous trip back from Europe in October 1998, and I felt even worse for three weeks after this trip, in spite of taking all the recommended precautions, I decided that it would be the last long overseas trip that I take.

The Book: *Smallpox and its Eradication*

Archiving WHO Records

The book on smallpox eradication could not be written unless there was an adequate archive of all records in WHO Headquarters in Geneva, so, in February 1980, Isao Arita initiated discussions to establish this. With advice from Miss Julia Sheppard, Archivist at the Wellcome Institute for the History of Medicine, and discussions with WHO administrators, Henderson and myself, Mr R. E. Manning was appointed Archives Consultant, Smallpox Eradication Unit, in November 1980 to carry out this task. He began work in December 1980, and submitted progress reports to the Meeting on Implementation of the Post-Smallpox Eradication Policy Committee, in February 1981, and to the Committee on Orthopoxvirus Infections, in March 1982. At the latter meeting it was noted that the records of the smallpox eradication program had been established in an archival mode and that there was a comprehensive inventory and guide indicating what was in each of the files.

Early Plans

As early as March 1978, I wrote to Arita, then Chief of the Smallpox Eradication Unit, explaining that in January 1980 I would have retired from my university job and would like to make the production of the smallpox book my principal post-retirement activity. At the time I thought that I could produce a book with 12 chapters in about three years, and at his suggestion I set out a budget covering the salary for a typist (this was before the days of personal computers) and funds for travelling to Geneva and to Baltimore (to see Henderson). By December 1978 this had evolved into a plan whereby I was the author and Henderson and Arita were designated 'general editors'.

Two years later this concept had developed into a book of 800 to 1000 pages, with 14 chapters and Fenner, Arita and Henderson as the authors. Dr I. D. Ladnyi, then Assistant Director-General of WHO and responsible for its work on communicable diseases, was then included in all further discussions. Subsequently Henderson, Arita and I agreed that Ladnyi should be asked to be an author, as the most appropriate person to symbolize the important contributions of the Soviet Union to the program. We did not expect him to be involved in the writing. By October 1980, after discussions with Henderson during a visit to Baltimore, the plan had expanded to a book of about 1,400 pages, in one volume, arranged in 16 chapters, for each of which a designated author (Fenner, Henderson or Arita) would take primary responsibility. While these discussions were going on I produced two draft chapters, on Virology, and Pathogenesis and Epidemiology.

Final Approval

On 31 March, 1981, the Director-General officially approved the proposal that Arita, as Chief of the Smallpox Eradication Unit, should undertake the preparation of a comprehensive book on smallpox and its eradication under the authorship of Drs Arita, Fenner, Henderson and Ladnyi, with these four authors and a publications officer serving as the WHO Editorial Board. The Head of WHO Publications, Dominic Loveday, undertook to do this work. It was also agreed that Dr Zdeno Jezek, a medical officer with the Smallpox Eradication Unit, and two secretaries in Geneva, would work full time on the project, and a comprehensive budget for three years was approved.

Reports of the Editorial Board

Although the Director-General's approval was dated March 1983, an Editorial Board, comprising Arita, Fenner, Henderson, Ladnyi and Loveday, with Ladnyi as Chairman, had already held meetings in March and September 1982, and subsequently met on 14 March and 11 October, 1983, 30 March and 13 December, 1984, and 26 March and 7 November, 1986. At the early meetings discussion was focussed on the structure and length of the book and the authorship of the various chapters. Subsequently it reviewed progress and looked at such matters as consultation with WHO Regional offices, publishers, estimated cost, formal distribution and promotion of the publication.

There were a few additional matters that warrant mention here. At the fifth meeting, in October 1983, the Board recommended to the Director-General that Jezek should be designated as an additional author, in view of his great assistance in assembling the materials and preparing the first drafts of several chapters. At an early meeting I was able to inform the Board that I had arranged for an excellent draughtsman in the Geography Department of the ANU to prepare all of the maps and diagrams. After a good deal of informal discussion it was decided that WHO should be the publisher, and in March 1982 Keith Wynn, who was in charge of printing at WHO, produced three full-sized mock-up books, on different types of paper and containing part of the text of the clinical chapter, which I had prepared, in 15 different fonts, from which the Board selected one. Arrangements were also made for each of the authors, and selected specialists in the fields covered in various chapters, to read and comment on draft copies of each chapter. Loveday informed the meeting, in March 1984, that Mrs Sheila Deck had been appointed editor. When he retired in 1986, Loveday took over from her. In 1984, John Wickett, an administrative officer during the eradication program, was taken on to supervise the acquisition and preparation of illustrations (other than maps and drawings). Also, in March 1984, I took responsibility for preparing name and subject indexes.

Personal Experiences in the Production

I had travelled a lot overseas during the eradication campaign, often combining smallpox work with other activities, but now I travelled even more frequently, especially to Geneva (13 trips) but also to Baltimore, to see Henderson (five trips) and elsewhere. I also corresponded very frequently with Henderson and Arita, and, as the book progressed, with Loveday (who was an excellent editor and became a very good friend), and I have in my archives copies of letters to 24 other scientists with whom I exchanged letters about the book.

As early as July 1978, Arita had agreed that WHO should make a grant to the John Curtin School of enough money to cover the salary and related expenses of a secretary/typist, an allowance for overseas travel and sundry expenses. The arrangements were finalized in 1980, with a grant of $US22,500 (then worth $A19,800) annually, for two years. Since typists in Australia were paid much less than they were in Geneva, I saved over $A8,000 in 1980 and used this plus savings in 1981 to purchase a Wang Computer (as used in WHO Headquarters) for $A13,219. WHO continued to send me enough money to pay the cartographer, maintenance on the Wang computer and my travel expenses to Geneva until 1986. From 1982 WHO unexpectedly granted me an honorarium of $US15,000, rising to $25,000 in 1983 and 1984 and $US20,000 for 1985, and finally, in 1987 a payment of $US6,000 for preparing the index.

I wrote the first six chapters, which provided background information, as soon as I moved into the John Curtin School in January 1980, starting with the first chapter on the clinical features of smallpox. I had finished the first draft of this chapter in March 1981, sent it to Arita and Henderson for comment, and then to Jezek (who was not then a co-author), A. R. Rao and J. K. Sarkar (India), A. W. Downie and A. B. Christie (England) and G. Meiklelohn and C. H. Kempe (USA) for comment. In October-November 1981, I visited all these people. In my Basser Library Archives, their comments are filed or stapled into the bound draft chapter. The most detailed comments were from A. R. Rao, the author of an excellent book on smallpox in India (Rao, 1972), with whom I spent three days in Madras, going over the draft page by page.

I did not need so much assistance with other chapters, but with every draft I enlisted the help of colleagues and usually made a point of visiting them and discussing the work after they had sent me their comments.

Besides getting critical help from scientists on draft chapters, from mid-1983 I sought help from Loveday on a variety of topics, initially on some technicalities of indexing and from 1986 on matters like permission letters for illustrations, the preface and acknowledgements, reference lists and the like. Fortunately, with the advent of facsimile machines in the mid-1980s, international communication became much quicker and it was possible to exchange drafts with hand-written comments overnight.

The Problem of the 'Whitepox' Viruses

The usual method of isolation of variola and other poxviruses throughout the campaign was inoculation on the chorioallantoic membrane of developing chick embryos. Variola virus produced small white pocks, standard strains of vaccinia virus produced large white pocks but some strains ('rabbitpox' and 'neurovaccinia' viruses) produced pocks with a haemorrhagic, ulcerated centre, and monkeypox virus produced small pocks with an ulcerated centre. In 1958, I had shown that a large number of white pocks, each different in size, could be isolated from rabbitpox virus and that pairs of these would recombine to yield wild-type rabbitpox virus when inoculated into single Hela cells (Fenner, 1959).

One problem that had worried us during the eradication program, but was not finally solved until after the declaration of eradication, was the isolation of what were called 'whitepox' or 'wild whitepox' viruses from several species of African wild animals, and later from stocks of monkeypox virus. Three sets of isolations of 'whitepox' viruses have been made. In 1964–65, four orthopoxvirus isolations were made from cynomolgus kidney cells, at the time being used for virus isolation in the National Institute of Public Health in Bilthoven (Gipsen and Kapsenburg, 1966); one was eventually recognized to be monkeypox virus, one was vaccinia, and two others were indistinguishable from variola virus (Gispen and Brand-Saathof, 1972). On two occasions in 1964, material from smallpox patients from Vellore, in India, had been handled on the same bench in the same laboratory, and skilful detective work by Kapsenburg identified the passage transfer of cell cultures during which contamination could have occurred.

The second set comprised four strains of virus, isolated by workers at the Moscow Research Institute for Viral Preparations in Moscow, one of the two WHO Collaborating Centres for Viral Diagnosis, from the tissues of a chimpanzee, a monkey, a squirrel and a multimammate rat shot in Zaire between 1971 and 1975 (Marennikova et al., 1972). The third set were viruses that produced white pocks which had been recovered from stocks of monkeypox virus that had been maintained for several years in the Moscow laboratories (Marennikova et al., 1979); all these isolates were identical and indistinguishable from those of the second set.

Seeking an explanation, Marennikova postulated that the third set were white variants of monkeypox virus, a virus which had a wide host range and was known to occur in Zaire. However, investigations by Keith Dumbell in London showed that all the 'whitepox' viruses isolated in the Moscow laboratories resembled Indian and not African strains of variola virus, and studies by Jim Nakano at CDC, Atlanta, confirmed these results. In relation to the third set, other studies showed that monkeypox virus, like other orthopoxviruses that produced pocks with a haemorrhagic centre, did produce white pock mutants, but, as with rabbitpox, these differed from one another and resembled monkeypox and not variola virus in most biological tests.

Clearly, it was important to clarify the situation, so one of Marennikova's staff was despatched to the Atlanta laboratory to see if she could reproduce her results there. She was unable to do this using material supplied by the Atlanta laboratory, so she was given permission to use one of the 'monkeypox' strains that she had brought from Moscow, and with this she was able to reproduce her earlier results (Esposito et al., 1985). These investigations demonstrated that all the 'whitepox' viruses, whether isolated from animals from Zaire or recovered from monkeypox virus strains from the Moscow laboratories, were laboratory contaminants, a conclusion with which Marennikova finally agreed.

Book Launch

Eventually, on 18 January, 1988, came the launching ceremony, held as part of the 81st session of the Executive Board of WHO. There were four speakers: an introduction by Dr A. Grech, Chairman of the Executive Board; an introduction to the book by myself, as senior author; then a speech by Dr Mahler, Director-General of WHO; and closing remarks by Dr Grech. This ceremony was followed by a press conference, at which the speakers at the launch and the other authors were available to answer questions. Unfortunately, Dr Ladnyi, one of the authors, had died in 1987 as the result of a car accident.

The launch was followed that evening by a dinner arranged by the staff of the former Smallpox Eradication Unit, the demise of which, on 1 January, had been announced by the Director-General that afternoon. They had chosen one of the very pleasant restaurants that abound in the villages around Geneva. It was a Monday evening and there were few others than our group in the restaurant. I remember noticing a pile of cardboard boxes in one corner of the restaurant, near our table, but I thought nothing of it. Then, as the very pleasant and relaxed dinner drew to a close, Keith Wynn, who was in charge of production of WHO publications, started to open the boxes and pass around special leather-bound copies of the 'big red book'. Everyone at the dinner who had been involved with its production signed every copy and all received a copy. Years later, in 1997, I donated my copy to the Rare Books Collection in the ANU Library.

Two Other Books Related to the Smallpox Program

There were always times in the 1980s when I could not work on *Smallpox and its Eradication,* and during these periods I produced two other books that expanded on various topics mentioned in that book. The chapter, entitled 'Variola virus and other orthopoxviruses', had been designed to provide background information for public health workers, and I thought that it was timely to expand that chapter into a book for virologists. The result was another book, *The Orthopoxviruses,* with Keith Dumbell and Riccardo Wittek as co-authors and published by Academic Press in 1989. Zdeno Jezek, one of the co-authors of *Smallpox and its Eradication,* had been involved in the survey of the human monkeypox set up in Zaire, with assistance from WHO, between 1980 and 1986. He and I collaborated in producing a short review of the results of this survey in another book, entitled *Human Monkeypox* and published in 1988. Wittek lived in Lausanne, so I was able to consult him and Jezek whenever I went to Geneva.

The Orthopoxviruses

Since the chapter on virology was one of the first I wrote for *Smallpox and its Eradication,* by 1982 I had enlisted the help of two co-authors, Samuel Dales and Keith Dumbell, for this book, and, in 1981, I arranged for the book, with those authors, to be published by Academic Press, publishers of my virology textbooks. The proposed publication date was October 1983. We decided that we needed another poxvirus expert who was a molecular biologist and, in 1984, I persuaded Ricardo Wittek to be a co-author. Then, in 1986, Dales had to pull out, and the Memorandum of Agreement had to be altered accordingly. However, by late 1987 it went to press and was published in 1988 (Fenner et al., 1988). It received excellent reviews, including one from a poxvirus expert , who concludes with these remarks: 'This book is highly recommended...and is worth every cent of its price. The chapters are extremely well written, and the content is superb throughout. Collectively, this work presents the best up-to-date summary of the features of these viruses anyone, novice or specialist, is likely to need for a long time to come.'

Human Monkeypox

Human monkeypox had been a problem in the smallpox eradication campaign ever since its discovery in 1970. From 1980 to 1986, WHO supervised institution-based surveillance of monkeypox in selected areas representing about 15 per cent of the land area of Zaire and containing about 5 million inhabitants. In 1984–86, WHO arranged for Dr Lev Khodakevich to supervise surveys in this area to determine the most important animal hosts, which turned out to be several species of squirrels.

Over a period from early 1985 to mid-1986 I corresponded with Zdeno Jezek, who had succeeded Arita as Chief of the Smallpox Eradication Unit, editing his articles on this program. In early 1987, we decided to publish a short book based on the work of the WHO program and I wrote to Joe Melnick, a friend of mine who was the editor of *Monographs in Virology*, published by S. Karger AG, suggesting that we should publish a monograph on human monkeypox in that series, with Jezek as senior author. He agreed, and Zdeno and I worked together from 30 July, 1987, until it was published in October 1988 (Jezek and Fenner, 1988). It is recognized as the definitive text on human monkeypox.

Award of the Japan Prize to Henderson, Arita and Fenner

In 1985, Macfarlane Burnet nominated D. A. Henderson, I. Arita and me for the Nobel Prize in Physiology or Medicine; the nominations were repeated in 1986 and 1987, from both Australia and the United States. In 1988, the World Health Organization was nominated for the Nobel Peace Prize for the eradication of smallpox, which unlike the other Nobel Prizes had been awarded to organizations (UNICEF, UNHCR) as well as to individuals. This nomination was submitted from Australia, the United States and Sweden, but like the earlier nominations, it was unsuccessful.

The Japan Prize, initiated in 1985 by the Science and Technology Foundation of Japan, with one award in the physical and one in the biological sciences each year, but each year in different designated topics, was regarded as the equivalent, for applied science, of the Nobel Prize for creative science. In 1988, the designated topic in the biological sciences was 'preventative medicine'. We were nominated, and for the first and only time so far, it was awarded for two subjects: to Henderson, Arita and me, for the eradication of smallpox; and to Drs L. Montagnier and R. C. Gallo, for the discovery of and diagnostic methods for the AIDS-causing virus. We and our wives all went to Japan for the presentation ceremony, which was held in Japan Prize week, 4–10 April, 1988. Details of this are set out later in this chapter.

The Threat of Smallpox Virus as a Bioterrorism Weapon

In an earlier section of this chapter, I have referred to the prolonged discussions about the destruction of stocks of variola virus. With hindsight, these discussions were irrelevant, as became apparent when Ken Alibek (1999) announced to the world that, in spite of its signature of the Chemical and Biological Warfare Convention in 1975, the Soviet Union had taken that virus into its massive biowarfare program in 1980. When this program was dismantled by President Yeltsin in the early 1990s, it was realized that some of the unemployed scientists would be very likely to offer their services, and supplies of freeze-dried viruses and bacteria, to other countries or organizations.

By 1994, it became known outside the Soviet Union that the Russians had moved their variola virus samples from Moscow to Koltsevo, in Siberia, where one of their major biowarfare installations was located. After the events of 11 September, 2001, the industrialized nations became concerned with the risk of smallpox as a bioterrorism weapon. Especially in the United States, but also in the United Kingdom and Australia, access to vaccine stocks was arranged and plans for responding to various levels of risk developed, ranging from a single case of smallpox anywhere in the world to outbreaks in the country concerned. In Australia, the responsibility for making these plans lay with Professor Richard Smallwood, who was then Chief Medical Officer of the Commonwealth and a close personal friend of mine. I was a member of a team he assembled, with representatives of all the states and territories, to produce such a plan; this was published in January 2004 (Australian Government Department of Health and Ageing, 2004) and made widely available from Commonwealth, State and Territory health departments. Over the past few years, from 1999 onwards, I have given a number of lectures on this topic to a wide variety of audiences, and helped with several newspaper articles on smallpox as a bioterrorism threat.

More Books

Portraits of Viruses. A History of Virology

The journal *Intervirology* was established in 1973 as the official journal of the Virology Section of the International Association of Microbiological Societies, and continued in that role until 1990. The publisher was S. Karger, of Basel, the Editor-in-Chief was J. L. Melnick, and there were several Section Editors, among whom A. J. Gibbs was Section Editor for Plant Virology for many years, and I was Section Editor for Taxonomy from 1973 to 1977. Gibbs had joined the Department of Microbiology in JCSMR as a Research Fellow in 1966 and was promoted to Senior Research Fellow in 1967. He returned to England in 1969, but came out to Canberra again a few years later, to a senior post in the Research School of Biological Sciences in ANU. We were always good friends and in 1978 we suggested to Melnick that we would like to produce a book on the history of virology by asking selected experts to contribute articles to *Intervirology* on the history of viruses or virus groups in which they were recognized experts. We undertook to select and persuade the authors of these articles, and also to act as editors of the book in which these essays would be collated. Between 1979 and 1986, 15 such articles were published in *Intervirology,* and in 1988 assembled in Fenner and Gibbs (eds) *Portraits of Viruses: A History of Virology*. Nine of the chapters dealt with animal viruses, four with plant viruses and two with bacterial viruses. Each chapter also provided interesting portraits of the authors, each a leader in his field.

White and Fenner, *Medical Virology*, Third Edition

The second edition of *Medical Virology* had been published in 1976 and was very popular as a university textbook. I started writing to Academic Press about a third edition in 1980, saying that David White and I thought that we might be able to get this to them by 1983. However, completing *Smallpox and its Eradication* was more demanding on my time than I had anticipated, and the third edition was not published until 1986. During that 10 years, virology had made great advances, and the book, in the same format as the second edition, was enlarged from 487 to 665 pages. Since I was much older than David, and at the age of 70 getting out of touch with molecular virology, I suggested that David should be first author, and he agreed. As with previous editions, the reviews were complimentary; different reviewers commented: 'Like its predecessor, this edition is beautifully illustrated...The single most appealing feature of this book is its extreme readability', and, 'The authors succeeded in presenting a fast growing branch of biology and medicine in a complete, condensed and at the same time comprehensive way.' Subsequently, Spanish, Japanese and Chinese editions were published and we received numerous requests for permission to use illustrations and/or tables in other books.

White and Fenner, *Medical Virology*, Fourth Edition

We signed the Memorandum of Agreement for the fourth edition in November 1990, but this time it was David who was overcommitted and the manuscript was not completed until 1993. After discussion with Academic Press, we used a different format, with pages that included text within a rectangle measuring 20.5 x 12 cm, compared with the 17.5 x 12 cm used for earlier editions. As a result, the additional information we had to cover (including a much longer section on Retroviruses) was included within 601 pages. Once again, it received excellent reviews. David had to retire from his university post in 1994, for medical reasons, and I had my 80th birthday that year, so neither of us could help with a fifth edition. Academic Press is very anxious to publish one, but so far they haven't been able to find suitable authors. The need is clear, since 253 copies of the 10-year-old fourth edition were sold in 2004; total sales of this edition, from 1994, were 13,279.

Fenner et al., Veterinary Virology

In January 1978, having found that several veterinary schools used *Medical Virology* as a textbook, I wrote to Dr J. Barsky, Vice-President of Academic Press, suggesting the possibility of producing a companion volume, *Veterinary Virology*. Academic Press made enquiries from veterinary schools in the United States, which convinced them that there was a market for such a book. Between 1979 and 1983 I wrote to a number of veterinary virologists and eventually asked Paul Gibbs, of the University of Florida at Gainesville, Fred Murphy at the Centers for Disease Control in Atlanta and Michael Studdert of the University of Melbourne, to join David White and me as co-authors. The plan we adopted was to use the scheme that had been so successful with *Medical Virology*, namely Part I, with 16 chapters on the Principles of Animal Virology, using examples from diseases of veterinary importance rather than human diseases, and Part II, with 19 chapters on viruses of domestic animals grouped taxonomically. An agreement with four of the authors was signed in June 1983, and with advice from Fred Murphy I persuaded Peter Bachman, of the Ludwig-Maximilians-Universität in Munich, to join us. Fortunately, I was able to visit Munich and see Peter in April 1984, during one of my many trips to Geneva. We allocated each chapter in Part II to one of these authors. When drafted, this was examined by all authors and, finally, it would go through my word processor, so that we maintained a unity of style. Royalties were shared equally between the six authors.

There was, of course, much more correspondence about this book than with *Medical Virology*, by letter and fax (email had not been invented). It was finally published in 1987, and was dedicated to Peter Bachman, who had died, unexpectedly, on 26 May, 1985. It received uniformly excellent reviews in veterinary journals, the only problem most of them mentioned was its high price.

In 1989, arrangements were made for it to be translated into Spanish; the Spanish translation was published in 1992.

Fenner et al., *Veterinary Virology,* Second Edition

By 1989, Academic Press decided that a second edition of *Veterinary Virology* should be published and a Memorandum of Agreement was signed in October 1990. With the agreement of all authors, Bachman's place was taken by Rudolf Rott, of the Institute of Virology in Giessen, Germany. The second edition was published in 1993, and although it covered the many advances in molecular virology made since 1987, both books were almost the same length.

Murphy et al., *Veterinary Virology,* Third Edition

David White and I ceased to be involved with the textbooks in 1994, but Fred Murphy, Paul Gibbs and Mike Studdert, authors of previous editions, and Marian Horzinek, of Utrecht University, produced a third edition in 1999, in the same improved format as that used for the fourth edition of *Medical Virology*. They dedicated it to David and me. The dedication, beneath portraits of each of us, reads:

> This book is dedicated to our dear friends, Frank J. Fenner and David O. White, the founders of a series of books that now includes three editions of this book and four editions of *Medical Virology*. They set a standard of scholarship that is impossible to match and a *joie de vivre* that made the writing and editing almost fun. They taught us that the subject of virology must be seen within the context of society as a whole as well as within the context of science. They envisioned virology as being so broad as to extend from its roots as a microbiological science, a molecular and cell biological science, an infectious disease science, to become a major contributor to the overall advance of human and animal wellbeing. All this as a single seamless cloth. We hope our students will come to understand the 'big picture' of veterinary and medical virology as well as Frank and David have throughout their amazing careers.

History of Microbiology in Australia

I had been a member of the Australian Society for Microbiology (ASM) since its foundation in 1960, President in 1964–65, and Honorary Life Member since 1975. In the November 1986 edition of its journal, *Australian Microbiologist*, Carolyn Beaton, the assistant editor, suggested that Australia's Bicentennial year, 1988, should be marked by a special issue chronicling 200 years of microbiology in Australia. Tragically, Carolyn died in April 1987. David White, then President of the Society, persuaded me to act as editor of a book on the topic and he acted as Chairman of an Editorial Advisory Panel to assist in the task. The result was a book of 610 pages, with 11 chapters:

1. The early days of microbiology in Australia
2. Teaching institutions
3. Research Institutes and CSIRO
4. Diagnostic laboratories and the Commonwealth Serum Laboratories
5. Australian contributions to bacteriology
6. Australian contributions to virology
7. Australian contributions to mycology
8. Australian contributions to protozoology
9. Industrial microbiology
10. National activities concerned with microbiology
11. International activities concerned with microbiology

To cover this broad field, I enlisted the help of 'Coordinators' for various sections of each chapter and, in addition, some 320 microbiologists selected by the appropriate coordinator supplied information which I organized and collated. Since some of them provided more material than could be accommodated within the book, I deposited any additional material in my file (MS 143) in the Basser Library Archives, along with glossy prints and if possible negatives of the 250 photographs of distinguished microbiologists. One device I had not used before but found useful here and in subsequent books on the history of science was to place a 'potted biography' beneath or alongside each portrait.

The book was produced by a friend of mine, Robert Kirk, who had taken on desktop publishing when he retired as Head of the Department of Human Biology in the John Curtin School of Medical Research. Glaxo Australia shared the cost of production with the Society, with a grant of $20,000. The ASM purchased 5,000 copies at a cost of $58,160, i.e., about $12 a copy, which has allowed it to provide every member with a copy; every new member receives one on election. To help pay for the printing, copies were sold for $75 (prepublication $50) to most of the institutions mentioned in the book.

The Australian Academy of Science: the First Forty Years

In 1993, I suggested to Peter Vallee, the Executive Secretary of the Australian Academy of Science, that I should update the extensive Appendices of *The Australian Academy of Science: The First Twenty-Five Years,* in preparation of publication of a history celebrating the Academy's jubilee in 2004. However, noting that *The First Twenty-five Years* was then out of print, he suggested that I should update and revise the whole book. The co-editor of *The First Twenty-five Years,* Lloyd Rees, had died in August 1989, and I agreed to do this. With the guidance of a small Advisory Committee (Fellows R. W. Crompton, L. T. Evans, N. H. Fletcher and the Editor of *Historical Records of Australian Science,* R. W. Home), the book was published in 1995. In the same format as the previous book, it grew from 286 to 503 pages.

The Australian Academy of Science: the First Fifty Years

Having finished the history of the first 50 years of the John Curtin School in November 2001 (see below), I started updating *The Australian Academy of Science: the First Forty Years* in 2002, with the intention in getting it published early in 2005. The task was quite different from the earlier versions, because all records were by then electronic. Once again, it was a matter of getting help from the secretariat and Fellows of the Academy. I decided to adopt a different format, using the fourth edition of *Medical Virology* as a model, both in relation to page size and details of page headings, etc. With special help from the Publications Manager, Maureen Swanage, and the Librarian, Rosanne Walker, the book was published in February 2005.

Biological Control of Vertebrate Pests: the History of Myxomatosis—an Experiment in Evolution

Late in 1991, I was invited to give a paper on the history of smallpox at a conference, 'Emerging Infectious Diseases: Historical Perspectives', held at Merieux Conference Centre at Annecy, in France, in April 1992. At lunch one day, I sat next to the symposium organizer, Professor Bernardino Fantini, Director of the Louis Jeantet Institute for the History of Medicine at the University of Geneva. I accompanied him on the trip back to Geneva by taxi and he suggested that we should collaborate on a series of papers on the history of myxomatosis. During our correspondence between June 1992 and October 1993, the idea of a series of papers evolved into the concept of a book on the biological control of vertebrate pests (of which the only two successful agents were myxoma virus and rabbit haemorrhagic disease virus). By January 1994 we had developed a plan for the book and I promised to start writing as soon as I had finished work on the fourth edition of *Medical Virology*. My frequent trips to WHO in Geneva made it easy to discuss problems face to face, and we also corresponded a lot by email. By January 1996, *The History of Myxomatosis* had evolved into a more general book: *Biological Control of Vertebrate Pests: the History of Myxomatosis—an Experiment in Evolution*. I did most of the writing, but I wouldn't have started the book, nor worked out the best structure for it, without Fantini's help. We arranged for publication by CABI International, Wallingford, in February 1998 and the book appeared in July 1999. It is a book of 339 pages, arranged in 14 chapters, with numerous diagrams and many photographs, especially of persons involved in relevant research. I regard it as one of my best books.

The John Curtin School of Medical Research: the First Fifty Years, 1948–98

In January 1997, I had a letter from the publishers, Allen and Unwin, who in 1996 had published *The Making of the Australian National University: 1946 to*

1996, saying that they would like me to consider writing a history of the JCSMR. At the time, I said that I was too busy with other books and the preparation of my archives for the Basser Library. However, by mid-1998 I had changed my mind and started to plan a book to be produced by me and David Curtis as co-authors. I chose to ask David because he had been Eccles' first PhD student, graduating in 1957, then rising through the academic ranks to be a Professor in 1966 and was Director of the School from 1989–92. However, I had been so impressed with the high quality and low price of *History of Microbiology in Australia,* which had been published in 1990 by Brolga Press, a desk-top publishing operation managed by Bob Kirk, that we decided to get him to publish this book. Late in December 1998, I wrote a letter to the Director of JCSMR for consideration by Faculty Board, canvassing this idea. Faculty Board approved and during 1999 David and I discussed its structure. We finally decided that it should consist of three parts: Part I, Development and Change, which was a chronological account of how the School developed and the gradual increase in coverage, by department and with later consolidation into Divisions; Part II, Highlights of Research, which consisted of 89 essays, arranged in 10 fields, according to discipline and Part III, Statistical Information, which included full details of all Academic Staff, Visiting Fellows, PhD students, Service by staff to organizations outside the School, and External Grants, in those days a minor source of funding.

We included many photographs, with long legends outlining the careers of all the major actors, and several illustrations of the development of the buildings. At this time Bob Kirk operated from Gunderoo, a village in New South Wales, some 35 km from Canberra. Publication was funded from a donation of $40,000, which I made over a period of two financial years. The School ordered 5,000 copies, and is able to give them to all new staff and students and to visitors. The book was launched on 15 November, 2001, by Dr Barry Jones.

Overseas Trips Other than Those Concerned with Smallpox

China, 1980, Australian Development Assistance Bureau

This trip was my first and only experience of travel with members of the Diplomatic Corps. In October 1980, I was asked to be part of a mission to China by representatives of the Australian Development Assistance Bureau (ADAB), to investigate the potential for Australian aid to China. The members of the group were J. C. Ingram, Director of ADAB (leader), T. Terrell and W. Newton (both ADAB), R. Dun, NSW Department of Agriculture, D. Little, Director-General of Public Works, Victoria, and me. It was most interesting, although I felt that I could not contribute much to the discussion. What I remember most vividly is flying to Lan-Chou, capital of Gansu province, over miles of heavily eroded loess mountains. On the ground, there was always a heavy mist of dust; this was the reason for the Huangho River being called 'the Yellow River'. We visited the Desert Research Institute, where they were trying to control the desert sandhills north of Lan-Chou by planting poplars on the ridges. We were given splendid meals and visited all the temples etc. around Beijing. I met up again with Dr Jiang Yu Tu, who had guided Joel Breman and me around on our visit to certify smallpox eradication in 1978.

Washington, 1982 to 1983, Fogarty Scholar

I had spent three months as a Fogarty Scholar in the National Institutes of Health (NIH) in 1971–72, and Bobbie and I had spent three months there in 1973–74. In August 1979, anticipating my retirement at the end of the year, I wrote to NIH to follow up a suggestion from the official then in charge of the scheme, Dr Peter Condliffe, suggesting that since I expected to be deeply involved in writing books on virology, I would like to spend several months at NIH, preferably as a Fogarty Scholar, in 1982–83. Condliffe came out to Canberra in January 1980 and we arranged to take up the Scholarship again, with Bobbie, between September 1982 and February 1983. I copied all correspondence to my good friend Bob Chanock. Previously we had lived at Stone House; this arrangement was no longer possible, but I had an office there. We lived in a flat opposite the Clinical Center at NIH.

I came over from Geneva, where I had been working on the smallpox book. Bobbie flew directly from Canberra. I spent a lot of time in the National Library of Medicine, which was on the NIH campus and had long runs of most medical journals, which were very useful for historical material for the smallpox book. I also worked on a possible third edition of *The Biology of Animal Viruses* (unfortunately Joe Sambrook, who was a very able molecular virologist and had agreed to be a co-author, pulled out and this was never published).

My colleague, David White, was spending a few months on study leave at this time. He had hired a car and we visited a number of the interesting places near Washington, DC. We saw a great deal of Bob and Beth Chanock, who often took us to concerts at the Kennedy Center and introduced us to the wonderful galleries and museums in central Washington.

In December, we went up to the Rockefeller University in New York to attend a memorial service for René Dubos, who had died earlier that month, and met many old friends of the 1940s. Then I went to Atlanta, Georgia, to the CDC and talked with Jim Nakano, Joe Esposito and other poxvirologists. Later I went to Gainesville, Florida to discuss the projected *Veterinary Virology* with Paul Gibbs, and a couple of weeks later to Baltimore to see D. A. Henderson in his office at Green Mansions, a Hopkins University building where he worked on the operational chapters of the smallpox book each weekend. Finally we attended a ceremony at Stone House, where the Acting Director of the Fogarty

International Centre presented me with a Fogarty Medal and accompanying Certificate.

Japan, 4–11 April 1988, Japan Prize Award Ceremony

Bobbie and I arrived in Tokyo on Monday 4 April, 1988, and were met at the airport, taken to the Akasaka Prince Hotel and given a press kit and a detailed program for the week. We met the Press on Tuesday, and the Governor of Tokyo and the Prime Minister on Wednesday morning, followed by a Japan Prize Lecture by each of us in a large hall, then an 'academic debate'. That evening, the Hendersons, the Aritas, Bobbie and I had dinner at the Australian Embassy, other guests included the Director-General of the Institutes of Health in Japan and his wife. The next morning, we met the Science Council of Japan and rehearsed the award ceremony in the National Theatre of Japan. The actual Award Ceremony occurred that afternoon, followed by a superb banquet in the Akasaka Prince Hotel, during which we met Crown Prince Akimoto and presented him with a copy of *Smallpox and its Eradication*. On Friday morning, we met Emperor Hirohito (who at that time was very frail and rarely met anyone) at the Imperial Palace. In the afternoon, we travelled by the bullet train to Osaka, where we gave the Japan Prize Lectures again. The next day, after meeting the Governor of Osaka in the morning, we spent the rest of the day sight-seeing in Osaka and Kyoto, followed by a splendid Japanese style dinner (including sitting at the table without chairs) at what we were told was the most exclusive restaurant in Japan. Our host was Mr Matsushito, 93 years old, who set up and largely endowed the Japan Foundation for Science and Technology, which operates the Japan Prizes.

On the Sunday, the smallpox group went to Kumamoto, Arita's home town. After dropping our baggage at the New Sky Hotel we visited several gardens, then D. A. and I went to the Kumamoto Hospital, where Arita had gathered an audience. D. A. spoke about international health perspectives and I about viral diseases in Australia. Next afternoon we visited Mt Aso, an impressive volcano in the middle of the island of Kyushu, then to the airport and back to Narita airport, before flying to Hong Kong next morning. After arranging for storage of most of our luggage, Bobbie and I flew to Guilin in China. We booked in at a hotel there and we went for a trip down the Liyuan River. Unfortunately it was very misty, and we could only catch glimpses of the famous mountains. Then back to Hong Kong and home, with a stop-off in Bangkok for Bobbie to buy some dresses.

China, 18–25 September 1988, review of an ACIAR Project

The Australian Centre for International Agricultural Research (ACIAR) was established by Prime Minister Fraser in 1980, at the suggestion of Sir John Crawford, then Vice Chancellor of ANU. Among its projects was the utilization of entomopathogenic nematodes to control insect pests in China, which had been initiated in October 1985, the principal investigator being Dr Robin Bedding, of the CSIRO Division of Entomology. In June 1988, I was invited to assist in reviewing the project, both in China and Hobart, Tasmania, where Bedding was producing nematodes on a large scale. I visited Hobart on 30–31 August, 1988, and China on 18–25 September, 1988. The other reviewer was Professor Qiu Shi-Bang, of the People's Republic of China. In China we visited the Guangdong Entomological Institute, where large scale production of nematodes was based, and the Biological Control Laboratory in Beijing, where it was intended that larger scale production could be carried out for use in the control of *Carposia* in apple orchards. We were very favorably impressed with the project and reported our opinions to Dr McWilliam, Director of ACIAR, in October 1988.

Japan, 1991 to 1996, Agency for Cooperation in International Health

In 1991, just before he retired from his position as Director of the Kumamoto National Hospital, Isao Arita set up a new organization, the Agency for International Cooperation in Health (ACIH), which was based in his home town, Kumamoto, and planned to provide support for preventive medicine in developing countries. From 1993, he assumed chairmanship of ACIH. Among its varied activities were international conferences on such matters as vaccine supply to the poorest countries. At his invitation, I participated in several of these conferences, usually acting as rapporteur, and I edited the reports of the meetings, which were usually held in Kumamoto.

Meeting on Global Vaccine Supply, 23–26 May 1991

11 international experts met in Kumamoto for two days to discuss such matters as vaccine supply for WHO's Expanded Program for Immunization, including both the production of new vaccines in developing countries and ways of assuring vaccine quality. On 26 May a symposium was held on 'Japan as seen by Foreigners'. After an introduction by Dr Arita, Dr Mark Radford, an Australian who had worked in Japan since 1985, and I discussed the topic from our different angles. After the speeches there was a vigorous discussion, and subsequently everything was published in both Japanese and English.

Second Meeting on Global Vaccine Supply, 3–5 August 1992

This meeting included 10 overseas and 17 Japanese experts and met in Tokyo. The topic was the same as for the first meeting, but this time experts from Indonesia, Brazil and India spoke of the situation in their countries, and one from the United States Agency for International Development (USAID) described their long experience with these problems. A very useful graph was produced in the report of the meeting which indicated, for all countries in the world, where they stood in relation to wealth, population and capacity to produce bacterial and viral vaccines. A series of recommendations was produced on global planning and coordination, achievement of vaccine self-sufficiency and assurance of vaccine quality, and, as before, I acted as rapporteur.

Third Meeting of the Consultative Group, Children's Vaccine Initiative, Kyoto, 7–9 November 1993

The Children's Vaccine Initiative (CVI) was a global coalition of governments, UN organizations, non-governmental and private organizations, industry and research groups focused on the goal of bringing new and improved vaccine into national immunization programs. ACIH had played a major role in organizing this meeting in Japan and Arita was the keynote speaker. It was attended by 212 participants, representatives from the Rockefeller Foundation, UNICEF, UNDP and the World Bank, and 13 members of the WHO Secretariat. At the conclusion of the three-day meeting, it produced the Declaration of Kyoto, essentially a document emphasising the importance of promoting vaccination world wide and the need for the broadest possible support to be mobilized among developing and developed countries to catalyze the priority activities described in the CVI strategic plan. I helped produce an abridged report for the ACIH.

Children's Vaccine Initiative and Jenner Commemoration, Kumamoto, 25–26 November 1996

There were 36 speakers at this meeting, which as well as celebrating the 200th anniversary of Edward Jenner's discovery, tried to work out ways of accelerating the process from laboratory research to the production of vaccines and their use in the field. For the first time, there was a detailed discussion of the problems in producing a vaccine against HIV, in which my colleague from the John Curtin School, Gordon Ada, participated. I prepared an abbreviated report for ACIH. This was the last meeting arranged by ACIH in which I was able to participate.

Bozeman and Yellowstone National Park, 11–25 July 1997

I had been invited to give the Edwin H. Lennette Memorial Lecture at the annual meeting of the American Society for Virology in Bozeman, Montana, in July 1997. Fred Murphy got in touch with me well before the meeting and suggested that I accompany their family on a week's trip through the Yellowstone National Park. He had a new RV (recreational vehicle), which had beds for four, shower, toilet, stove and refrigerator. Fred, Irene and their son, Ric, his two young boys and I travelled in the RV; his other sons had brought their bicycles. We had a wonderful trip all around Yellowstone, not only the first national park in USA, but one of the most wonderful in the world, with entrancing hot springs, geysers and a great trip down the Snake River. Then we went to Bozeman, where the meeting was very interesting. I gave my lecture and saw a lot of old friends, including Joe Esposito, Grant McFadden, Olin Kew, Dick Moyer and Mary Estes.

Home Life and Bobbie's Death

There is brief mention of my marriage to Bobbie Roberts in Chapter 3 and of the children in Chapter 5, but she was such an important factor in my life and career that I must say a bit more about our life together, and her death. Throughout my career as a Professor of Microbiology, and even more when I was Director of the John Curtin School and of CRES, Bobbie was a tremendous support. With her help, we often entertained staff and visitors, with dinners at home and in summer, parties in our spacious garden. I travelled overseas a great deal. Most of these trips were short, and especially in Geneva I worked all the time. She did not come on these trips and did not want to. But, especially after my retirement, she often came as well, as outlined in the previous section.

Bobbie had always been a moderate cigarette smoker and when I was head of the Department of Microbiology I used always bring her back duty-free cigarettes when I had been overseas. But as soon as I became Director, she gave up smoking and told me to get rid of all the cigarettes in the house. Her initial distress, hunting everywhere for one more cigarette, brought home to me how addictive smoking is for some people (I had never smoked). Bobbie was very active in a range of community activities. Almost immediately after our arrival in Canberra, she was invited to become a Councillor of the Canberra Mothercraft Society. As a Triple Certificated Nursing Sister, she served with distinction, representing the Society as a delegate to the National Council of Women of the ACT and for many years was a member of the Executive. She supervised the monthly clothing sales for the National Council of Women and helped with the teas that were given every 'Pension Thursday' to the early pioneers of Canberra, before the advent of Senior Citizens Clubs. She was a member of the Pan-Pacific and South-East Asia Women's Association, helping many people from those areas settle into life here, and as one of the first members of the Ex-Servicewomen's Sub-Branch of the Returned Services League of the ACT, she represented the RSL on the Services Trust Welfare Fund, on which she served with distinction for many years. Among the many charities that Bobbie helped regularly were the Guide Dogs for the Blind, the Smith Family, the Knitting Guild, the Save the Children Fund and UNICEF, which in 1995 recognized her many years of service with an award.

As a close friend has said, 'It was Bobbie's way to "say it with flowers"' and the garden at 8 Monaro Crescent was the venue for many fetes and garden stalls, for many organizations, including the Canberra YWCA Annual Garden Sale, which raised thousands of dollars annually, and she gave help with flowers and plants to fetes held by Legacy and the Boys' and Girls' Grammar Schools.' Besides gardening, her hobbies included tennis, golf and bridge, and she was a member of the University Ladies Drawing Room Committee, the Commonwealth Club and Friends of the National Gallery. In January 1980, she was decorated with a Medal of the Order of Australia (OAM) at a ceremony at Government house, for community service.

In 1989 she was found to have colon cancer, and had a colonectomy, with good results for several years. Then, in 1994 she was found to have extensive secondaries in the lungs, which progressed in spite of radium treatment and chemotherapy. For some months she was confined to bed, at home, but in October 1995 she was sent to the Respite Care Facility, on the shores of Lake Burley Griffin. She gradually got worse, but insisted that I should go to London in early December to receive the Copley Medal of The Royal Society, although at that stage she was at death's door. She died on 28 December, 1995.

To many, she will be long remembered as a loving friend. This is well encapsulated in a letter from Kunang Helmi, the eldest daughter of Indonesian Ambassador Helmi, who was a near neighbour in the late 1950s: 'In fact what I really want to say is how much I love you Aunt Bobbie, for what you are and what you did. You set me a shining example of what kindness and generosity are about—I often think of you in my prayers, as do Rana and Rio [Kunang's younger siblings].'

Although it was long expected, I was devastated by her death. Company at the John Curtin School each week day was a great help, and after there had been time to repaint the interiors of the main house and the extension that we had built in 1981–82, I moved into the extension and Marilyn and her family moved into the main house, which has proved an excellent arrangement for both of us. Even so, it took about three years before I could adjust to Bobbie's absence and, only then, I told myself: 'I see so many women who have lost their husbands and adjust to life as widows; I must adjust, as a widower.'

Celebrations of the Lives of My Mentors

I was deeply involved in celebrating the centenaries of the births of two of the three scientists (other than my father) who had the greatest influence on my career, Lord Florey (1898–1968) and Sir Macfarlane Burnet (1899–1985). I had written biographical memoirs and obituary notices for both of them, I was involved in the unveiling of the Florey Stone in Westminster Abbey in 1982, and in the celebrations of the Florey Centenary in 1998 and the Burnet Centenary in 1999.

Howard Florey

I wrote the obituary notice in the *Australian Journal of Science* (Fenner, 1968), and entries on Florey for *Scribner's Dictionary of Scientific Biography* (Fenner,

1972), the Roll of the Royal Australasian College of Physicians (Fenner, 1988) and the *Australian Dictionary of Biography* (Fenner, 1996).

I was also involved in the dedication of the Florey Stone in Westminster Abbey. The idea that there should be a record of Florey in one of Britain's great cathedrals was due to Dr Cecil Hackett. Lady Fairley, the widow of Sir Neil Hamilton Fairley and a close friend of the Hacketts, had noticed that there was a plaque commemorating Sir Alexander Fleming in St Paul's Cathedral, in London, but nothing commemorating Florey there, or in Westminster Abbey. Mrs Beattie Hackett was a voluntary helper in Westminster Abbey of many years standing. Cecil Hackett took it upon himself to get something done, and obtained support from The Royal Society and the South Australian government and agreement from the Dean of Westminster Abbey for the placement of a commemorative stone on the Abbey floor, adjacent to those of Herschel and Charles Darwin. Because I was an Australian, a Fellow of The Royal Society, had worked with Florey on the establishment of the John Curtin School, and was a close friend of the Hacketts, I was asked to give the address at the ceremony on All Souls Day, 1982, when the Florey Stone was unveiled by Lady Florey. Subsequently, I signed the register of those who had given 'sermons' at Westminster Abbey, something that I had never expected!

Florey was deeply involved in the establishment of the JCSMR and was Chancellor of the ANU at the time of his death. In 1969, The Royal Society and the ANU jointly sponsored an appeal to provide an endowment for Florey Memorial Fellowships, to be awarded to postdoctoral students for advanced work in Britain or Australia. I raised over $6,000 from the staff of the JCSMR. In 1973, just before I resigned from the Directorship, I persuaded Council of the ANU to name the chair held by the Director 'The Howard Florey Professor'. My immediate successor, Colin Courtice, was the first to use this title. In 1981, The Royal Society established Howard Florey Lectures to allow eminent scientists to give lectures, in alternate years, in Australia and Britain. The first Lecturer was Sir Andrew Huxley, then President of The Royal Society, who visited Australia in 1982, and I was the second, visiting England in 1983.

Florey Centenary Celebrations

Thirty years after his death, in 1998, the centenary of Florey's birth was celebrated in Australia and Britain. I was involved in the celebrations in both places.

Celebrations in Australia

These were organized by Dr John Best, Chairman of the Australian Institute of Political Science and a close associate of Dr Michael Wooldridge, then Minister for Health and Family Services. In 1996, Best set up a national committee and state committees in Adelaide and Melbourne that organized a wide range of activities: a two-day scientific meeting in Canberra, hosted by the JCSMR; a Florey Medal to be given to an outstanding Australian scientist, the initial presentation to be made at a Florey Day Dinner in Adelaide on 24 September, 1998; and a variety of arrangements with schools to celebrate Florey's achievements. I was a member of the National Committee and was the opening speaker at the Florey Centenary Symposium on *Helicobacter pylori*, held in Canberra, I launched the inaugural Medibank Private Teachers Awards at Florey Primary School in Canberra. I was a speaker at the Howard Florey Centenary Symposium 'Infectious Disease in Humans', organized by the Nature and Society Forum and held at the Academy of Science Dome in March 1998 and I helped organize the July 1998 issue of *Microbiology Australia,* the journal of the Australian Society for Microbiology, titled 'Howard Florey; a man of many parts' and wrote the article on Florey. I also wrote the biographical note on Florey for the 'Portrait of Howard Florey' exhibition held in the National Portrait Gallery, Canberra, 21 September to 15 November, 1998.

Celebrations in England

John Best arranged for a small group of Australians, of whom I was one, to go the England at the end of September for the British celebrations, which were held in Oxford at Queens College, where Florey was Provost at the time of his death and at the Science Park, where Bob May, then Chief Scientist, opened Florey Hall, and at The Royal Society in London. It was all very interesting and enjoyable. My only contribution was to give a short speech at The Royal Society reception to outline Florey's contributions to the establishment Memorial Fellows (see above) at The Royal Society reception.

Frank Macfarlane Burnet

I wrote long and detailed Biographical Memoirs of Burnet for the Australian Academy of Science and The Royal Society, which differed in some details, and a short one for the American Philosophical Society (Fenner, 1987a, 1987b, 1987c).

Burnet Centenary Celebrations

Pleased by the success of the Florey Centenary celebrations, Health Minister Michael Wooldridge and John Best took responsibility for organizing celebrations for the centenary of Burnet's birth. Understandably, because of his long association with the Walter and Eliza Hall Institute, these were held in Melbourne and both the Hall Institute and the Macfarlane Burnet Centre for Medical Research held their own functions. I was a member of John Best's committee, a speaker at the Macfarlane Burnet Centenary Symposium on Immunology and Virology, held at the Hall Institute on 4–5 August 1999, I helped organize the July 1999 number of *Microbiology Australia,* the journal of the Australian Society for Microbiology, and wrote the article 'Burnet, the virologist, 1925–40', and I wrote the biographical note on Burnet for 'A Broader Vision; Celebrations of the Centenaries of the births of Macfarlane Burnet, Jean Macnamara and Ian Clunies Ross', held at the National Portrait Gallery between 2 September and 24 October, 1999. I also wrote a short article on 'The scientific achievements and legacy of Frank Macfarlane Burnet' for the 1989/1999 Annual Report of the Walter and Eliza Hall Institute, and I was on the Organizing Committee and gave a lecture at the Burnet Centenary Symposium: 'Q Fever', held at the Queensland Institute of Medical Research on 12–14 October, 1999.

Prizes and Awards

Probably because I was still working, with advancing age I received a number of prestigious prizes and awards. I will list these chronologically, with a few remarks on some of them, most of them under four headings: national awards, fellowships and the like, named lectures, and prizes.

National Awards

On Australia Day 1989, it was announced that I had been awarded the highest honour in Australia, Companion of the Order of Australia (AC), for service to medical science, to public health and to the environment.

In 2002, I was approached concerning nomination for the 2003 Australian of the Year Award. I said that I did not want to be nominated for that award because, if selected, it carried too many responsibilities for a person my age, but I would agree to nomination in the Senior Australian of the Year category. The nominator agreed, but the local Australia Day Committee ignored her proposal and selected me to be ACT Australian of the Year, 2003, and I am proud of that award; fortunately I was not chosen for the national award.

In November 2005, I received a letter stating that I was one of four finalists for the ACT Senior Australian of the Year, 2006, and later that month I was selected for that award. At a ceremony in Canberra on 25 January, 2006, the representative of Queensland, a nurse of Aboriginal descent, was chosen from the seven State and Territory nominees to be Senior Australian of the Year, 2006.

Fellowships and the Like

1980	Honorary Life Member, Australasian Society for Infectious Diseases
1991	Emeritus Member, American Society for Virology
1991	Honorary Fellow, Indian Virological Society
1993	Honorary Member, Australian Veterinary Association
1996	Fellow, American Academy of Microbiology
1998	Patron, Nature and Society Forum
1999	Honorary Fellow, Australasian College of Tropical Medicine

2000	Patron, Australasian Society for Infectious Diseases
2001	Patron, Sustainable Population Australia
2002	Honorary Fellow, University House, ANU
2004	Honorary Life Member, Australian Conservation Foundation

Named Lectures

1980	Sir Macfarlane Burnet Address, Australasian Society for Infectious Diseases
1983	Florey Lecture, The Royal Society of London
1985	Burnet Lecture, Australian Academy of Science
1987	John Murtagh Macrossan Lecture, University of Queensland
1987	A. W. T. Edwards Memorial Oration, The Australian Society for Medical Research
1989	Wallace Rowe Lecture
1990	Gordon Meiklejohn Lecture
1997	Edwin H. Lennette Lecture, American Society for Virology
1999	Derrick–Mackerras Lecture, Queensland Institute of Medical Research

Prizes

1980	ANZAAS Medal, Australian and New Zealand Association for the Advancement of Science
1980	ANZAC Peace Prize
1982	Fogarty Medal, US National Institutes of Health
1986	Mudd Award, International Union of Microbiological Societies
1988	WHO Medal
1988	Japan Prize (Preventative Medicine)
1989	Advance Australia Award
1995	Copley Medal, The Royal Society of London
1999	Senior Australian Achiever of the Year Award
2000	Albert Einstein World Award of Science
2002	Clunies Ross Science and Technology Lifetime Award
2002	Prime Minister's Prize for Science
2003	Centenary Medal

Figure 11.1. Frank Fenner receiving the Prime Minister's Prize for Science, from Prime Minister John Howard, August 2002

Marks of Recognition other than Prizes and Awards

Since my retirement, I have received a number of marks of recognition other than the prizes and awards mentioned in the previous section. I will list them chronologically.

1982	On 8–12 February, 1982 a former PhD student, John Mackenzie, then Associate Professor of Microbiology in the University of Western Australia, organized an International Conference, 'Viral Diseases in South-East Asia and the Western Pacific', held in Canberra and dedicated to me. The proceedings were published as a book of the same title, by Academic Press, Sydney. I was presented with a leather-bound copy, signed by most of the contributors, which is now housed in the Basser Library.
1986	The Frank Fenner Visiting Fellowship at the Australian Animal Health Laboratory.
1990	The Frank Fenner Research Award of the Australian Society for Microbiology (ASM). This is presented at the Annual Meeting of the Society, at which the recipient delivers the Fenner Lecture. I attend most meetings of the ASM and usually present the Award.
1992	In 1991, the ANU purchased a multi-storey building on Commonwealth Avenue, formerly known as the Gowrie Hostel and used for the accommodation of public servants. It was refurbished and converted into a residential college, with rooms for about 500 students. The ANU Council, meeting in March 1992, renamed it Fenner Hall. Over the years since then it has developed into a vibrant

student community, and I have enjoyed participating in their activities.

1992 Docteur *honoris causa,* Université de Liège. On the occasion of the 175th anniversary of the University of Liège in 1992, the Veterinary Faculty of the University arranged that I should receive an honorary doctorate. The celebrations were elaborate. There was also a symposium in the Veterinary School, in which I participated and gave a lecture.

1994 In April 1994 the ANU Council approved a prize—the Frank Fenner Medal, to be awarded for the outstanding PhD thesis in the JCSMR each year.

1994 In November 1994 the JCSMR opened a 'Fenner display' in the lower level foyer of its building.

1995 Honorary Doctor of Science, Oxford Brookes University.

1996 In 1996 the JCSMR set up Fenner Merit Scholarships, offering not less than $5,000 to attract Australian students of high calibre to pursue postgraduate research at the School.

1996 Honorary Doctor of Science, The Australian National University.

1999 In 1999, Steve Redman, while Director pending the arrival of Judith Whitworth, commissioned artist Mathew Lynn to paint a portrait of me, which now hangs in the foyer of the building. It is a fine portrait, and I enjoyed sitting for it and observing the way in which Mathew talked to me as he painted, in order to draw out my personality.

2002 In 2002, I was made a Life Member of the ANU Endowment for Excellence and presented with a silver tray by the President of the Board, inscribed 'The Australian National University has pleasure in admitting Frank Fenner as a Life member of the ANU Endowment for Excellence in recognition and grateful appreciation of the generous support of the University and in particular for establishing The Frank and Bobbie Fenner Fund.'

2003 In May 2003, the ANU honoured me by naming the building which houses the administrations of both the Faculty of Science and the new ANU Medical School the Frank Fenner Building. It was opened by Science Minister, Mr Peter McGauran, on 21 May, 2003.

2003 In August 2003, the Australasian Society for Infectious Diseases, of which I had been Patron since 2000, named its most prestigious award the Frank Fenner Award for Advanced Research in Infectious Diseases. I have had the pleasure of presenting the award at the ASID Meetings in Alice Springs in 2004, Busselton in 2005 and Wellington in 2006.

2005 On 12 April, the Chief Minister of the ACT, Mr Jon Stanhope, unveiled the first 17 plaques installed in the ACT Honour Walk, on the Pedestrian Walk, Ainslie Avenue, Canberra. I was unable to attend because I was in Adelaide, but later I saw the plaque outlining my contributions to science; it was the only plaque recognizing a scientist.

Lectures and Newspaper Interviews

Omitting those listed under 'Prizes and Awards' and those given at meetings of the Frank and Bobbie Fenner Conferences on Medical Research and the Fenner Conferences on the Environment, between 1980 and 2006, I gave 107 lectures, 25 to overseas audiences and the rest within Australia. The majority of the lectures covered general topics, 14 were on smallpox, six on monkeypox, and four on mousepox. Over the same period I also gave 38 newspaper interviews, always in response to enquiries by journalists.

Donations

In addition to minor donations to various charities and the like, and an early donation of $1,000 to the National Gallery of Australia in 1983 (as a Founding Donor), I have made substantial donations to the Australian Academy of Science, The Australian National University and an NGO, the Nature and Society Forum.

Donations to the John Curtin School of Medical Research

In 1984, five years after my retirement, Bobbie and I considered our financial situation and what we needed for our own and our daughter's and grandchildren's futures, and decided to make annual donations to support the two scientific interests that had been central to my life: medical research and concern for the environment. Stimulated by the example of Sir Frederick and Lady White, who had made substantial donations to the Australian Academy of Science to establish an endowment fund, the income from which could be used to sponsor Academy conferences, we agreed that such conferences were the most cost-effective way of using relatively small amounts of money. From 1984 onwards we therefore made annual donations to the Australian Academy of Science, to support conferences on current environmental and conservation problems in Australia (see below), and to The Australian National University, to build up an endowment fund the income from which could be used to support annual conferences on aspects of medical research of interest to the John Curtin School of Medical Research. Initially (1984), the annual donations to each institution were small ($5,000), but in 1986 we increased them substantially (to between $15,000 to $30,000 a year). In 1997 these reached $250,000. With income and additions from the University's Endowment for Excellence Fund this reached $386,220 by the end of 2005; the University hopes to increase this to $500,000, which will be adequate for the support of annual conferences in perpetuity. Between 1999 and 2006, I made other donations to the School, totalling $87,000, for a variety of purposes.

This year, 2006, realising that I would soon really retire and cease coming to the School each day, I gave thought to disposal of my published books and my medals. The Basser Library already had all my books and bound reprints of my scientific papers. After consultation with my daughter and the Director of JCSMR, Judith Whitworth, I decided to give the medals and a set of my books to the JCSMR; the value of these, under the Australian Government's Cultural Gifts Program, was $38,025.

Donations to the Australian Academy of Science

Beginning in 1970 and continuing annually until 1983, I donated some of the royalties received from published books, particularly *Medical Virology*, to the Australian Academy of Science, to set up an Environment Fund, the proceeds of which were to be available for use of various Academy initiatives relating to the environment, often proposed by the Standing Committee (later National Committee) for the Environment. My donations to this fund totalled $19,050.

As noted earlier, from 1984, Bobbie and I increased the size of our donations to the Academy of Science as well as the JCSMR. By 1988, the Academy decided that sufficient funds were available to start the conferences, and, with increasing donations, by 1997 the Environment Conference Fund reached $230,000. This was judged by Council to be enough, given the success of the Conferences in attracting sponsorship, to maintain them indefinitely at a rate of one Conference annually.

Because the funding for conferences on medical research in the John Curtin School were initiated at the same time, there were discussions with Bob Porter, Director of JCSMR, and Ralph Slatyer, Chairman of the Environment Conference Committee, about the naming of the two conferences, resulting in the names used here: 'Fenner Conferences on the Environment' and 'Frank and Bobbie Fenner Conferences on Medical Research'. The first Fenner Conference on the Environment was in held 1988 and there have been meetings each year since then. The proceedings of most conferences have been published and copies are available in the Basser Library.

I was executor of the will of Alfred Gottschalk, a Fellow of the Australian Academy of Science and a close friend of mine, who died in 1973, and at his request I arranged for $35,000 to be donated to the Australian Academy of Science as an endowment to support the annual award of a medal to a young scientist, who was not a Fellow, for distinguished research in medical or biological science. The first award was made in 1979; it and the majority of subsequent awards have gone to biomedical scientists. I saw the need for another similar award, in biology other than medical research. Given Gottschalk's background in biomedical research, and my broad interests in the conservation of biological diversity as well as preventive medicine, I decided, after consultation with the

Academy Council, to set up a fund for a medal for the work of a young scientist in biological science other than biomedical research. Having donated enough to support the Environment Conferences indefinitely, in 1997 I started to make donations for such a medal, which Council, following precedent, named the Fenner Medal. The first award was made in 2000, and the endowment reached the required total of $100,000 in 2001.

I had always been interested in the history of science, and hence in the Basser Library, and in 1996 and 1997, I donated $10,000 to the Academy for use by the Basser Library, initially for cataloging and to support the Video History Project. In 2000, I made a donation of $20,000 to the Library Fund, the bulk of which was used to put in place a series of cupboards for housing archive boxes, at the rear of the top floor of the Shine Dome. In 2003–04 I donated $25,000 to the Academy to help with the production of video histories of Fellows and in 2004–06, another $25,000 for the Basser Library, a total of $80,000 between 1996 and July 2006.

Other Donations to The Australian National University

From 1996, I began making donations to the ANU to be used by Fenner Hall. These amounted to $118,000 by the end of 2002, but after I was awarded the Prime Minister's Prize for Science in 2002, I decided to use the tax-free $300,000 that accompanied that Prize for donations. In 2003, I donated $100,000 from that source to establish Bobbie Fenner Scholarships at Fenner Hall, which was augmented in 2004 and again in 2005 by donations of $23,000, plus $38,000 for improving facilities in the Hall, reaching a total for Fenner Hall, by July 2006, of $293,000.

In March 2005, I donated $5,000 to the Judith Wright Award Fund, and in 2005–06 I initiated donations of $30,000 for scholarships for students in the Centre for Resource and Environmental Studies (CRES). I intend to maintain the latter donations as long as I am alive.

Donations to an NGO, the Nature and Society Forum

I have always felt guilty about my inability to become involved in community activities (my wife Bobbie more than made up for my deficiencies here), but in 1992 I had joined the Nature and Society Forum (NSF), a brainchild of my friend Stephen Boyden. In the 1990s, I took an active part in its activities, and in 1997 I was invited to become its Patron. In 1998, I decided to make annual donations to NSF to assist their work, usually of $20,000 annually. After receiving the Prime Minister's Science Prize, I donated $50,000 to NSF, and by June 2006 the donations totalled $214,045.

Two Personal Celebrations

During the 25 years covered in this chapter, two events were staged marking milestones in my career at the ANU: my 80th birthday in 1994 and the 50th anniversary of my appointment to the ANU in 1999.

1994, The Sixth Frank and Bobbie Fenner Conference

Celebrating my 80th birthday, this conference, organized by Gordon Ada and entitled 'Viruses, Vaccines and Vectors', was attended by a number of distinguished overseas colleagues and long-time friends of mine: Isao Arita, Bob Chanock, Ciro de Quadros, Paul Gibbs, D. A. Henderson, Dick Johnson, Fred Murphy, Bernie Moss, Parker Small and Rob Webster, as well as leading Australian scientists. On the first day, the Chief Minister of the ACT, Rosemary Follett, opened a Frank Fenner exhibit in the lower level foyer of the JCSMR, and on the evening of the second day, a birthday banquet was held in the Dining Room of Old Parliament House, at which the audience included, as well as the speakers, many old students and research staff and their wives. D. A. Henderson, Bob Chanock and David White spoke, and D. A. presented me with a unique award, a sculpture in brass symbolizing my election as Grand Master of the Order of the Bifurcated Needle. The occasion was all the more pleasant for me because my wife Bobbie, my daughter Marilyn and her husband and children were all there. Bobbie died just over a year later.

1999, Fifty Years in the ANU

On 29 July, 1999, was the 50th anniversary of my appointment as foundation Professor of Microbiology in the ANU. I celebrated it with a dinner in the Common Room at University House, which was attended by as many of my past students, research staff and general staff as could come. My daughter Marilyn acted as Master of Ceremonies, and the Vice-Chancellor, Deane Terrell, Ian Marshall, my first PhD student, Stephen Boyden and Gordon Ada spoke. Among other items on display was a book containing photographs of all academic staff and PhD students from the Department of Microbiology, 1952–67, at the time of their appointment.

References

Alibek, K. with Handelman, K. 1999, *Biohazard. The Chilling True Story of the Largest Covert Biological Weapons Program in the World,* Hutchinson, London.

Arita, I. and Gromyko, A. 1982, Surveillance of orthopoxvirus infections, and associated research, in the period after smallpox eradication, *Bulletin of the World Health Organization,* vol. 60(3), pp. 367–75.

Australian Government Department of Health and Ageing 2004, *Guidelines for Smallpox Outbreak, Preparedness, Response and Management*. Australian Department of Health and Ageing.

Blanden, R. V. (ed) 1989, *Immunology of Virus Diseases,* produced by Brolga Press for the John Curtin School of Medical Research.

Dumbell, K. R. and Huq, F. 1986, Epidemiological implications of the typing of variola isolates, *Transactions of the Royal Society of Tropical Medicine and Hygiene,* vol. 69, pp. 303–6.

Esposito, J. J., Nakano, J. H. and Obejeski, J. F. 1985, Can variola-like viruses be derived from monkeypox virus? An investigation based on DNA mapping, *Bulletin of the World Health Organization,* vol. 63, pp. 695–703.

Fenner, F. 1959, Genetic studies with mammalian poxviruses. II. Recombination between two strains of vaccinia virus in single Hela cells, *Virology,* vol. 8, pp. 499–507.

Fenner, F. 1968, Howard Walter Florey: Baron of Adelaide and Marston, *Australian Journal of Science,* vol. 31(1), pp. 37–9.

Fenner, F. 1972, Howard Walter Florey, *Dictionary of Scientific Biography,* vol. 5, pp. 41–4.

Fenner, F. J. 1988, Florey, Lord Howard Walter, Baron Florey of Adelaide and Marston, in *Roll of the Royal Australasian College of Physicians,* G. L. McDonald, (ed), vol. I, pp. 95–7, The Royal Australasian College of Physicians, Sydney.

Fenner, F. 1996, Howard Walter Florey, *Australian Dictionary of Biography,* vol. 14, pp. 188–90.

Fenner, F. 1987a, Frank Macfarlane Burnet, 1899–1985, *Historical Records of Australian Science,* vol. 7(1), pp. 39–77.

Fenner, F. 1987b, Frank Macfarlane Burnet, 1899–1985, *Biographical Memoirs of Fellows of The Royal Society,* vol. 33, pp. 101–62.

Fenner, F. 1987c, Sir Frank Macfarlane Burnet, Biographical Memoir, *Year Book of the American Philosophical Society,* 1987, pp. 90–5.

Fenner, F. (ed), 1990, *History of Microbiology in Australia,* Brolga Press, ACT, for the Australian Society for Microbiology.

Fenner, F. and Fantini, B. 1999, *Biological Control of Vertebrate Pests. The History of Myxomatosis, an Experiment in Evolution,* CAB International, Wallingford.

Fenner, F. and Gibbs, A. J. (eds) 1988, *Portraits of Viruses. A History of Virology,* S. Karger AG, Basel.

Fenner, F., Wittek, R. and Dumbell, K. R. 1988, *The Orthopoxviruses,* Academic Press, San Diego.

Fenner, F. J., Bachman, P. A., Gibbs, E. P. J., Murphy, F. A., Studdert, M. J. and White, D. O 1987, *Veterinary Virology.* Academic Press, Orlando, USA.

Fenner, F.J., Gibbs, E. P .J., Murphy, F. A., Rott, R., Studdert, M .J., and White, D. O. 1993, *Veterinary Virology,* Second Edition, Academic Press, San Diego, USA.

Gispen, R. and Brand-Saathof, B. 1972, 'White' poxvirus from monkeys, *Bulletin of the World Health Organization,* vol. 46, pp. 585–92.

Gispen, R. and Kapsenberg, J. G. 1966, Monkeypox virus-infectie in cultures van apeniercellen zonder duidelijk epizootisch verband met pokken, en in een kolonie van apen lijdende aan pokken, *Verslagen en mededelingen betreffende de volksgezondheid,* vol. 12, pp. 140–4.

Jezek, Z. and Fenner, F. 1988, *Human Monkeypox,* Monographs in Virology No. 17, S. Karger Publishers, Inc., New York.

Marennikova, S. S., Selukhina, E. M., Mal'ceva, N. N. and Ladnyi, I. D. 1972, Poxviruses from clinically ill and asymptomatically infected monkeys and a chimpanzee, *Bulletin of the World Health Organization,* vol. 46, pp. 613–20.

Marennikova, S. S., Selukhina, E. M., Mal'ceva, N. N. and Matsevich, G. R. 1979, Monkeypox virus as a source of whitepox viruses, *Intervirology,* vol. 11, pp. 333–40.

Murphy, F. A., Gibbs, E. P. J., Horzinek, M. C. and Studdert, M. J. 1999, *Veterinary Virology,* Third Edition, Academic Press, San Diego.

Rao, A. R. 1972, *Smallpox,* Bombay, The Kothari Book Depot.

White, D. O. and Fenner, F. J. 1986, *Medical Virology,* Third Edition, Academic Press, Orlando.

White, D. O. and Fenner, F. J. 1986, *Medical Virology,* Fourth Edition, Academic Press, San Diego.

WHO 1984, The current status of human monkeypox: Memorandum from a WHO Meeting, *Bulletin of the World Health Organization,* vol. 62(5), pp. 703–13.

WHO 1986, Committee on Orthopoxvirus Infections, *Weekly Epidemiological Records,* vol. 38, pp. 289–93.

Part II. The Life of Charles Fenner

Chapter 12. The Fenner Lineage

Introduction

During my childhood in Adelaide in the 1930s, Fenner was not a common name. In 1937, for example, the name occurred only twice in the Adelaide telephone book: C. A. E. Fenner (my father, whose forebears came from Germany, at 42 Alexandra Avenue), and A. G. Fenner, the head wool appraiser for Elder Smith and Co. Ltd. The latter family had migrated from England, and of course Fenner's cricket ground in Cambridge is familiar to many. Although I have little time for genealogical studies, since they ignore the female genes, my father had acquired a lot of information about the genealogy of the relevant Fenners, including from a substantial book (*Hessisches Geschlechterbuch*, 1971) so I begin an account of his life with a brief summary of that history.

My grandfather, Johannes Fenner, was born in the village of Niedergrenzebach, in the province of Hesse, in Germany, on 10 February, 1841, the fourth of seven children. In 1860, troubled by the prospect of being conscripted for military service after Hesse had come under Prussian control, and attracted by the idea of gold mining, he went to England and in Liverpool he embarked on the ship *King of Algiers*, bound for Victoria. He left the ship in Melbourne and went to Talbot, near Ballarat. He was naturalized there in 1874, and on 19 January, 1875, he married Mary Thomas, whose parents had come from Staffordshire, in England, and who had been born in Thebarton, a suburb of Adelaide, on 16 February, 1852. They had eight children, four boys and four girls, one of whom died as a baby and another at the age of four years. Charles Albert Edward, the fifth child, was born in 1884. The youngest member of the family, Thomas Richard, enlisted in the Australian army in 1915 and was killed by 'friendly fire' at Mouquet Farm in France on 29 August, 1916. Johannes Fenner died on 13 July 1923; his wife, Mary, died in Ballarat in 1939.

Some of the German Fenners have been very interested in their family history. During trips to Germany in 1931 and 1937, my father and mother went to the village of Niedergrenzebach, where my grandfather was born. Below I quote edited portions of Father's diaries that relate to these visits (C. Fenner, 1931, 1937).

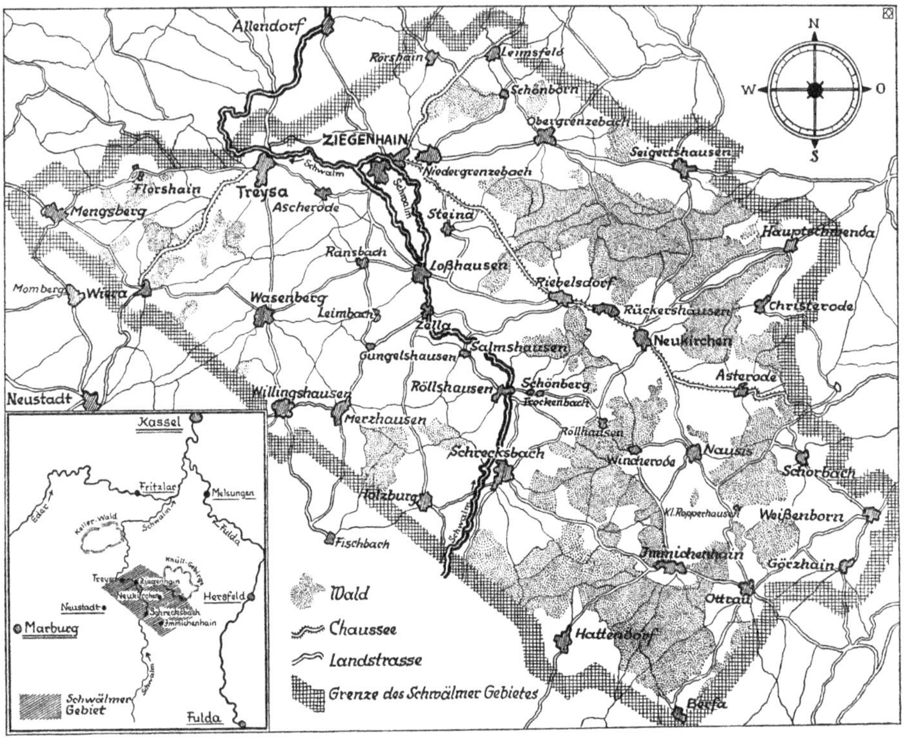

Abb. 1. Karte des Schwälmer Gebietes. Links unten Übersichtskarte

Figure 12.1. Map of Hesse and the Schwalm valley

Extracts from Charles Fenner's 1931 and 1937 Diaries

27 October 1931, Niedergrenzebach

After breakfast in Treysa [the nearest railway station], we left with Max Pirsch, the driver, to go to Niedergrenzebach, in search of my father's people and his birthplace. There were the hills and fields and woods where my father played, and where he went to school, and where he put the teacher's son's face in the mud of the Grenzebach. There were no fences between properties, just as he had said, and which looked so strange to us. There were fruit trees along the roads, and there were strange old many-storeyed houses, with stock (cows, horses, pigs) below and people above.

The day was superb. These Fenners turn out to be a peculiar people, very strong on tradition, fond of the past, and living in houses that the Fenners have occupied for centuries. Again, they are, as Dad used to tell me, Schwälmer people—a race apart, with their own religions and manners, and all distinctly set out in their different dress. The dress of the Schwälmer I knew from my grandfather's and uncle's and aunt's photographs (Figure 12.2). I did not know that they were still used, the men's curious embroidered cloaks, the brass-buckled shoes, the pill-box hat, which covers a knot of hair, as it has done for centuries, and as it does today, to her great pride, for my own rosy-cheeked second cousin, Anna Catharina Fenner!

Brautpaar und Brautjungfern

Der Bräutigam trägt auf dem Dreimaster „Die Lust", den hohen Strauß aus Glaskugeln, der mit Rosmarinzweigen umsteckt ist. Diese werden nach der Hochzeit eingepflanzt.

Figure 12.2. Schwälmer men's and women's attire

Yesterday at Marburg, we saw several people get on the train in the same curious, ancient, and (to me) quite beautiful costume. I said to Peggy, little believing that it was true: 'There may be a Fenner among those.' Sure enough, there was my own blood cousin, Anna Catharina Fenner.

Well, we got to Niedergrenzebach. It was a thrill. Such a quaint and ancient place. I can't describe it, with its packed series of high, curiously gabled buildings, hundreds of years old, and its crooked streets, and so on. There is nothing like it anywhere else, except perhaps the neighbouring village of Obergrenzebach. We went to the village schoolmaster; he could not speak English. We went to the farrier's. He again was most kind, but he had forgotten his English. Then we went to the home of the farmer, Johannes Adam Fenner. We met the man himself, in this four-square farmyard, all smiles and cheerfulness and goodwill. Yes, he was Johannes Fenner. Yes, he had married my father's cousin and inherited the family property! And he took us and showed us the carved stones, of the farm, going back, I think, for hundreds of years, where Johannes Fenners had owned and built, with *Gottes*' help, this farm. There were four such inscriptions, all of different dates, one carved in stone, one carved in the wood over the door and two painted in German on the great beams of the buildings.

Then we went into the house and had bread and hard but sweet butter and pork sausage and good white Schwalm wine. And we all sat down and drank and ate together. And chatted away very merrily in German-English with no interpreter but Peggy. And I cracked jokes in German, and all laughed at them. The daughter Anna Catharina was 31 that day, her *geburstag*! The older boy—the *Kronprinz*—is Johannes and the second is Heinrich Claus. Dad's father was Hans Claus.

Next day I saw the church Dad went to and the churchyard where his father and mother are buried, but could not visit their graves, for here in the crowded little churchyards of the Schwalm, one occupies a grave for 30 years only, then someone else is buried there. The 'house' is the holy place, the place of tradition. Just before we left, I went into the bedroom and my cousin John Adam Fenner, pointing to the smaller bed said: *'Hier deine vater schlafen'*. Then to the big bed: *'Hier deine vater geboren.'* So there it was. I had got to Dad's birthplace as I had set out to do.

16 July 1937, Rotenburg and Niedergrenzebach

I have met four members of the Fenner family here. At Spangenberg the burgomeister is Theobald Fenner, cousin to Dr Carl of Hannover. A big man, fine home, pleasant wife, two sons, one daughter, and a glorious country home in the woods. I have photographs of his town house, nestling below the great *schloss*-crowned rock of Spangenberg. So up there I had the longed-for pleasure of wandering in these woods, the beech and oak very beautiful, the pine and spruce woods ominously still and dark, and not easy to get along because of the lower dead branches of the trees. And we made pleasant family contacts, and learnt a little more German. Peg does quite well. And Dr Daniel Fenner and his very alert and charming *frau* were there to put us right.

And then off on a very long drive to Ziegenhain and Niedergrenzebach in the valley of the Schwalm, where the folk movement, and love of home, and retention of the old costumes, is so strong. And we went to the house where the first Fenner of record lived in the fifteenth century in the centre of the double-walled, moated, part of Ziegenhain. And we got Schwälmer books and cards. For each member of the family comments to us on my being an unmistakable member of the Fenner family—Spangenberg Fenner knew me as a Fenner before he knew who I was at all. So we are just transplanted Schwälmers—all the family.

And I went to my father's home, and I wandered in his church and churchyard, and saw the graves of Fenners, and on the church walls saw the honour rolls with the Fenner names for the war of 1814 (Waterloo etc.) and for the wars of the 1860s and 1870s, and for the war of 1914–18. In the Napoleonic wars three Fenners, two Johannes sen. and jun. In the last war three. And I thought of the honour rolls at home in Talbot and Dunach [in Victoria], with Fenners in them too. And I agreed with Adam Fenner, when he hoped that no more would members of our family and of those still in Germany be opposite to one another in the trenches. A nephew of his was in the German trenches opposite the place where my brother Tom was killed in August 1916, or thereabouts. And so it goes. Alongside two of the names of the 1814 honour roll were pinned the medals won on the field. There they have hung quiet and untouched for a century. And in the Fenner garden I brought a piece of stone for memory of Dad, who had loved it all so much, but not so much that when the land became Prussian he left, for he would not become a Prussian soldier. Or so they say. I do not know.

Dad asked me if I ever I visited Germany to go to his mother's grave—for him. For her death before he returned to see her was the bitterest thing in his life. And so I found the place of the Fenner family—an area—no stones, in the old churchyard. And we went and saw the old Fenner home of 1453—*Die Unser Liebe Frauen Kolben*—and saw the old lady Fenner, with 17 generations of occupancy—500 years—of that one house. And I more than ever was sorry I had not been able—with Peg who is of the Fenner family—to attend the *Sippentag* (kinship day). I hope that some day Frank will come and he will be able to speak German also, and that will be ever so much better. Whatever wars may happen in between, and God forbid they may be between Germany and us, he will find that these family bonds are stronger than anything that war may harm. And he will get a warm welcome from all. Dad did not tell me much of his home. But one thing I recall well. He often showed me the blue cornflower and revealed that he was always touched by the memory of this emblematic flower of his home.

We went and saw the homes of four other Fenners in Niedergrenzebach and the places where my father's sister lived, and much more interesting family history. And there was a long drive home, by basalt, and *muschelkalk*, and *rotsandstein*, and so on. For my cousin Dr Daniel is a geologist also. The name Rotenburg is from the reddish soils of the *Rotsandstein*? Upper Cretaceous.

17 July 1937, Rotenburg/Fulda

At 2pm we set off in the car. Took Dr Daniel Fenner and his wife Caroline and his *zwei söhne und eine tochter*, away up the valley of the Fulda, through villages old and new, to Bad Hersfeld. These *Bads* (watering and bathing places, like Bath) are all over Germany. Have visited or seen several. They are always very pleasant and restful places. It is no wonder folk take 'the cure' there.

First we went to the Kursaal. We got an English dictionary from the waiter who spoke good English. Then we sat down for a family gathering. There were ten Fenners, six of the direct line, and four *frauen*. All were descended directly from the Grebe *(grabe)* line of Fenners. Burgomeister Johannes Jost Fenner, who built the old home at Niedergrenzebach, or rebuilt it, in 1742 (I forget the date, but it is carved on the foundation stone and on the beam over the door and you may read it for yourself some day, *mes enfants)*. Of the six Fenners present, four were doctors of university standing. Dr Daniel, Dr Heinrich, Dr Emilie (the cleverest of the lot, from Berlin) and Dr Charles (the exotic member). Eckhard is a *pfarrer* (parson) of good standing, and Christian is a *Buchhandler* of degree. So the Fenner line, which in essentials was a farming line, has in our case blossomed out into other paths. There have been bishops and generals and engineers also, but they are farther removed.

Eckhard Fenner, a clever fellow, well learned in architecture and philology and history, but not so much so as Dr Emilie, wrote out (quite from memory) a lineal table to show that we are all descended from Johann Jost — also from Kurt 1648 (also from Oswald Fenner 1400 and something). And I have photos of the original crests of the family of 1555 (one flag) and of 1698 (two flags). And all this gives one an interest in heredity and a pride in the stability and sanity and soundness of one's ancestors, without any unnecessary and inappropriate swank about it. I am very sorry that I could not get to the Pfingsten celebration of the Fenner Family (*Sippentag,* May 16) at my father's own old home village of Niedergrenzebach, but this smaller *sippentag* at Hersfeld, where we and our *frauen* ate and drank and walked and talked together made up for it. First at the Kursaal, then to the Gardens where we 'took the waters' and had our photos taken. Peg wanted me to be taken drinking a glass of water as a surprise for the family. And we talked of all the other Fenners and who they were and how they were related, and what their characters were, and they all agreed that on face value alone I was a Fenner, with many family characteristics. Then when we had drunk from the Lullus spring (Lullus was a missionary who founded Hersfeld) we went over to the great ruins of the once magnificent church of Hersfeld, and discussed its arches and styles and carvings and history from AD 736, when it was first built (heathen temple there before) to 1100 and when it was rebuilt until Napoleon's time, when it was destroyed to its present magnificent ruins. After 1700 however it had been deserted because under Phillip of Grossemüttige (1504-67), who gave the first Fenner crest, in the 1500s the rise of Protestantism had been so great. Phillip was one of the two powerful friends of Martin Luther and helped to establish the Reformation. Phillip also had two wives. We were in the home of the second one yesterday. Then to the Rathskeller where we dined heartily on Kalbschnitzel and gherkins and rye *brot* and beer. And then *Aufwiedersehen.* And we left for home very late and a long way to drive. But it was another look at the beauty of the German landscapes by moonlight through the blue mists.

Hirt Family History

So-called 'genealogies', as commemorated in *Hessisches Geschlechterbuch* (1971) and the Fenner Sippentags, are lop-sided, since they trace the history of the male line only. It has been much more difficult to find out much about my mother's ancestors. As related below, Father married Emma Louise Hirt on 4 January, 1911. Mother was born at Narracan, in Gippsland, Victoria, on 19 August 1883. Her father, Johannes Gottlieb Hirt, was born in Silesia about 1825 and migrated to California and then Australia. He died in 1898. Silesia (German Schlesien) was an old Polish province which became a possession of the Bohemian crown in 1355, passed with that crown to the Austrian Habsburgs in 1526, was taken by Prussia in 1742 and returned to Poland in 1945. Her mother, whose surname was Kaiser, was born in Dreysa, a village near Bautzen, in Saxony, and migrated to Australia at the age of 15. They married in 1859 and she died in 1934. Some other female members of the Kaiser family had migrated to South Australia and married other Germans, including several Kleinigs, in the Angaston region in South Australia. As children, Lyell, Winn and I (Frank) used to stay on the farm of one of the Kleinigs at Ebenezer, near Tanunda, during school holidays. When father and mother visited Europe in 1931 it was not possible for them to visit Silesia but they made a point of finding the village (in Saxony) where my maternal grandmother was born. I have extracted the comments set out below from father's diary of their 1931 trip.

Extracts from Charles Fenner's 1931 Diary

29 October 1931, Drehsa

We got up at 7.20 this morning and by 9.20 were at the Haupt Bahnhof to catch our train for Bautzen. Got a fine detailed map of this and Dresden, including Drehsa, and this was most helpful in our quest for the birthplace of *grandmutter*. The way out was very beautiful, as always in this lovely, fertile, much-used and long settled land. Peg does not think it was as pretty as the Marburg country but I think it was. The fields were usually larger, there was a much greater area of woods, the tops of hills were mostly woods, forests, beautifully kept, here of fir and pine, with borders of beech and birch—a few larch and spruce. The fields are fenceless, with the usual small square stones as 'land-marks' between properties. A church on the cot or saddle of the low hills. Villages nestling along the ever-running streams or on the hillsides. Most beautiful. We also passed huge paper mills and great breweries.

At long last we got to Bautzen, a huge fine town. Castles, fortresses, river, market-places, museums, theatres, fine streets—a big town we should call it. There are scores and scores of villages around it. We got a taxi. The lad, who was not smart or shaven, did not impress us favourably at first. In fact, at one stage, out in the country, when both he and we thought we were lost, we were just a bit scared. And indeed, he had us quite at his mercy. He did not understand one word of English, and but little of our combined German. However, we got to the long-talked-of Drehsa. We had found, at last, the village that Peggy had sought, where her mother was born. A pretty little place, large houses of the *alte haus* type, crowded together along the stream, and dominated by the *Schloss* (castle) of some Duke. We drove to the schoolhouse, and spoke to several people. They were not so smiling and sunny as those in other parts. Perhaps in these outback villages those who speak English are still hated as enemies. But one woman was very nice. The women and men we saw everywhere as they worked in the fields. One got a fine idea of what is meant by a 'peasant people'. We went to the school. No-one there. It was not the school that *grandmutter* went to, but we were on the roads she trod, and the woods and the fields where she played.

Then the schoolteacher came. Some one had sent for her. Such a nice, rosy-cheeked, strong and hearty girl, from Berlin. She liked Drehsa for 'a little while'. She complimented Peggy on her German and the two chattered away like magpies. She took us to a man's house, and he told us where Peter Kaiser, the *Burgomeister* had lived, and where Georg Kaiser, *Burgomeister* also, had lived. But now the Kaisers do not live there. So we went to see it, and took photos of it—the house where *grandmutter* had been a girl with her brothers and sisters, four of whom had gone to Australia, and she the only one left. And I photographed the geese in the lane by the house and the stream and the street. There are no fences; one walks to one house through another, as at Niedergrenzebach. Then we went up to the *Schloss,* and we said goodbye to the little pleasant Berliner teacher and we set off back.

The Fenner Coat-of-Arms

On a subsequent trip, in 1937, my father had met a cousin, Dr Daniel Fenner, who was very interested in the family history and organized some of the early post-World War II *Sippentags*. Daniel Fenner had constructed an abbreviated history of the Fenners of Niedergrenzebach which goes back to 1465. My father belonged to the 17th generation; during this long period they had devised three versions of a 'coat-of-arms'; the final version is illustrated in Figure 12.3. Shortly after Fenner Hall had been named (after Frank Fenner) as a residential college of The Australian National University in Canberra in 1992, its governing council agreed to accept this as their emblem, used on the entrance to the building (Figure 12.4).

In July 1949, Bobbie and I came to England from the Rockefeller Institute in New York and bought a car, a Ford Prefect, which we collected within a week of arriving in England. After driving around Britain and much of Europe, in October we came to Germany, where we had already established links with some German relatives. We drove, with my second cousin Dr Otto Fenner, to Niedergrenzebach and saw the Fenner house for ourselves.

Figure 12.3. The Fenner coat-of-arms

'This escutcheon was given to Ekkhardt Fenner, Actuary of the Landgrave Philipp of Hessen in Marburg as a reward for his services on the occasion of the famous discourse between Luther and Zwingli in Marburg in 1525.' The original is in the castle archives in Marburg.

Figure 12.4. Fenner Hall, The Australian National University, Canberra

Notice at the entrance on Northbourne Avenue.

References

Fenner, C. 1931, MS178/6/3A5 (1931 diary) Basser Library, Australian Academy of Science. Also lodged in PRG 372: papers of Dr C. A. E. Fenner, held in the Mortlock Library of South Australiana.

Fenner, C. 1937, MS178/6/5B (1937 diary) Basser Library, Australian Academy of Science. Also lodged in PRG 372: papers of Dr C. A. E. Fenner, held in the Mortlock Library of South Australiana.

Hessisches Geschlechterbuch. Bearbeiter (revised) by Pfarrer I. R Gottfried Ruetz. Achtzehnter Band (1. Schwälmer Band). in *Deutsches Geschlechterbuch*, Band 157, 1971, C. A Starke Verlag, Limburg an der Lahn, pp. 161–380.

Chapter 13. Childhood, University, Marriage and Family

Primary School, 1888 to 1895

My father, Charles Albert Edward Fenner, was born on 18 May, 1884, in Dunach, a small village 35 km north of Ballarat and at the foot of a volcano, Mount Greenock. In December 1879, his father, Johannes Fenner, who had been living in the area and working in the local gold mine, had taken over the license of the Dunach Hotel. It was a low, rambling building, and apart from the rooms used by the family there were eight rooms for public use. He gave up the license in 1892 and became a poultry farmer, keeping turkeys which used to free-range over the plains. The Fenner family stayed there until the old building was burnt down in about 1910; then they lived in the school-house for some years. I remember my father telling me that when he was a boy one of his jobs was to take the turkeys out before he went to school and collect them after he came home.

As Father related to my brother, Bill Fenner, in 1954, when he was in bed in Adelaide with a stroke, he had always had a dog, although he said that his most vivid early memory was when the family bought him a pony, Dolly, who was part of his life, although he was thrown off and broke several ribs. He went across the road to the one-teacher school when he was four years old; he cried on the first day, as I did myself. When he was eight years old and had just finished Grade Four he received a certificate from the Victorian Education Department saying he had completed the necessary education and could leave school. However, he went on and at age 11 got his merit certificate.

His mother got Charles a job as an 'apprentice' at the printing office of the *Talbot Leader* (a local newspaper). He was paid two shillings and sixpence a week, which he gave to his mother. He couldn't think of anything else to do with it. At the time Talbot, seven kilometres from Dunach, had a population of about 5,000 compared with Dunach's 400. After five years as a 'printer's devil', when he was 17, he left the *Talbot Leader* and joined the Victorian Education Department as a pupil-teacher in local primary schools. Two years later, in 1903, he became 'principal' of two one-teacher bush schools. He supervised these two schools (I will call them 'A' and 'B') by teaching in school A for two days, setting enough work for the rest of the week, and then going to school B, teaching there and setting work for the first days of the next week. He would then have the weekend at home.

Figure 13.1. Students and teachers at Dunach School in 1890

At that time there were three Fenners at the school, Charles, Jack and May.

Melbourne Teachers' College and the University of Melbourne

In 1907, Charles Fenner was accepted for the two-year course to the Melbourne Teachers' College, then a residential college within the grounds of the University of Melbourne. During this period he gained university entrance qualifications and engaged in College activities, including editorship of the College magazine. From there it was a succession of brief appointments, in each of which the local inspector reported his performance as most promising. Then he departed from the normal course of teacher training by returning to the University of Melbourne as a Kernot Research Scholar, majoring in geology, under Professor E. W. Skeats, and biology, under Professor Baldwin Spencer. He won all the scholarships and prizes that were available to him at that time, graduated BSc with First Class Honours in 1912 and gained a Diploma of Education in 1913. His archives in the Basser Library of the Australian Academy of Science contain several large notebooks with drawings relating to first-year physics (1907), human anatomy (1910), comparative anatomy (1911), and two books of field notes on geology (1911 and 1912). These testify to his artistic abilities, which were also used in his early scientific papers, for which he was awarded the degree of DSc by the University of Melbourne in 1917.

Figure 13.2. Charles Fenner with his father Johannes Fenner, in 1903, when Charles was 'principal' of two one-teacher schools

Marriage to Emma Louise Hirt, 4 January 1911

Emma Louise Hirt was the youngest child of Johannes Gottlieb Hirt and Maria Hirt (née Kaiser). Her father migrated from Germany to California and then to Victoria. Starting as a gold-digger in Ballarat, he later became a successful boot-maker. Maria Kaiser was born in Dreysa, a village near Bautzen, in Saxony, and migrated to Australia at the age of 15. She and Johannes married in 1859 and she died in 1934.

When the gold ran out, they and their nine children moved to 101 Eyre Street, Ballarat. Some years later, Johannes and Maria moved to Thorpdale, in Gippsland, where he bought a cherry orchard. The older children who had not married stayed in Ballarat, acting as housekeepers and looking after the younger ones who were still at school. Much later, the older sisters, Lin and Anna, who were widows, and Crin and Paula, who were spinsters, looked after one or two of us children in the same house during our school holidays.

On 19 August, 1883, when Johannes was about 70 and Maria about 50 years old, they had another child, who was christened Emma Louise. After her husband's death in 1898, Maria moved back to 101 Eyre Street and Emma Louise (names she hated) went to primary school and then to Grenville College, a co-educational secondary school in Ballarat. In 1906 she began training as an infants' teacher at the Melbourne Teachers' College and met Charles Fenner, also a resident at the College. Charles responded to her dislike of her given names by calling her 'Pegasus', shortened to 'Peggy' or 'Peg'. After completing her training she taught at the primary school in Inglewood, northwest of Bendigo (Figure 13.3). Late in 1910 she bought her wedding dress in Bendigo. They were married in Ballarat on 4 January, 1911, and went to live in Fitzroy, a suburb of Melbourne. Charles taught and continued his studies at Melbourne University and their first child, Charles Lyell (named after the famous English geologist) was born in Fitzroy on 17 August, 1912.

Later Teaching Positions in Victoria

After a few months at Sale High School, in 1913 Charles was appointed Headmaster of the Mansfield Agricultural High School and the family moved to Mansfield. However, in November 1914 he was appointed Principal of the Science Departments in the Ballarat School of Mines and Industries, with teaching responsibilities in the Geological branch of the School. The family moved to a house on 2 Doveton Street, next door to 101 Eyre Street. The next two children, Frank Johannes (later John) and Winifred Joyce (Winn), were born in Ballarat on 21 December, 1914, and 26 August, 1916, respectively.

Figure 13.3. Photograph of Miss Emma L. Hirt

Miss Emma L. Hirt (the adult on the left) teaching at Inglewood Primary School.

Figure 13.4. Photograph of the married couple

From left: the bridesmaid, Emma's sister Paula; the bridegroom; the bride; the best man, W. M. Sullivan. Seated: Emma's nieces Joyce Love and Elva Hirt.

Field trips for the geology students and his own weekend studies focused on the geology of the Bacchus Marsh district, including Werribee Gorge, and it was for papers on this work that he was awarded the degree of DSc by the University of Melbourne in 1917. Scientific papers written while he was living in Victoria are listed in the Bibliography in Chapter 16. In 1915, he was appointed Principal of the Ballarat School of Mines, but he was soon to move on. When he left, the Council recorded the following testimonial:

> Mr Charles Fenner has held the position of Principal of the Ballarat School of Mines and Industries since November 1914. Throughout his appointment he has fulfilled his duties with a zeal and enthusiasm that merit the highest praise. He is an organized man of exceptional ability and his keen insight into the needs of technical education has enabled him to place the departments under his control on a higher level of efficiency than they have ever before reached.
>
> In his lecture work Mr Fenner has had charge of the Geological branch of the School and his labors therein have met with marked success. In the Technical Schools' examinations in Geology, Mining Geology, Petrology and Mineralogy, the examiners have on each occasion specially commented on the excellence of the work done by his students. He has made a special point of the field work, which has become a very popular part of the students' curriculum, and in all branches the economic bearing of the subjects taught has been continually stressed. Mr Fenner has also prepared students for the Geology Examinations (Science and Arts Courses) at the Melbourne University, and there, too, his students met with more than ordinary success. He has the power of inspiring in his students a love for the subjects he teaches. Mr Fenner's brilliant University record speaks for itself. He is a cultured gentleman, and combines with the qualities of decision and tact a charming personality that has made him exceedingly popular with the Staff and students.

Appointment as Superintendent of Technical Education

In 1915, the senior officers of the South Australian Department of Education were the Director, the Superintendent of Primary Education and the Superintendent of Secondary Education. On 5 November, 1916, Charles Fenner, after the passage of an Act to consolidate and amend the Law relating to Public Education, was appointed Superintendent of Technical Education, a newly created senior post in the Department. His subsequent career in the Education Department is described in the next chapter.

Retirement and Death

Charles Fenner, who had been appointed Director of Education in 1939, retired because of ill health in 1946. He then worked at home and as a volunteer at the

South Australian Museum until he had a stroke early in 1954. He died on 9 June, 1955. The house at 42 Alexandra Avenue (see Chapter 1) was sold in November 1956 and Mother moved to a smaller house at 10 Springbank Road, Panorama, where she lived until 1964, when she went into a nursing home. She died on 9 February, 1966. Lyell's eldest son Ted lived in the house for about a year and then it was sold.

The obituary notice in the *S.A. Teachers' Journal*, after recording my Father's career in detail, concludes with the following comments:

> A sense of humour and other human qualities which he possessed endeared him to those who worked with him. In conference his remarks were flavoured with wit and his serious pronouncements were received with respect. He always convinced his hearers that he knew what he was talking about, because he really did know. Many people have fine bodies but very ordinary minds. Dr Fenner had a remarkable mind but by no means a robust constitution. It may be said that for many years his mind had to fight his body and that a lesser man would have yielded to the infirmity of the flesh long before he did. Even after his retirement from the Department he continued to work and gave part time service to the South Australian Museum for a number of years. He wrote as long as he was able.
>
> He came to South Australia during World War I and served as Director of Education during the whole of World War II. In the foreword of one of his books he wrote: 'In a world of war, I have as far as possible not referred to war, for wars pass, and peace ultimately prevails.' Those who admired and respected him and held him in affection will like to think of him as at peace. He was a man of great attainments and he rendered outstanding service to his day and generation.

References

Fenner, C. 1931, MS178/6/3A5 (1931 diary) Basser Library, Australian Academy of Science. Also lodged in PRG 372: papers of Dr C. A. E. Fenner, held in the Mortlock Library of South Australiana.

Fenner, C. 1937, MS178/6/5B (1937 diary) Basser Library, Australian Academy of Science. Also lodged in PRG 372: papers of Dr C. A. E. Fenner, held in the Mortlock Library of South Australiana.

Chapter 14. Charles Fenner, the Educational Administrator

Introduction

Charles Fenner made his livelihood, for most of his life, as a senior administrator in the Education Department of South Australia, where he was Superintendent of Technical Education from 1916 to 1939 and Director of Education from 1939 to 1946. During this period he also made important contributions to science, both as a teacher and a science communicator, mainly in the fields of geomorphology and human geography (see Chapter 16). In describing his work as an educational administrator, I have included relevant parts of the entry on his life in the *Australian Dictionary of Biography* (Trethewey, 1981). I have also made extensive use of an essay by Hyams (1990) which provides a detailed account of his educational work. As mentioned in the introduction to this book, the diary that he wrote on his trip to North America and Europe in 1937 provides interesting information comparing various aspects of education in Europe and America with the situation in Australia. Chapter 15 consists of extracts selected from this diary. To balance Hyams' interpretation of his contributions to education in South Australia I have added some comments from other sources at the end of that essay.

Extracts from the Australian Dictionary of Biography

On his appointment on 5 November, 1916, a local newspaper, *The Register*, introduced him as 'a leading educational authority...a grand teacher and organizer' with 'fine powers of lucid expression'. However, as Trethewey (1981) points out:

> When Fenner was appointed, regret was expressed that no South Australian was considered suitable for the job. Although theoretically in an ideal position to influence the department's activities, he was frustrated by having to work within a pre-established framework, by the financial constraints of a world war and the Depression, by political procrastination, by opposition to his proposed reforms and by ill health. His plans for a unified technical education system were undermined by the autonomous South Australian School of Mines and Industries and by delays, until 1940, in expanding secondary technical education on the model of Thebarton Technical High School (opened 1924).
>
> Fenner encouraged innovation, an example being an individual freedom scheme of learning introduced in 1927 at Thebarton, details of which are outlined in a paper by Fenner and Paull (1930), which was read before the Education Section of the Australasian Association for the

> Advancement of Science in Brisbane. Fenner's criticisms of traditional schooling, summarized in his first essay after appointment as Director of Education (Fenner, 1940), were influential in liberalizing the primary school curriculum. He supported the teaching of technical subjects in high schools and liberal subjects in technical schools. He established a vocational guidance and placement scheme and argued for raising the school age. He was also responsible for technical training courses for unemployed youths during the Depression and for reconstruction schemes for troops returning from both world wars. He helped draft the 1917 Technical Education for Apprentices Act which, with its concept for compulsory, part-time, technical study for apprentices, set a precedent for other States.
>
> Fenner was a figure of his time in stressing the views of education for citizenship and technical education as a means of providing skilled labour to develop South Australia's industrial base. In other respects he was forward thinking: his proposals and educational articles were based on research on overseas and interstate trends, appraised in the light of local needs and conditions...His Directorship coincided with World War II. This exacerbated Fenner's impatience at being unable to effect changes that he had advocated for two decades; he increasingly sought refuge in writing. Overall, local prejudice no doubt contributed both to the difficulty in having his ideas accepted and to Adey's appointment to the Directorship in 1929, despite Fenner's broader and longer experience of high-level administration and his superior intellectual standing.

Essay by Hyams (1990): 'Superintendent of Technical Education, 1916 to 1939'

Charles Fenner was appointed as Superintendent of Technical Education in 1916, and was charged with the responsibility of developing Technical Education in South Australia. His initial work was to establish Apprentice Training, consolidate the Art School and develop special schools concerned with technical subjects (Jolly, 2001).

Hyams writes:

> In various ways Dr Charles Fenner was typical of the generality of Australian educational administrators: competent, intelligent, but without a distinctive and developed education doctrine. There is even doubt that he attained more than mediocrity in personnel leadership, as distinct from policy making. Yet in other ways he was unique. Most Directors of Education climbed to the bureaucratic summit in education either through the primary or secondary schools sector; Fenner came through the Cinderella division of technical education, and his record in that area

underlines some of the issues with which South Australia (and indeed Australia as a whole) grappled in the somewhat belated realization of potential links between education and the economy. Fenner was also different because he was a scholar; yet his scholarship was in science, not in pedagogy...The context of educational thinking [in 1916, when he was appointed superintendent of the technical education division of the South Australian Education Department] appeared opportune for Fenner in his new role. Technical education had been an increasing focus of the public record of educational leaders in Australia as part of growing economic nationalism and rivalry, a worldwide phenomenon of the late 19th and early 20th centuries. The spectacular rise of imperial Germany was often quoted as a model for investment in technical education, reinforced by the crisis of international war. Given the world perspective, there was also the special position of South Australia. Australia as a whole was still very much in the preindustrial phase of industrial development, with South Australia particularly dependent on rural activities. From time to time community leaders in the State had argued that its future economic welfare lay in developing a manufacturing base, and the chief ingredient of this, Fenner urged, was the local availability of a trained workforce. Since South Australia was deficient in the standard industrial resources of coal, iron and timber, there was all the more need to compensate with a vigorous policy to develop a local supply of a trained workforce. This need not be an idle dream: Fenner was able to quote the example of the American provincial town of Worcester, where in spite of a dearth of natural resources 'an atmosphere of mechanical skill creates a supply of good workmen, and this has brought factories of various kinds'. (South Australian Parliamentary Papers, 1917).

What he found in South Australia in 1916 was little more than an educational shell for contributing to any Worcester-type economy. The 1915 Act of the legislature had provided for the divisions of the State educational system into primary, secondary and technical sectors, and Fenner was the first superintendent of the new technical branch. In the following year Donald Clark, Chief Inspector of Technical Schools in Fenner's home state of Victoria, had on invitation visited South Australia and furnished a report recommending vocational training for the bulk of older school children, the establishment of Victorian-type junior technical schools for that purpose, and generally a technical education system fashioned largely on the Victorian scheme. With that foundation it was left to Fenner to build the edifice, and his efforts earned for him in the legislature, a quarter of a century later, acknowledgement as 'the father of our technical education system'. (South Australian Parliamentary Debates, 1940).

Numerous deficiencies confronted Fenner in the existing educational provision and he was quick to condemn them. Already established technical institutions, he believed, had been developed without any relation on the one hand to factory or workshop and on the other to the primary level of schooling. Subjects taught in technical school classes bore little reference to factory requirements, while there was a gap between the highest level of primary school and the lowest classes of the technical school. For children facing that gap there should be some form of junior technical school. He lamented that the proportion of Australian adolescents receiving any form of instruction (including part-time courses while being employed) was so low; whereas over half of the 15–18 age group in America was being schooled, the figure in Australia was only 12 per cent and for South Australia a mere nine per cent. In terms of international economic competition Fenner regarded this as a deplorable waste.

The answer did not lie, he believed, in a simple extension of general secondary education at all. Specialization was very much part of the Australian pedagogical ethos of the early twentieth century. Hence the existing high school was not to be the model of universalization; its academic curriculum and its orientation towards public examinations, university entrance and professional careers should remain the preserve of the privileged minority. For the bulk of schoolgoers Fenner advocated a route which equated modernization with specialization: 'schools must deal with the actualities and activities of life, and will be justified or condemned according to the service they perform for the betterment of the community'. This called for the development of both separate commercial schools for young people destined for clerical occupations and junior trade schools relevant to factory work; it included provision for girls gravitating towards domestic work.

But Fenner was addressing an Australian public which was not yet persuaded of the need for educating any but a minority beyond the age of 12. His was a vision of a network of junior technical schools catering for most young people of ages 12 to 16. Their program would be general education with a nonacademic basis, allowing identification of the pupil's preference and ability for a particular trade, for which there would then follow a further two years of intensive specialized trade training and then an apprenticeship in the selected and available occupation. In this way specialised training, vocational guidance and selection would form an articulated and unified process. Realistically, however, he acknowledged that the upward extension of popular secondary education and expansion of full-time vocational training was not yet politically and financially feasible.

> Yet even his concept of extensive part-time continuing education encountered popular resistance. Although the various classes enrolled apparently large numbers of recruits, the actual attendance was far less impressive, with many students drifting away before the year's work was completed. . . . Visits to classes in the country towns left him depressed: 'Boys and girls, freed from the influence of compulsory school attendance at the age of 14, are reluctant to take up evening studies, and are with the greatest difficulty brought to realize the value of such additional training.' Despairing of that arrangement, he was thus strongly inclined towards the notion of compulsory attendance.
>
> The compulsory ideal was not to be realized with part-time training, but it enjoyed some measure of success in the case of the apprenticeship scheme. Fenner, like other leading Australian educationists, clung to the British tradition of apprenticeship as a major element in technical education and was of course influenced by his own youthful experience as an apprentice. This he preferred to the European alternative of incorporating specific trade training in secondary schooling, believing that British practices enduring since Elizabethan times could, with adaptations to modern times, most suitably correlate workshop and schooling.

Comment by FF: The preceding paragraph refers to Charles Fenner's views in 1924. Perusal of the comments in his 1937 diaries, when he examined educational practices in the United States, England and Germany, suggests that when he had an opportunity to compare practices in these countries, he greatly preferred the patterns operating in the United States and Germany to those in England.

Hyams goes on:

> Notwithstanding the development of mechanization, mass production and repetitive processes, he believed that the supply of machine operators and skilled tradesmen depended on a sound basic training from ages 16 to 21. He challenged popular views that industrial modernization in fact required less skill, and returned to the theme of South Australian economic self-sufficiency. Even more than before, he argued, there was a need for capable workmen 'with ability in reckoning and in the reading of drawings and prints, and with some resourcefulness and adaptability'.
>
> Because he held that apprenticeship should properly include some formal technical education as well as workshop guidance, Fenner strongly urged the introduction of compulsory courses for apprentices. Not surprisingly, therefore, he was influential in the drafting of legislation which in the Apprentice Act of 1917 gave the government the power to compel technical education in proclaimed trades. Indentured apprentices could

thereby be required to attend classes for six hours per week, 40 hours per year, partly during working time and partly in their free time. As chairman of the Apprentices Advisory Board, Fenner was strategically placed to encourage the development of apprenticeship. But he had to proceed cautiously, finding that the application of compulsory classes had to be made gradually and according to varying conditions of the market and differences between individual trades.

By the end of the 1920s there were patent signs of economic recession; in the 1930s the demand for apprentices slumped, employers attempted to extricate themselves from indenture-committed formal courses of instruction. The system, he submitted, had survived four centuries of developing industrial conditions in British countries and the economic depression was just one more crisis which it could be expected to survive. The shortage of supply of skilled young workers and the industrial pressure of international war brought the realization of Fenner's goal and in 1940 the Technical Education Bill was passed, extending the provisions of the Apprentice Act to unindentured learners.

Apprenticeship had been one of the linchpins of Fenner's concept of a comprehensive technical education system; the other was the junior technical school. This took much longer to develop. A single Education Department junior technical school was established in the Adelaide inner suburb of Thebarton in 1924 (initially known as the Thebarton Technical High School) but it was not the immediate forerunner of a system of such institutions. A nonacademic schooling network began in the following year in the form of Central Schools, in reality a number of post-primary annexes in existing large primary schools and offering commercial, junior technical or homemaking courses. This was not in effect the Fenner concept of junior technical education, but something of a compromise between the utilitarianism of vocational training and the academic emphasis of the contemporary Australian high school.

Comment by FF: Father encouraged innovation, an example being an individual-freedom scheme of learning (Dalton Plan) introduced in 1927 at Thebarton, details of which are outlined in a paper by Fenner and Paull (1930). His confidence in this school was demonstrated by the fact that the three sons who were then eligible for secondary education, Frank, Tom and Bill, all attended that school. I obtained both my Intermediate and Leaving certificates there, and found the independence provided by the Dalton Plan excellent as an introduction to study at the university.

W. J. Adey, who succeeded McCoy as Director of Education in 1929, held fast against early specialization, quoting both American practices and the renowned 1926 Hadow Report in Britain in supporting universal secondary schooling, but commencing with a general education. Fenner's moment, however, was near at hand. By 1936 he was able to report that the boys' Central schools (which were not separate schools, but upper divisions of large primary schools) were waning. Some improvement in employment conditions as the worst of the depression receded meant a drift of adolescents into unskilled, deadend, but now more obtainable jobs. One solution to the situation would be to raise the school-leaving age from 14 to 15; this extended schooling, Fenner believed, should be in the direction of a junior technical type of education. Yet as time progressed he appears to have modified his stance on the nature and timing of specialization; by the end of the 1930s he was calling for a junior technical school which would simply have a general education program with some manual craft and science work.

It had long been Fenner's belief that vocational training carried with it an obligation to provide vocational guidance. The idea was never encouraged during McCoy's regime in the department, and it was only at the end of his directorship that he was inclined to concede the need for specialist guidance services to assist school-leavers with employment matters. Shortly before his death in office in 1929 he allowed the establishment of a modest system of guidance, generally under Fenner's supervision. Here Fenner was in his element, with his usual propensity for surveys, statistics and carefully constructed graphs, looking at the relationship between schooling and the job market. Through interviews of pupils and a system of stored information he arranged for the compilation of profiles of school leavers to indicate the employment potential of each of them. But even when Director, Fenner still had to counter traditional prejudices against professionalism in these areas of concern, and secured the appointment of the first vocational guidance officer for departmental schools in 1940, 'only after much representation' on his part.

There was, of course, more than a single dimension to his view of the role of technical education. Just as he saw it as the mainspring for economic advancement for the community, he also shared with other champions of technical education that fervent belief in its pedagogical value. Such faith was largely expressed in his emphasis on manual training as part of the primary school curriculum. It was not, as he had asserted in 1917, when he first joined the department, just something which happened to be useful, such as skill in carpentry, ironwork or bookbinding. It also had mental and even moral qualities, allowing scope for developing habits of neatness, accuracy, and perseverance as well as taste. He was still extolling these virtues of handicraft activity in general schooling 20 years later. He reiterated its contribution to good habit formation—order, exactness, cleanliness and neatness. And he lauded its role in inducing respect for work of all kinds and in developing the quality of self-reliance in the individual child.

Fenner may have been ahead of many of his contemporaries in his thinking on technical education and his efforts to force the pace of its development. But in one other respect, that of the vocational education of women, he does not emerge as a progressive. Here he reflected current middle-class conservative views; his view of the female half of the community was one mainly of married women and few females in careers. Evincing no desire to change that situation, he postulated a form of technical education for girls that was essentially domestic training; it was to take account of the fact that most young women would cease paid employment before the age of 30 and would establish homes of their own. The type of training he had in mind involved not merely cooking and cleaning, but also such matters as household accounts, hygiene, child care, needlework and the 'power of taste and expression' as exemplified by language and drawing. But competence and motivation for women's chores were also the means to a further goal with which Fenner identified himself—the supply of paid domestic service. The chronic shortage of domestic labor was a continuing and even increasing problem for middle-class Australia and Fenner was sympathetic to its cause. He saw the shortage of supply as a revolt against harsh and unpleasant conditions in domestic service, but he believed that a system of education in household 'arts' would increase the skill, reduce the hours of employment needed and hence enhance the status of domestic help. What Fenner failed to appreciate sufficiently was the way in which the issue reflected both changing labor market conditions and the dialectic of social class. Women were increasingly diverted from paid domestic service by a gradual expansion of alternative employment opportunities, including those in the former domains of men. Calls for domestic training were often interpreted by working-class representatives as part of middle-class labor exploitation.

Fenner's identification with technical education in Australia was in part due to his long administrative association with that field. He remained for 23 years as superintendent before promotion to Director of Education

The Bulletin, 24 April 1929

DR. C. E. FENNER, Superintendent of Technical Education in South Australia.

The Bulletin, 29 October 1930

Dr. CHARLES FENNER.

The Bulletin, 25 December 1935

This isn't a picture of Charles Fenner, S.A. Director of Technical Education, but an assay. The divided chin which goes with imagination is shown clearly; the contemplative, slightly-ironic mouth has been separated from the heartlike nose with which Coventry has sought to emphasise that his subject's good nature is written on his face. And nobody gazing on that thatch could doubt that its owner had earned the right to wear bay leaves. The only falsity suggested is that there is a classic mind under that ample cover, instead of a brilliant scientific one which made its owner Kernot scholar, a first-class honorman in biology and geology and a D.Sc. of Melbourne University. Dr. Fenner was an Australian delegate to the Centenary Science Congress in London in 1931. He has written widely in the press, apart from his purely technical efforts.

Assay of a D.Sc.

Figure 14.1. Cartoons of Charles Fenner published in *The Bulletin*, 1929, 1930 and 1935

in South Australia. For a decade he had served McCoy well and with considerable admiration, even though his Director had not been an easy man to deal with and at times had appeared to have little concern for Fenner and for technical education. On McCoy's death in 1929 Fenner was the senior superintendent in the department and might well have expected the top post, but the succession went instead to William James Adey, Superintendent of Secondary Education. Fenner's wife did not hide her bitter disappointment at the result, and the family could not help but note that Adey was a cousin of the Chief Secretary, one of the government ministers, and that he was a South Australian, and not like Fenner an import from another State. But it was not in Fenner's nature to make an issue of the affair. In any case nepotism and favouritism would have been difficult charges to substantiate; certainly extant Cabinet records do very little to confirm the family's views. The recommendation of the Public Service Commission quite deliberately asserted in 1929 that 'it is known that there is no person available in the service who is as capable of filling the position in question as the person recommended'.

Figure 14.2. Charles Fenner, Director of Education, at his desk

Director of Education, 1939 to 1946

In May 1939, W. J. Adey retired after 52 years of service in the department. Fenner was appointed Acting Director and his appointment as Director was confirmed on 1 June, 1939. He retired, because of ill-health, on 16 May, 1946.

Hyams' essay continues:

> His previous appointment had been at the time of World War I and now elevation to the directorship coincided with the outbreak of World War II, a conflict which ended less than a year before his premature retirement in 1946. He was able to proceed immediately with a scheme he had long favoured, replacement of the boys' Central Schools by junior technical schools. The Central Schools...were not physically separate schools, but merely the upper divisions of large primary schools and under the mixed supervision of the superintendents of technical, secondary and primary divisions of the State Education Department. What was needed were separate entities, and these emerged in 1940 as junior technical schools, acknowledged secondary establishments. To help them along, Fenner secured government acceptance of the principle of housing them in separate buildings, of inclusion of machine shops in their facilities and

of an eventual extension of schooling, compulsory and non-fee paying, to the age of 16.

However, his final years of high office were scarcely opportune for the fulfilment of any comprehensive educational philosophy that he might have developed. As Director he was immediately beset by shortages of physical resources for education, by loss of personnel to the military services and by a necessary preoccupation with patriotic exercises and voluntary fund-raising campaigns in the schools. Yet even given such extenuating circumstances, it cannot be said with confidence that he had developed any distinctive educational credo, apart from his strong bias towards technical education. In effect, then, his stance and role on educational directions and educational reforms was not remarkable; he appears to have been neither much ahead nor much behind most Australian educational administrators of that era.

For example, leading educationalists across Australia had been pleading for an upward extension of the school age, and Fenner was one of a number of them. As a consistently enthusiastic advocate of that reform, he was not particularly influential in its local realization. Late in 1939 he urged Shirley Jeffries, South Australian Minister of Education, to make a gesture towards raising the school-leaving age by dropping fees at the end of the term in which children reached the age of 15. This, he felt, would particularly stimulate attendance at the vocational type schools. The ministerial reply came promptly and in terms of the prevailing philosophy of a conservative government: 'To approve of the above recommendation would involve a complete reversal of all that I have stood for in connection with these matters: namely that where State services may be availed of by citizens and they can afford to pay some portion of the cost and take advantage of the service, they should be required to do so according to their financial resources.'

Fenner let the matter rest there. Nearly six years later he was able to secure acceptance in principle of 16 as the school-leaving age. But this was a pipe dream; the government had decided in 1942 to raise it to 15 in two stages, and even this modest goal would take years to realize. The move was simply part of a nationwide drive towards extended schooling, and South Australia in this respect was in advance of some States and behind others.

Talk of educational reform in general was very much in the air at that stage, especially in terms of its anticipated pivotal role in the forthcoming era of postwar reconstruction. Fenner moved very much with the times by arguing in 1941 that the moment had arrived for South Australia to adjust to the needs of an increasingly complex civilization, especially by overhauling its old educational structure. His sense of the intellectual and political climate for such an issue was accurate; in the next year as a result of some pressures within the legislature the government established a commission of enquiry into the State's educational system. Although it yielded few immediate important and practical results, the enquiry canvassed vital educational issues in its lengthy existence to 1949, most of them addressed in later times. Asked in 1945 to respond to some of the recommendations of the committee, Fenner lent support to many of its arguments directed towards raising the quality of the State's teaching service, and especially the case for improved salary structures. But he demurred on some of the proposals affecting criteria for recruitment. These included the notion of intelligence tests for entrants and the suggestion that all teacher recruits should reach the level of university matriculation in their general education. Both provisions, he countered, would unduly emphasize intellectual aspects at the expense of the natural talent for teaching. In this stance he reflected the conventional thinking of the education hierarchy. Such a conservative policy on teacher recruitment was consistent with his earlier view on the appropriate priorities in teacher preparation. In his disapproval of some of the training directions favoured by the Principal of the Adelaide Teachers College, in 1941 he had pronounced that too much emphasis was being placed on the university aspect of student teacher training and ordered removal of that emphasis from the departmental circular.

He strongly supported Area Schools. Not only were they the product of the now much favoured rural consolidation of schools, but for him they had the added attraction of offering secondary education not based on the traditional model of the state high schools. Even these later institutions were the target of his attention and he was able to secure for them curricular modifications to temper their academic emphasis; their programs were to include art, craft and manual training 'without loss of their present prestige as gathering grounds for potential university students'. While diluting a little the academic schools, he was able to raise the status of the junior technical schools by supporting the claim to bring their teachers' salaries into line with the levels paid to high school staff.

Staunch technical education advocates such as Fenner were obliged in Australia to come to terms with the increasingly fashionable case against early specialization. His attitude was put to the test in 1941 in the concern which had arisen over secondary school students commencing commercial courses in their first year. Fenner argued that sound educational practice required the delay of specialization until after the first two years of secondary schooling. If this was not practicably possible in all cases, then the department should stand firm at least on delaying subjects such as shorthand, typing and bookkeeping until after the first year. These subjects he regarded as purely vocational training. Such a category did not include, in his opinion, domestic arts, manual work and art and craft, all of which, he had always maintained, possessed pedagogical rather than merely vocational merit.

The decision accordingly to drop the avowedly vocational subjects at the early secondary school stage met with opposition. The Superintendent of Secondary Education argued against the change and representatives of school parent committees gave it a hostile reaction. Their principal objection to the move was that it would tempt parents to transfer them to non-government commercial colleges to secure the two years of vocational training they wanted. The alternative, to spend more than two years at secondary schools to include such training, would impose an unfair burden on poorer families. Fenner stood firm in resisting all counter contentions. Nor did he flinch at the irony of rejecting the argument of the Superintendent of Secondary Education that the commercial subjects in question could be regarded as having intrinsic educational value—a case strikingly similar to his own much reiterated view of technical subjects. In his stance he cited the worldwide trend to postpone specialization in education. But he also addressed the socioeconomic issue underlying the dispute. The financial burden on parents, he insisted, would now be greatly alleviated by the federal government's scheme of child endowment, thus making it economically possible for parents to keep children longer at school. Moreover, the children themselves had to be protected; it was 'little more than exploitation on the part of parents and employers to subject children of such tender age to vocational subjects in order that they may commence to earn money at an early date'. In his resolve not to be diverted from the change Fenner had the strongest encouragement from his minister, who had frequently discussed the policy with him, had independent and firm views on the matter, and took the sole responsibility for its implementation.

Such victories were not frequent during Fenner's directorship. As far as administrative leadership there was added to successes a mix of frustration and failure. In terms of organizational affairs he made some advances, especially with the inspectorate, a body which was the key element in operating all the Australian State educational bureaucracies. Fenner concentrated on ushering his school inspectors into a role of inspirers rather than examiners, as in the past. He led them on a deeper and wider thrust into professional reading to suit this preferred emphasis, and utilized their talents to a unprecedented extent by including them on newly created departmental curriculum boards. Yet there were flaws and contradictions in the changes. He involved inspectors in examination of school financial accounts, thus emphasizing their police role in another direction. He also failed to delegate to them sufficient authority to assume effectively the function of local district leadership in education.

Obviously he did not greatly relish much of the administrative burden thrust upon him. He was oppressed by the multiplicity of routine duties which could not be delegated even though they encroached severely on his scope for planning and broader policy making. He was also restricted in doing what he liked most in the job, visiting schools and discussing with teachers matters of everyday conduct of their affairs; only half a dozen such visits could be managed in his first two years as Director. Compared with his counterparts in other States and countries, Fenner felt clearly disadvantaged; they had, he complained, considerable freedom for reading, discussion, visiting schools and planning developments. This was because many school systems, unlike that in South Australia, had the added position of a Deputy Director, to handle in general numerous administrative chores and to manage affairs during the Director's overseas visits. At first the government was not prepared to accede to Fenner's plea for help. It finally listened when the Public Service Commissioner commented that the burden on the Director was too great to enable him sufficient opportunity to study educational developments, especially those occurring overseas, and that this handicap would best be eliminated by creating a Deputy Director's post. The government acquiesced and made provision for it in an Education Act in 1946, unfortunately too late to be of any benefit to Fenner.

Fenner had complained that the burden was so onerous as to absorb much of his time at home, both evenings and weekends. This, it was claimed, was also detrimental to his health. Indeed he had never been particularly robust, and his doctor was able to certify that he had suffered from a 'deficiency' since childhood [now, in 2006, I (FF) believe that his illness was due to infection with *Helicobacter pylori*]. The strains of office from 1940 onwards had adversely affected his nerves and his blood pressure. In 1945 he had required hospitalization and a three-month period of leave from work. Hence it was hardly surprising that in May 1946 he should request and receive early retirement. Ill health was a major determinant in his decision to quit, but unhappiness also figured in the resignation. The cordial relationship with his political master which Fenner had enjoyed during the Jeffries regime was not to be repeated with R. J. Rudall as Minister of Education at the end of Fenner's administration. Rudall was inclined to want a more direct hand in running the Education Department and asserted his inclination strongly. The result was that the two men often crossed swords. Fenner finally concluded that it was better to retire at 62 than to carry on in that way for a further three years.

In summary, this brief account of the career of Charles Fenner [as an educational administrator] tells us a little about public life in an era in Australian history. In the dichotomous control of government departments shared by the political head and chief professional officer, there was always considerable potential for the professional to exert an important influence on the evolution of policy...He occupied the top position only briefly and during an international war; he was neither shrewd nor charismatic; and this was not a period in which South Australia could be characterised as a progressive State in matters of public schooling. More than in administration, more than in technical education policy, Fenner enjoyed success and satisfaction in the scholarly world of science. Perhaps the final judgement should be that fate had dealt him an unkind hand; he was a public figure who missed his true vocation.

Other Aspects of Charles Fenner's Career in the Education Department

Comments on Secondary Education on Return from Europe, 1932

In an interview with *The Mail*, Fenner commented:

> The extent of our debt to England and Scotland for the system [of secondary education] we have in this State was apparent on a visit to schools in those countries. We have not been slow in learning, and it will be readily admitted that we have done our educational work even better that those from whom we have learned. It is difficult to compare the systems of Britain with those of our own State, because of essential differences in organization. In Britain each school is practically independent except for small monetary grants and a system of Government inspection which is not very rigid, and quite in contrast to the centralized system in South Australia. There are some advantages in the British system, but there are more both from the point of view of efficiency and of promotion in ours. Vocational guidance is facing the same setbacks as it is here. More is being done in the way of part-time education on the Continent than elsewhere. There one finds lift boys and hotel attendants, 16 to 18 years of age, going to school two or three days a week and working for the rest of the time.

The South Australian School of Arts and Crafts

The School of Arts and Crafts, which had been long established, was taken over by the Education Department in 1909, and suffered from poor teaching facilities, which were made worse by the fact that the Exhibition Building, where it was housed, was taken over in 1919 as a temporary hospital to cope with the influenza epidemic. The situation did not return to normal until September 1920. Fenner stressed the need for a permanent and more suitable building, a dream which did not eventuate until it moved to Stanley Street, North Adelaide, in the late 1950s. In 1924, he established a Girls Central Arts School as a branch of the School of Arts and Crafts, as a school where the general education of girls along with their arts subjects, a school that, with Thebarton Technical High School, receives extensive coverage in *A Broader Vision* (Jolly, 2001). Father had a special interest in art and was a close friend of Hans Heysen, and he had a special interest in promoting the work of Ivor Hele when he was a student at the School of Arts. As G. S. Macdonald said in obituary notes in the Education Gazette: 'Hans Heysen was one of his oldest friends, Ivor Hele one of his most loved young protegés.' Ivor Hele painted a portrait of him in his DSc gown, which is currently in the foyer of the State Education Office. Ivor also gave him several fine paintings, one of which I inherited.

Appreciation by Australian Broadcasting Commission (S.A. Division)

> It is impossible to think of the progress of Educational Broadcasts in South Australia without becoming deeply appreciative of the efforts and interest of the its first liaison officer, Dr C. Fenner. Taking office towards the end of 1933, Dr Fenner gathered together a small team of eight enthusiasts and inaugurated a modest series of experimental broadcasts, the details of which are recorded in the first educational broadcast pamphlet ever published in Australia.
>
> Most of the original broadcasters are still actively engaged, and the findings they derived from a variety of experiments tried in the early stages have been confirmed and consolidated into the present high standard performance. Difficulties were encountered and criticism was sometimes vigorous, but the enthusiasm of our leader was unfaltering, and his cheery optimism never failed. He was a constant source of encouragement to everyone concerned, and throughout the whole of his six years of office he has been a tower of strength, fathering the growing activity to its present breadth and complexity.
>
> There are now 13 different topics organized by South Australian leaders in addition to those undertaken by Federal authorities, and about 50 teachers and other speakers and three dramatic companies are engaged

every term broadcasting the fruits of their experience in schools. This booklet, indicative of the general progress, has steadily increased in both size and value, and is now distributed to more than 4,000 teachers and scholars in approximately 300 listening schools.

An Appreciation by J. S. Walker, Director-General of Education, 1967

Talking about notable events reminds me of something that happened 60 years ago which I believe was of outstanding importance and significance for our State—the establishment of Government High Schools. Adelaide High was the first, and W. J. Adey was its founder and the architect of the State system of secondary education...A second landmark in the history of secondary education in South Australia was the establishment of Junior Technical Schools (now called Technical High Schools) and Area Schools by a very forward-looking Director of Education, Dr Charles Fenner. That was in 1940, more than a quarter of a century ago. Dr Fenner foresaw the time when nearly every boy and girl would go to secondary school because primary school would no longer be a sufficient preparation for life. He also realized that children differ widely in interests, needs and aptitudes as well as academic ability, and only a fraction of those who passed through secondary schools would go on to tertiary study.

At the time our High School courses were directed towards gaining a Public Examination Board certificate at the Intermediate or Leaving level. The Leaving was the matriculation for University entrance so the courses were essentially a preparation for University study. Dr Fenner thought it was unreasonable to suppose that these University-oriented courses would be appropriate for all secondary students. And so the curriculum in the Junior Technical and Area Schools was custom-built to suit the needs of the majority who would leave school after the third year.

The courses in English, mathematics, science, social studies, art and craft, music and physical education in these new secondary schools were not soft options. Every student was stretched to do his best so that with few exceptions each of them left school with a sense of achievement and fulfilment; and surely every boy and girl is entitled to this. The examinations were conducted by the schools themselves with guidance from senior officers in the Education Department. After some years of trial, most employers were glad to accept boys and girls from these schools on the Head's recommendation. They judged the schools by their reputation—and by the worth of the young people who passed through them (reproduced in Jolly, 2001).

Tribute from the Hon S. W. Jeffries, *S.A. Hansard,* July 1947

Members will realize that I am very interested in education, having held the portfolio of Minister for Education for 11 years. I am pleased indeed to see the progress of education since I vacated office and that the prospects for further progress are so bright. During my absence from this House the late Director of Education, Dr Chas Fenner, resigned on account of ill-health. I should like to say how greatly I appreciated him as Director of Education. He possessed outstanding educational qualifications. He had great foresight and was a man of ideals, and from the day he came from Victoria, 20 or 25 years ago to inaugurate technical education in South Australia, to the day of his resignation, he gave himself unreservedly to the cause of education. Dr Fenner was respected by the Ministers and Directors of Education in other States because of his knowledge and character. When I attended interstate meetings of Directors and Ministers I observed that they looked to him for guidance and I think that it is only fair that I should pay tribute to him for the work he has rendered to education for so many years. His foresight in the matter of technical education proved to be of inestimable value to this State when war broke out, because, but for the technical schools and the men trained in those subjects, we would have been in a sorry state. I do not suggest that he alone did all this. He was ably seconded by his superintendents and inspectors and by the great majority of teachers, most of whom give their services without stint for the welfare of the boys and girls.

Comments by Thiele and Gibbs (1975)

W. J. Adey's successor, as Director of Education, was Charles Fenner. The appointment was announced on 1 June, 1939, first as an acting brief and then permanently when Adey's long service leave expired. As a whole the appointment was expected and approved. Fenner's background was quite exceptional; a science degree with first class honours at the University of Melbourne, followed later by a doctorate, an impressive list of publications and scientific research papers on geography, geology, anthropology and natural history, a reputation as a world authority on glass meteorites, overseas experience as a delegate at scientific conferences and as an educational observer and guest speaker, ten years experience as a part-time lecturer at the University of Adelaide and a constant stream of newspaper articles, reports, conference addresses and public comment...

23 years in the position of Superintendent of Technical Education had made Fenner very well known throughout the State. He had pioneered

the compulsory education of apprentices, supervised the post-war reconstruction scheme, developed technical courses of many kinds and established widespread contacts with industry. He was also keenly interested in art, in the development of art and craft in schools, in the work of the Art School and in audio-visual education. If ever there was a Director with personal experience and interests at all levels of education, from small outback schools to the University, it was Charles Fenner.

The remaining months of 1939 were very busy ones, the most important single action being the decision to close down the Central Schools. Their attachment to primary schools had always mitigated against public recognition of their work; Fenner therefore announced that they would 'reappear in 1940 as junior technical boys schools and junior technical girls schools, each with a separate entity, and with the status of a secondary school within the technical division.' His penchant for good public relations was immediately apparent: well-attended meetings of parents were called to explain the move, exhibitions of student work were held and illustrative graphs and diagrams began to appear in official documents and newspapers, as they had been doing for two decades in his own reports.

In September the war that had been looming for so long finally broke out and the Schools Patriotic Fund (SPF) was established to coordinate the efforts of all schools. It proved to be an astonishing enterprise...The impact of the war on technical education was enormous. From early 1940 G. S. McDonald, the newly appointed Superintendent, and J. S. Walker, the youngest inspector in the State, were busy organizing facilities and services for wartime use. By May more than 400 RAAF fitters were being given courses at the Grenfell Street Trades School, and training in machining, fitting, welding and tool-making were getting under way for munitions and aircraft construction...By September the Grenfell Street Trades School, the Adelaide Technical College and the School of Mines went on to a three-shift roster, working 24 hours a day. Specialist teachers and instructors assumed heavy workloads...

Despite the war, the struggle for professional improvement went on...Fenner himself was constantly involved, writing on the cultural value of technical education, on the school leaving age, and on the 'stuffing versus stimulating' approach to teaching...In Adelaide in 1942 there was angry criticism of the Education Department and the Civil Defence authorities because the schools were utterly unprepared for war, there were no air raid shelters, no medical supplies and no emergency plans...And so the city became pockmarked with mounds of earth and zig-zag trenches, black-out hoods and paint went on windows and car head-lights and adhesive strips crossed bare panes of glass to give shatter-proof protection.

In November 1942 Executive Council approved a new set of Regulations which revised the scholarship system, reduced the age of entry into secondary schools, required students to take three year's study before attempting the Intermediate examination, enlarged the authority of inspectors...Above all, machinery was introduced for the continuous revision of the curriculum. There was to be a Curriculum Board for each of the main areas—primary, secondary, technical and rural—with the superintendent as chairman in each case and places for male and female nominees of the teachers themselves. The Minister said the new regulations were 'a great step forward' and praised the Director and his officers for their organizing skill...Meanwhile the squeeze in the schoolrooms had begun, and there was only one solution, temporary accommodation. In 1943 Fenner announced the Department's large scale commitment to 'portable schools'. These, he said, consisted of a single room about 24 feet square, and were being turned out as fast as the Architect-in Chief's Department could build them...

On 17 April 1946 C. L. Abbott resigned from the Ministry, giving as his reasons the extreme workload of three portfolios...Exactly a month later, on 16 May 1946, Fenner retired because of ill health at the age of 62. He, too, had fallen victim to what Abbott called 'the almost insupportable burden...of administering this huge undertaking', which was what the Education Department had become. In casting about for a successor, the commentators were at no pains to disguise the burden.

References

Fenner, C. 1932, Interview with *The Mail*, 13 January, 1932.

Fenner, C. and Paull, A. G. 1930, Individual Education. An account of an experiment in operation at Thebarton Technical High School, South Australia. *Educational Research Series No. 1*. pp. 7–40. Melbourne University Press. Also reproduced in Jolly, 2001, pp. 104–12.

Hyams, B. K. 1990, Charles Fenner: Scientist who would be administrator. *Biography, an Interdisciplinary Quarterly,* vol. 13, (1), pp. 57-75.

Jolly, E. 2001, *A Broader Vision. Voices of Vocational Education in Twentieth Century South Australia.* Michael Deves Publishing, Adelaide.

South Australian Parliamentary Papers (1917).

Thiele, C. and Gibbs, R. 1975 *Grains of Mustard Seed,* Education Department, South Australia.

Trethewey, L. 1981, Fenner, Charles Albert Edward, *Australian Dictionary of Biography*, vol. 8, pp. 481–82.

Bibliography of Charles Fenner's other Publications on Education

Fenner, C. 1919, Apprentice education, *Science and Industry*, pp. 500–4.

Fenner, C. 1924, Apprentice training. An experiment in compulsory specialized adolescent education. *S.A. Education Department Bulletin* No. 1.

Fenner, C. E. ed. 1934, *Occupations for Boys and Girls*. S.A. Education Department.

Fenner, C. 1934, *An Intermediate Geography of South Australia*. Whitcomb and Tombs Pty. Ltd. (Written for secondary school students, this book went through seven editions.)

Fenner, C. 1935–6, *One Hundred Years of Education: Part 5, Technical Schools and Colleges; Part 7, Special Developments from the Education Act of 1915*. Modern Ideas in Printing, vol. X, Printers' Trade School, Adelaide.

Fenner, C. 1936, The value and importance of apprenticeship, *S.A. Education Department Education Gazette*, January 1936.

Fenner, C. 1937, Some notes on technical education: what is being done elsewhere. *S.A. School of Mines, Annual Report*, pp. 45–51.

Fenner, C. 1939, *Some Aspects of Transition from School to Workshop*. Australian Council for Educational Research, Melbourne University Press.

Fenner, C. 1939, The cultural value of technical education. *The Education Gazette*, No. 649, vol. LVI, p. 176.

Trends in education. Address given by Dr Fenner at the Teachers' College, published in the *S.A. Teachers' Journal*, 23 December, 1939.

Fenner, C. 1940, Individual educational requirements for modern citizenship. *S.A. Education Department Bulletin* No. 8.

Fenner, C. 1941, *Education (Simple Views on a Complex Problem)*. Printers' Trade School, Adelaide.

Fenner, C. 1942, *Education in South Australia: Present Tendencies and Post-war Possibilities*. S.A. Education Department.

Fenner, C. 1945, South Australian schools. *S.A.Teachers' Journal*, for celebration of the fiftieth year of the South Australian Public Teachers' Union.

Fenner, C. 1946, Tendencies in education in South Australia. *Handbook of the Twenty-Fifth Meeting of the Australian and New Zealand Association for the Advancement of Science*, Adelaide.

The *Education Gazette*, between 1917 and 1946, had short articles on *Adventures in Art, Ernabella—a Freedom School, What is the Aim of Our Schools?, Village Surveys,* and *the Australia Language*. There were also Introductory Notes to a number of educational works, extracts from several addresses, and reviews of 23 books covering many aspects of education.

Reports to the Minister of Education were submitted by Charles Fenner as Superintendent of Technical Education from 1917 to 1938, and as Director of Education from 1939 to 1946. All were published in *South Australian Parliamentary Papers* each year.

Chapter 15. Overseas Trips, Diary Extracts on Education

Introduction

The Introduction to this biography explains the background to the material selected from the diaries produced on Charles Fenner's two overseas trips with his wife Peggy in 1931 and 1937 (C. Fenner, 1931, 1937). The 1931 trip was undertaken when he was chosen as one of the Australian delegates to the Centenary Meeting of the British Association for the Advancement of Science (BAAS); the others were: Sir Hubert Murray, President-elect of the Australasian Association for the Advancement of Science (AAAS); Professors Kerr Grant and Chapman, of the University of Adelaide; Professors Ewart, Hartung and Skeats, of the University of Melbourne; and Dr Clive Lord of Tasmania, with funding from the British Association. Fenner was also instructed by the Director of Education 'to enquire into the question of broadcasting in schools, visual education and the operation of the Hadow Report in England and Scotland'. They travelled there, via the Suez Strait, on the *SS Balranald* and back, around South Africa, on the *SS Jervis Bay*.

The 1937 trip to North America and Europe was funded by a grant from the Carnegie Foundation and was focused on educational matters. They were accompanied on this trip by Dr Draper Campbell, Head of the Dental School in Adelaide and an active member of the Royal Society of South Australia, and his wife Elizabeth. On this occasion they went first across the Pacific Ocean to California, then across the United States and to Canada, and then to the United Kingdom and continental Europe and home via South Africa.

The extracts below come from diaries of both trips. They represent only a small fraction of the daily entries containing information about educational matters and have been severely edited. Throughout both trips, Father made diary entries every day. A few have been selected for his comments on educational matters; each entry is distinguished by the place and date.

Some Emotional Reactions to Travel

I (Frank Fenner) recently watched a television documentary on the Australian SBS of the fish markets in Bombay, which showed the fantastic crowds of men and women, many carrying baskets of fish on their heads. This reminded me that Father and Mother made their overseas trips long before television showed us how people in different parts of the world lived. Even the glimpses of London that Australians received were in magazines such as *The Illustrated London News*. This needs to be borne in mind when reading their initial reactions to Colombo and London, set out below; there were similar reactions to Aden in the 1931 diaries.

5 August 1931, the First Day at Sea

I am a bad hand at recalling dates, but I shall never forget 5/8/31. It seems to me that for years I have been preparing and arranging for August 5 as for the end of everything; the end of the world. And I can still remember how I, how both Peggy and me, just hung out to that date, so run down and so dead tired that we thought we should never get untired anymore.

And now August 5 has come and has gone, 'tis ever so far away in time, just as our dear home and family and friends are distant in space. I am scribbling this on the afternoon of August 6 lying on my back in the cabin and we are out upon the Great Australian Bight somewhere in the Roaring Forties and the little bit of sea I can see through the porthole is most unquiet and the decks are deserted. The sea and a high proportion of the passengers are heaving away most industriously. Poor old Peggy is very ill, and hasn't eaten or drunken anything all day. She is still in bed and is patiently waiting for whatever may be the end. Our greatest anxiety, the five dear kids and home, will I am sure be in safe keeping with Miss Hawson and Stella and with all the friends who have volunteered to keep an eye here and there.

19 August 1931, Colombo

Peggy's birthday. Many happy returns of the day. Today has been like no other day of the 17,520 days which I have lived. My mind reels before the idea of putting anything on paper about it that could be at all adequate. It would really be much wiser to make no entry at all. But Hartung says 'Go on. Write it up.' So I shall put down the inadequate words, and leave most that really matters unsaid.

Kerr Grant thinks that to an Australian, or to anyone from the colder less peopled lands, the first day at Colombo gives something that can never be repeated. I quite believe it, though the whole world lies ahead, and only Colombo is behind, and so little behind at that, that the smell of Colombo is still with me. That smell! The most powerful and all-pervading thing in all Colombo. It may be incense, or it may be the frangipani flowers and the innumerable other incredibly brilliant flowers, or it may be from trees, or damp earth and mould, or it may be a human emanation from these teeming myriads of people, a few of whom and a glimpse of whose life we have seen today.

It is perhaps the most close-packed series of experiences that I shall ever enjoy as new things in any one day. For Peg it has been the most outstanding of birthdays. It has overcome her. She dropped on the bunk in her clothes after dinner and there she is now, fast asleep. This seething, selling, begging, smiling, pleasant people, this mass of humanity spawned by a fertile soil, a hot sun and abundant rains, this race of courteous thieves and robbers, takers and getters. These altogether charming people, raucous robbers and all, in all their indescribable variety of form and colour, of clothes and caste, and race and religion. Men, hundreds of them, like beautiful bronze Gods; most shapely and beautifully coloured muscular bodies. The sweetest little girls, and the loveliest brown-eyed babies. Hartung and I were two whites in a crowd of jabbering slim-bodied brown men. They didn't gather round us, but we had to steadily ignore them or we should have been surrounded out of pure curiosity. Little kiddies crowded round us with hands held out for a penny—'Master', 'Sahib' and so on, to attract you. I think the outstanding human thing that remains in my mind is a myriad of outstretched pink palms, upturned for whatever 'Master' might be pleased to give. Hundreds of people have written all this down in their diaries and in published books before, with far more knowledge and art and insight than I have. Nothing I can write will really state the case in reality for those for whom this diary is being written. So I give it up.

From the point of view of economic and human geography, the experience of today has become a part of me and will never cease to colour and to affect what I say and do regarding man and his reaction to his environment.

12–13 September, 1931, London

It is seven o'clock Sunday evening in this great city and I sit down before our gas fire in our comfortable and cheap room at Endsleigh Street, WC1 and prepare to make a few inadequate entries in this diary. However, I must try to keep up daily entries, for the sake of preserving memories of this trip and for those for whom this is written, and to make up for the long letters we are not writing, and are not going to write.

Having been only thrilled by Colombo, and Suez, and Aden and Cairo, and Port Said and Malta and Granada and Gibraltar, I casually remarked to Peggy 'I expect a thrill also awaits us in London.' But I really did not think that cold, foggy, stodgy London could have much that was thrilling. For did we not already know Melbourne and Sydney! Ye Gods! Well, as I have said, it is useless to attempt description. It can't be done. Traffic! Strewth. It's marvellous. We ride everywhere we can on the motor buses, on the tops of the buses. And you whiz along at 30 mph and you get where you're going very cheaply, and see the street at the same time. Traffic jams are quite common. But they clear up in no time once the way is open. The drivers are miracles.

The London policemen and the London bus conductors are as wise and as pleasant and obliging as everyone has said they were. We ask scores of questions every day and scores of others we don't ask, and try to find our way about in this stupendous maze and whirl of life by means of things learnt from the guide books—often with disastrous results in long walks.

Home. We slept till eight. Up and about. Off to church at St Clement Danes. This was a rich and wonderful experience. Old Samuel Johnson always attended there. He was a bigoted, dogmatic old Londoner. I am more sympathetic with his opinion than I was. It must be a great thing for Londoners to have this great city as their own. It is a most stimulating place—stimulating thought and memory, and the climate is bracing and stimulating too. I have walked miles and miles today, and seen hosts of things, but am not tired. The evening meal, which is here called 'supper' on Sundays, is at eight o'clock.

13 May 1937, London

I sit here in the Imperial Hotel, at six o'clock, bright sunlight, wishing it was dinner time as I am very hungry, not having had anything since breakfast. But I see that dinner is not till 6.30 while lunch ceases at 2.30. In America, how different. Not only, from San Francisco to New York, is the food much more abundant, more varied, and more perfectly cooked and served, but the eating houses are always open and the eater is always welcomed and well treated. In this hotel management business, including lighting, plumbing, food, and service, the States (and Canada too) just leave this country standing. Compared with what we generally saw and enjoyed in the States, from California to Washington, the general conditions of London hotels is not far beyond the Noah's Ark stage.

24 July 1937, Geneva to London

We have wandered on the continent for seven weeks, from capital to capital, from school to school, museum to museum, crossing frontiers and passing terrifying customs, packing, unpacking, and repacking, changing our currency and customs. And we had never really had a REST. We left London seven weeks ago! Dieppe, Paris, Basle, Zurich, Ragatz, Leichtenstein, Innsbruck, Salzburg, Linz, Vienna, Budapest, Balaton, Graz, Villach, Cortina d'Ampezzo, Venice, Lake Garda, Verona, Padua, Milan, Chartreuse, Genoa, Nice, Monte Carlo, Grenoble, Geneva, Dijon, Paris, Liège, Aachen, Hannover, Hildesheim, Braunschweig,

Hameln, Berlin, Prague, Nuremberg, Rotenburg, Spangenberg, Hersfeld, Ziegenhain, Frankfurt, Geneva, Montreaux, Interlaken, Rhone Glacier, Lucerne, Basle, Calais, London.

Edited Extracts of Some of the Comments on Education

24 March 1937, San Francisco

Rang Dr Kemp, Head of Education in the University of California at Berkeley. Got from him the phone number of the City Superintendent, Mr J. P. Nourse. Took a taxi up to the Civic Auditorium and there had a great time with Mr Nourse, his assistant Mr Schmidt, and his director of publications Mr Mullaney. Most illuminating and interesting. They were questioning as much or more than I was. But we all four enjoyed the afternoon very much. The Easter holidays rather interfere with my visits to schools here.

They envy us our tenure of Director's and Superintendent's positions, our control of the Teachers' College, our powers of dismissal of insubordinate teachers. They said our Directors were 'entrenched autocrats' and that in these matters we were 100 years ahead of them. Am not sure whether I wouldn't be willing to see much more of those things go if we could in its place get such an enthusiastic public support for education as provides the school salaries, and the magnificent buildings, grounds, and equipment here. Our Australian education, much more so South Australian education, is miserly and parsimonious in costs. The Americans stoutly and doubtless rightly resent the idea that they are wasteful, but they do spend a much more generous amount per child, in all grades of education, than we do. This would not be a welcome statement to my lords and masters in South Australia, but it is true.

2 April 1937, Los Angeles

It seemed a never-ending road to the University (UCLA), a great place, building after building, thronged with the young life of America, a few Japs and a Negro occasionally. Japs very small and capable, Negroes rarely up to average. Dress seemingly careless and independent. Boys with jerseys and pullovers. 14,000 students. A vast staff. Visited the Geography, Geology and Education departments, and met and talked with many very interesting folk. Dr Frederick Leonard, Astronomy, authority on meteorites. President of the Society for Research in Meteorites. Wants me to write a paper for their June meeting on Australites. Recommended me as a Fellow of his Society for Research into Meteorites, only one of its kind in the world.

Met and talked with Dr Marvin Darsie, very alert and interesting, Dean of the Teachers' College, about their courses, selection, training, teaching, and placing. Most of their teachers placed in this county. Some in others. May get married without resigning. Dr Seagoe and her offsider, head of the statistical branch of the Education Department of the University. Very interesting talk also. Have notes and literature on all these educational things. Dr Kazuo Kawai, alert Jap, Oriental geography and history. The list of staff, showing just when they are lecturing and when they are free for interview (a very common thing) is like a department store catalogue.

3 April 1937, Pasadena

Taxi to Pasadena. Mr George Henck, who was hospitable and full of enthusiasm. We clicked at once, long talk on the Civilian Conservation Corps (CCC), education of adolescents, of apprentices, and of apprentice schools. Visited Pasadena Junior College, 18–20 years, first two quarters University, but more practical. No fees. Everyone here looks forward to 20–21 as the school leaving age. The fitting, turning, testing, electrical and other workshops excellent. Quite different from anything I had anticipated.

There had seemed to me to be little technical training in California but this was an illusion. They do not call it technical education, but in their secondary schools there is a most important, widespread, and valuable extension of what we would call technical training. For instance, the aeronautical courses in the Pasadena Junior College (18–21 years) are remarkably well equipped, very closely connected with the aircraft industry, and have the best instructors the craft can provide. They are making a full size all-metal (ALCLAD) plane to fly, tested and tried throughout, every stress designed and calculated here by the boys and masters, dies, rigs, moulds, patterns, everything. Far better workshops in this junior college than in the University, the School of Mines and all the other technical colleges in South Australia rolled into one. Had a long talk and inspection of these workshops and facilities and talked to the lads working there.

9 April 1937, Santa Fé, New Mexico

An extraordinary position can exist in an American State education department as a result of the appointment of State and County superintendents (directors) being a matter of politics. In a long and interesting conversation with Mr C. de Baca, Assistant Superintendent here in New Mexico, who has legally exactly the same powers as Mr Rogers the Superintendent, I learnt how he got his job. He was of an influential Spanish family (dating back to the conquistadores of these parts). He has never been a teacher or administrator. Like Mr Rogers, he supported a Democrat, Mr Solid Smith, and when the Democrat representatives got in last election, he was appointed to his job, also Mr Rogers. Thus it may be that superintendents are more interested in politics than in education. For instance, there were no real biennial superintendent's reports issued until four years ago. Señor C. de Baca is strong on the statistics and he is anchoring the department in that way.

I found that this State has much in common with ours, but their problems are vastly different. I find, too, that States vary. As much as California is ahead of us in nearly every way educationally, so is New Mexico behind us in very many ways. In California the leaving age tends towards 20–21. Here in New Mexico, according to C. de Baca, the child is compelled to come to school at 6. 'Beyond that we have not much power.' It seems that there is no upper compulsory age. The Indian villages have their schools, as I saw at Taos (Towse) but are controlled from the Department of the Interior in Washington. The Mexican, some call them Spanish, and the Penitente folk, have schools. In all the primary language must be English, primary and high. There is no separate thing called technical education, though even at so small a place as Taos there was an Industrial Arts School. I imagine that in larger places like Albuquerque, Santa Fé, Gallup, and Raton the American boys and girls have just about as good educational opportunities as they have in California, but in the rural districts it is far different. The law of certification of teachers gives them some hold but does not remove teacher appointments from political control. Here, as in California, they were entranced by the scheme in operation in South Australia, and I was much cross-questioned as to our attitude and position during or consequent upon political changes. I admire almost everything American. I think, however, that the matter of political patronage is as utterly rotten as it can possibly be.

14 April 1937, Chicago, Illinois

Up early. Set out to go to the University of Chicago, a vast and beautiful series of buildings, away to the southern part of this vast city. Got straight to Professor Judd's rooms. The girl clerks are always very pleasant, intelligent, and helpful. Went to his office. The usual question, 'What aspect of education interests you?' is always embarrassing, because the whole gamut of matters interest me. However, I was ready for this question this time, and fired it at him. All my special queries: tendencies of secondary education, attitudes towards vocational education, place of cinema and radio in modern school, technical schools, apprenticeship, rural education, the part education is playing in the field of unemployed youth, Civilian Conservation Corps activities, the effect of the depression on educational activities, and of the 'passing' of the depression. By the way, in this vortex of abounding wealth, they still consider they are not out of the depression period. Perhaps they regard the boom days of the pre-depression as normal and await their return.

It appears that the Army authorities were given the organization of the Civilian Conservation Corps, provisions, camps, food, etc. Then authorities from the labour department were put in charge of the general organization. Neither of these cared for the educational aspect. And indeed they urged that a large number of the Civilian Conservation Corps lads were at the best poor educational material. However, the educational authorities pressed for some part, and an educational adviser is appointed in each case, but he has no authority except to deal with boys who voluntarily come to him and ask for educational help. Chicago is the centre of the Fourth Army Corps Area, and the head of six or seven Civilian Conservation Corps camps. The officer in charge is an old educator. So matters educational may be a little better off here.

Teachers' colleges are State institutions but not under the Education Department. Re political matters, there are no less than 32 Superintendents elected by popular vote! But in advanced places such as New York the matter has been overcome; there the Superintendent (Commissioner) practically has a life tenure. He is appointed by a Board of Regents, representative educational men of high probity and no political colour that need obtrude. Possibly the more backward States will slowly come up to something like this.

Visited two very fine schools, an elementary and a high school, both laboratory schools of the University of Chicago. Wholly within the control of the faculty, each with 400–450 scholars. Children selected by the fact that there is a fee rising from $150 for the kindergarten. Children come aged four, two years kindergarten, six years elementary, three years high, and two years Junior College (senior High) completing at age 16+. Saw through the whole of both schools. They are certainly superior places. The buildings are well lit, warmed, and ventilated. The classrooms are quite unlike ours, more 'atmosphere' in them, more equipment, more furniture. The science, botany, zoology rooms were like nothing I've seen except at Harrow (in England). The kindergarten was delightful, and I feel that there was more real joyousness, earnestness, friendliness, and freedom than in most of our schools, perhaps more than in other American schools. Headmaster of elementary school, Mr Gillette, very charming and was stopped and spoken to by children in classrooms and corridors.

Then the High School. Most impressive were the dancing rooms, the gymnasiums, the gorgeous natatorium and the girls' class therein. Shower feet etc before entering. Water continually replenished and twice treated. Clear as glass. The school libraries remarkably good. Also the main library, a revelation of earnest application to reference and study books. They read much more than we do. The class room libraries are larger than the main library in some of our schools. The fact is we haven't begun to learn the elements of library education in our home country. The manual shops are less standardised than ours. For instance, there is a'power unit shop' in the high school, a large place with all kinds of engines, electrical, gas, internal combustion, aeroplane, motor, and so on, merely so that they may each have an adequate understanding of the motor car they must drive. Pottery very good. Excellent kiln also. Smithing, cabinet making, wood turning, benchwork, fitting, turning, but they explained that their shops are not well equipped. They are far beyond Thebarton Tech., but our folk wouldn't stand for this, wouldn't send their boys up to 5 years high school because we have too poor an ideal of education.

In the days of the depression and since, the taxpayers' associations have violently attacked education. McCormick, editor or owner or both of the *Chicago Tribune*, is a violent antagonist of education. So that position is much like ours. But here the middle wealthy folk and the poor folk have a healthy belief in education and they sway the matter. The 200,000 Negroes get fine new schools, both here and in other Northern cities.

16 April 1937, Detroit, Michigan

Re the political control of public service appointments, which is so general here, and so dangerous, one wonders how a nation that grew largely on England's ideals could have adopted a scheme whereby Superintendents, Police, Judges, and other folk in public authority should be appointed according to their political party. Professor Judd explains it thus: About the time most of the States came into being and when their constitutions were being written, there was much mistrust and suspicion of public servants. The farmers, in particular, did not believe in them. So, in their excessive idea of the value and importance of the popular vote, they placed all these matters in the hands of the parties or people they (the voters) elected. So there you are. Leading and important centres such as New York have got away from this.

Took a taxi to the Cass Technical High School, recommended by Eltham, where Mr Allen was the Principal. I had heard of million dollar schools, and had seen the Frank Higgins carefully and others less so, which I think are in this class. I have never seen anything so fine and well-organised and smooth running as this school. It was most delightful. Mr Allen took me to the top floor and worked down. Building cost $4,000,000. Equipment $1,000,000. Total = £1,000,000 English! They got their value. It was all planned out by this very wise gentleman, Mr Allen. Had a delightful time. The highlights were the general set-up, the locker system, the spaciousness, the originality, the senior students, the art classes, the electrical, aeronautical, motor, woodwork, mechanic, etc, so with physics, chemistry. We could not have such a school in Australia. It is a part of a social, economic, industrial, and financial system which we have not got. I have an excellent handbook, and I made some additional notes. Indentured apprenticeship here but little, mostly unindentured. 'Beginners' are taken from Cass. They place 300 every semester. These boys (and some girls) rise to the zone of skill and pay between the craftsman and the engineer, such as foreman, laboratory assistants, salesmen, and shop executives. If only our employers at home would realise that salesmen and directors should have a personal knowledge of processes! For every one university man, they need ten Cass-style men, and an army of skilled craftsmen and operatives. This Detroit area is one of high technological efficiency, no guesses, no trial and error, they must know.

26 April 1937, New York

We talked with Mr John Russell at Carnegie Corporation offices, 552 Fifth Avenue. He was pleased to see Peggy, thought maybe that I didn't have a wife at all. We had an interesting chat about schools in America, and about American life and homes and children, snobbery and 'snootishness'. They have private secondary schools that ape the 'great public schools' of England, classic and snobby. But all evens up when they get to the University. He suggested many things (not schools) that we should see in New York.

28 April 1937, New York

Shown in to Dr Lee. He is a Berkeley professor, first lent to San Francisco for Superintendent of Education, then lent to New York for this National Occupational Conference (NOC). Fine man. Had good long talk with him. Found out all I could about the NOC, also the Civilian Conservation Corps (CCC, the US Youth Unemployment scheme), and the Works Progress

Administration (WPA), all doing so much for the unemployed and the unskilled.

The NOC was set up four years ago, by the Carnegie Corporation, the reason being that American youth was in a dilemma. Something had to be done. Seventy men and women form the conference, which meet for three days or so once a year, with an executive committee of six or eight. These folk serve as a clearinghouse for all experiments, research and publications dealing with occupational adjustment, so that teachers, and vocational guides and counsellors might have systematic and authentic material to help them. They do not teach. They coordinate and correlate. 'If the governments would do what they should, the NOC need not exist.' They stimulate the government and educate public opinion.

There is a Technical Committee in addition to the Executive Committee. One project is to take 14 leading school superintendents (collected from Seattle to Providence), have a two-weeks tour of selected places, with an implied obligation that they will put into operation the best of what they see. Twelve days travel, then two days discussion and conference in some quiet town. With this there are on four occasions a National Hook-up, that is, a USA-wide series of addresses on these problems, to enlighten public opinion. He also told me more of the CCC and gave me a letter of introduction to Mr Oxley, the educational officer for CCC camps, whom I shall see in Washington.

29 April 1937, New York

Got a taxi to the Metropolitan Vocational High School, which is a secondary school, free, compulsory, and secular. Mostly Italians, for the school is in a crowded Italian residential quarter down the East Side, just near the foot of Brooklyn Bridge. I spoke to two Negro boys, one from Harlem, one from Brooklyn, miles and miles away. In the later years boys and girls may specialise in printing, auto mechanics, cafeteria work, grocery (have a regular shop, two cafés, etc), plumbing, dentistry (saw boys 15+, first year, making crowns on sets of teeth and so on), building trades, fitting and turning, radio and electrical.

Then they may, in their last years, go to a separate division, the Printing School, the Maritime occupations, Needlework, Homecraft, or Fine Arts. In the evening the same classrooms are used for adult and apprentice evening classes, from 5 to 9.30 pm, also free. Noted again the freedom of these schools, the interest, the friendliness between instructors and scholars (much above ours in average age), and the variety of set-ups in the various workshops and craft rooms.

3 May 1937, New York

Having seen much of American schools and schooling, and the young women and men in large numbers, 80–90 per cent at school until 18 to 20 years old, I think I understand a number of American ideas better. They have a different background from us. They are more emotional and more friendly and kind. At General Motors today, with their young staffs of executives, I wondered to what extent their outlook was different, as men, because for generations these young people have lived longer under school methods of discussion, of emulation, reward, acclamation, etc. Perhaps they continue to evaluate things in those terms in their adult life, rather than get the different points of view that we think are proper to adults. Perhaps we're wrong, and they're right. Are these young men and women essentially different and more schoolboyish (or girlish) and so on than folk in our normal adult workaday world?

Another remarkable fact is the large part the Universities play in things. Executives everywhere are sticking close to their Universities and thinking in terms of academic men, books, etc. A final thought from my readings and wanderings in New York City and in USA is the enormous influence upon American life, reaching almost everywhere in visionary and idealistic matters, of those two men, Rockefeller and Carnegie.

26 May 1937, London

Went to Broadcasting House and met Mr McGregor of the Empire Broadcasts. Learnt of his schemes and methods. Felt that we should have a short-wave at home. Told him of what we could do so far as Australian schools are concerned, the Federal Schools Broadcast Committee, means of overcoming time difficulties, etc. Took several notes that will be helpful. Down to Mr Cameron, Secretary to the Schools Broadcast Committee. A past Director of Education. Very agreeable and helpful. Explained how the whole show works here, number of listening schools had increased enormously during past three terms. Explained how the schools cooperate with the BBC; the Committee decides on the personnel, the talks, the talk content is supplied by the broadcaster, but is edited and improved quite a lot.

Their scheme seems to be just the opposite of ours. No teachers. Get experts of all sorts, edit and arrange their stuff, and encourage schools to get this additional glimpse of things outside the curriculum. Aim to supplement the teacher's work by something the teacher can't do himself. Must have the expert. Initially there was much apathy, and much difficulty re reception. Now increasingly good.

27 May 1937, London

Thanet Street School, Mr White Headmaster. The school is a small one; kindergarten to senior, only 109 scholars all told. Headmaster and five teachers, though the staffing basis is nominally 40. Went round the classes with Mr White. Met and spoke with the teachers and spoke to one grade of boys on geography. The class I heard and saw, Grade A of Senior School, listened to a broadcast in the Regional Geography Series by Dudley Stamp. The pamphlets were excellent. That's where the lecturers put in their work. These are the charm and the success of English school broadcasts. Teachers prefer not teachers, but experts, folk who know something they don't know. Dudley Stamp spoke for 20 minutes, very slowly and simply, as a good teacher, of his and his wife's journey to Buenos Aires and across the Andes. Simple facts, references to the maps and photos in the book (which the boys in the class turned up quickly and readily as the talk went on). No statistics or anything heavy to remember. The teacher made no preparation, beyond handing out the books and atlases. Nor any close, except to ask the boys to write a composition on South America. We try to get our teachers to do much more, alert the whole lesson, with maps and chalk and blackboard. And I think that's a good idea.

4 June 1937, London

Strand School, Brixton Hill, with Mr Dawe, Headmaster. Strand School grew up in the Strand, from casual evening commercial classes, and was transferred away down to Brixton Hill, when all about there were open fields 30 years ago. Seems to me nothing is ever planned in England. Everything just 'happens', grows up, like Topsy, haphazard and undefined. A complete contrast to the more expensive, more scientific, more systematic work in USA (Civilian Conservation Corps etc). Something to be said for both. I think, if funds can be secured, the correct compromise is in the direction of something that meets a definite known demand (English scheme) and has been proved in practice (English scheme) but is then systematised, properly organised, and extended for all those who come within its definitions (American plan, which is more scientific, more democratic, and more costly).

To return to secondary education. I saw something of English secondary schools in 1931, during a very thorough look around Harrow, and a more superficial visit to Eton. And I have seen all types of elementary, central, and technical schools here. But today for the first time I saw a State secondary school. It was a picked school, rather good locality, quite a decent building. The Headmaster is an alert and doubtless very capable man. But the school itself is dead, stuffy and smelling of the ghastly tradition of classicism and the 'great public schools' with all their frowsiness. The secondary schools I saw in America were immensely ahead of what I saw this morning. Strand School was proud of its library, which had many books, and was a lending library. But they haven't the slightest idea of using a library as a school activity, as it is every minute of the day at the Horace Mann (New York) and the University High School of Chicago. They are absolute back numbers, losing valuable possibilities and they don't know it. I feel sure England would scorn to learn anything from America. Bad as we are in Australia, we have that one advantage over England.

The Headmaster agrees that the school is bound hand and foot by the conditions of the public examinations of the University of London, which he regards as an absolute blight upon the whole field of secondary education in England. All I have seen, both at home in Australia, and in America and Europe, incline me to a primary education 5–11 or 12, two years general preliminary secondary education of the Junior Technical type, and then a selection into high school and technical schools of various kinds, not according to a literary examination, but on a basis of desire, school record, and potentialities of employment. Strand School has good standard labs, chemistry, physics, and nature study. A dead museum, a moribund library, unattractive classrooms, a good assembly hall and organ, fine playground with gymnasium being built. If it is general average of good State secondary education in England, then it is 50 years behind the same kinds of schools in America or Victoria.

2 July 1937, Hannover

I had two letters that were of the utmost value in opening doors in Germany. One I had obtained by my request through the Consul General in Australia and was a permission from the *Reichminister* to visit schools. The other and more powerful was a letter from the Chief Finance Minister of the Reich, Dr Hjalmar Schacht, approving of my visiting Labour Camps and *Landjahr* camps.

Today I visited three schools. First the Hindenburg Schule, Anderteusche Weise 26. Magnificent hall. Hobby rooms, planes, gliders, zoology and biology laboratories, physics, chemistry, gymnasiums and classrooms for over 100 students. An *Ober Realschule,* for 10–18 year-olds, a very fine institution. Then to an apprentice school. *Fach Schule,* Kleine Duven Strasse, an old primary school made into a trade school. Here we saw the workshops for the cooks, for the woodworkers, and for the electrical workers. Cooks get two days off, one for school, in their employer's time. Ditto others. The lads were young but very earnest. The teachers were good tradesmen, also very earnest and anxious to show and tell all they could, even to the foreigner from Australia.

3 July 1937, Hannover

To the *Fachschule* again. The *Ober Direktor* took us around. These men speak only German, but when they speak of work and equipment and tools and operations there is but one language. And with Wolfgang Fenner [a relative] to interpret here and there, I got on famously and collected valuable information, the more so here where apprenticeship is so highly considered, where technical education is so long established, and where there is such a tradition. They do not call these schools trade schools, but professional schools. This school was for smiths, motor car mechanics, hairdressers, and dentists. I was most impressed by several things that I have not seen in technical schools in Australia, America, France or England.

There are always planes overhead here, and there are always numerous men about in uniforms; Army, Navy, Air, Labour Corps, Storm Troops, Motor Corps; many honorary, many official.

5 July 1937, Hannover

I was taken to see young men at work in four or five places, draining the moors, preparing the land, changing and widening the courses of streams, and then to four separate camps at widely different districts and countrysides, Neustadt, Hildesheim, and Braunschweig. I was shown every detail of the camps, and every query was answered, and I met all the Field officers, and saw the men at work, at play, in hospital, eating their camp food, and saw them at their drill with spades, and the changing of sentries, and the vegetable gardens, and the sanitary arrangements, and the issue and repair of clothes and tools, and the library and books, and the kitchens and cooks, and every bedroom.

Every boy in Germany who is not medically prevented must attend these camps for six months. Wolfgang [Fenner, a relative] did and enjoyed it. I had been told it was a remedy for unemployed youth. But it is not. It is totally different from the Civilian Conservation Corps of USA. There is throughout all these camps a strong military discipline, but everyone is happy and every room is named and decorated in a way that shows individual pride in their work and joy in life. The badge and symbol of the *Arbeitsdienst* is well chosen: the spade and two ears of wheat, digging, draining, clearing, planting, to promote and increase fertility. Germany, they will explain, has no colonies. Therefore she must make complete and full use of all the land she has. There must be no moor. There must be no flooded areas.

To the Australian mind, and more so to the American mind, there would be the mental reaction that this is militaristic. So it is. But these camps are not to make soldiers, for every man later serves in the army. They are not to relieve unemployment, although in their birth and origin that was one of the reasons for their establishment, but that exists no more. They are a part of the education of German youth. I questioned some of them, they were men 18–20 years old of all degrees of wealth and occupation, merchants, mechanics, teachers, students, labourers, wood cutters, clerks, every type and condition. For six months they eat, work, sleep, sing and drill together. They also have instruction in the principles of National Socialism, that also is a part of their education.

Although I have said that Americans and Australians might think these camps military, it is true also, I am convinced, that the real spirit and effect of all these camps is in accord with the most treasured ideals of both those countries, namely to teach equality, to teach the labourer to know and to respect the merchant and the student, and vice versa. The whole impression is one of thoroughness, efficiency, unity, and joy in work, also of Service. The movement is embodied in its name, *Arbeitsdienst,* Service through labour.

Another impression I got was that to me, a foreigner and a stranger, everything, from the headquarter depot to the last point in work or play or housing or administration or equipment, everything was openly and freely and frankly shown and discussed. The officers and men believe deeply in the movement, are intensely proud of their organisation and achievement, and are anxious that anyone who is interested should have every opportunity of knowing all about it.

8 July 1937, Berlin

Handwerkerschule der Reichshauptstadt, Charlottenburg. This great five-storeyed brick building was specially built in 1900 or thereabouts for the'professional' (that means trade in our language) school work for apprentices. The great forward move in concentrated technical education in Germany appears to have come in the late 1870s after the nation had settled down into peace following several wars.

First I went to the Technical School. There are departments for Architecture and Room decoration, Theatre scene painting, Painting generally, Masonry and sculpture, Gold and silversmiths, Pottery and ceramics and Artistic smithing. These German technical schools are very massively and 'deliberately' built; they don't look like schools, and have but a small sign by the door. The equipment in all cases is excellent, not halfhearted. The instructors, whether in art or trade (profession they call it) are of very high standing and the heads of departments carry the title of Professor.

In the schools and workshops all the materials are insistently German, for, they say, we have no money to buy goods from other lands. Also, I think, there is the belief, pretty well founded, that anything they want can be found or made in the country. There is nothing unpleasant or belligerent in their attitude. It just comes casually now and then into the conversation. The quality of the work in this school was remarkably high. They work with good well-prepared material. Also the long hours, seven hours a day for six days a week, 42 hours compared with our 25 or less.

It was clear to me before I left Australia that everyone who wished to visit German schools must first inform the authorities of the Reich, and must get permission. That suggested that they were reluctant to show their schools, etc. This is quite a wrong impression. Everywhere they ask for, and receive, and welcome foreign students (there is a Technical School Guide, printed in English). Berlin is not a weaving or textile centre. But the school here is wonderfully complete, and has students from all provinces of Germany and from overseas. So I saw all through the *Textil und Moden Schule*. And that alone was an education and an inspiration. One idea is: Get teachers and leaders that know their jobs, pay them well, and give them the freedom and the equipment they need. They will deliver the goods. I recall particularly the remarkable equipment, three enormous workshops, for weaving, all manner of machines and looms. The history of textiles. The making of ersatz fabrics, *vistra* more like cotton, and *wolstra*, more like wool, and, of course, artificial silk. There they all are, and all their mixtures, also with wool and cotton in some cases, for the *vistra* and *wolstra* are not yet perfect. But they certainly do make beautiful fabrics.

One marked also the insistence on hand-work, in this machine age. They have the machines, but they do insist upon the importance of individuality and upon personal handmade work; smiths, masons, printers and weavers. The fashion design was enough to make one's mouth water, the thoroughness, the abundance of materials, the research in old designs, the school of theatrical design, the going back to Nature, copying and conventionalising native flowers, the beautiful delicate designs, the methods of working them up, the phosphor bronze plaques, the making of the materials, the mannequins, the fittings, the designs themselves, the cutting and fitting, all these in separate departments of the school, but all the girls (or boys) go through them, and all get work in the various factories of the textile industries.

In both these great schools there is the full time day school. In the same rooms there is also the evening school for those at work. The textile school with all its equipment and teachers and maintenance has but 250 or 350 evening students! But nothing seems to be spared to do the job well. Technical education in Germany is quite a different thing from technical education in England, again from the same thing in America.

9 July 1937, Berlin

Höhere Graphische Fachschule. The usual fine equipment, independent of the number of students, the chief point being that whatever is taught must be taught well. There were departments of typography, lithography, photography and reproduction, book-binding—the most beautiful books I have ever seen so far as covers, binding, and lettering are concerned. Also advertising art and cartography.

To indicate the thoroughness of the courses: cartographers, who are to be mere drawers of maps, must also study physiography, and they must go in the field and actually survey an area, make model relief maps, and so on, so that they thoroughly understand just what they are doing. Everything else was just as thorough. There are day-time courses, full-time, for about 250 students, to become commercial artists etc. And there are evening courses, four hours per evening, for tradesmen and others, about 350 of them. The day students, many of them girls, are from 16 to 20 years old.

Here a tradesman is first a *lehrling* (a learner), for four years, then a *gehilfe* (helper), for five or six years, and then, if he passes a stiff exam, a *meister* (master). These *meister* exams are hard, says the *Herr Direktor*, and will be harder. In the first year there is, each week: drawing 22 hours, typography 8 hours, geometrical drawing four hours, and lettering eight hours. Total 42 hours per week. In the second year much the same, up to 45 hours per week. The third and fourth years are stiffer in their requirements, but not longer in time.

16 July 1937, Rotenburg

Every German is not the devout worshipper of every aspect of the New Germany, as is generally thought outside Germany. I met and talked with several people who were very plainly spoken about things. In their logical German way they say of certain laws and movements, 'It is good, but it is bad, too.' Or 'That is bad, but it is also good in some ways.' I think the chief resenters are church folk. I talked with Jews there and with rebellious-minded people, but with them all is a strong love of Germany, and no hate or smouldering dislike of things. Just opinion.

19–21 July 1937, Geneva, International Education Conference

Sixth International Conference on Public Instruction, under the auspices of the International Bureau of Education, Geneva. I was given a seat at the top of the table. On the right is Mohammed Haidar Khan, who represents Afghanistan, then three Germans, including Dr Grafe, whom I had met in Berlin. Opposite me is the representative of Ireland, and on left was M. Piaget, the head of the Bureau. And there were representatives from Iran and Equador, and Colombia and Venezuela, Poland and Argentine, France and Iceland, Latvia and Esthonia, Spain and Portugal (though they are next door to being war enemies at the moment) and many more. The speeches are in English or French, and the Secretary, Miss Butts, translates them in a truly remarkable way.

The reports were very interesting; Afghanistan, Spain, Denmark, Poland, Turkey, and so on, with troubles just like ours, and some different. The Spaniards for instance, who presented their report much better than anyone else to date, are engaged in a fierce war. Yet there were three delegates telling how, during the year, the amount of money spent on education had been increased, and the number of schools also increased. China was interesting with its account of 13,000,000 school children! Judging by what I see I have more notes than any others. Walking up with the British delegates this morning, they remarked on the international tensions that were around the table, not shown, but sometimes suggested. There were Bolsheviks and Nazis, Japs and Chinese, Spanish and Portuguese and so on. And all on the edge of great dangerous potentialities.

It has been most enlightening and stimulating to me to meet and to hear all these diverse peoples, with their varied problems, their differing geographical environments, and their political points of view; all races, political creeds and religions. Even to hear that another such man's problems were the same as one's own was very interesting. And how they were handling it. And how many countries, long settled in educational work, whom one would think would be stable, all engaged in 'reforming' their secondary and middle schools.

27 July 1937, London

This evening I sat down and wrote two rather long *Australasian* articles. It is very strenuous, after a long day. But it must be done, as the *Australasian* articles must be nearly run out, and I do not want to break the unbroken record of continuity of over 21 years.

Interviewed Mr St John Wilson, of the Instructional Centres of the Labour Ministry. Got some more literature, and have arranged to go to a city (Park Royal) and a country (Culford) centre, the first a Government Training Center (GTC), and the second an Instructional Centre (IC). Talked with Mr Wilson the whole afternoon about these things. The interesting point is the extraordinarily different way that almost the same problem is tackled in USA, Germany and England. Here (because of prosperity and abundance of employment, I think, especially since the re-armament phase opened) the scheme is much less than in either of the two other countries. Here are only 8,000 total, compared with several hundred thousand in each of the other two. But it is very interesting, the more so as it is likely to be the way that will appeal to the taxpayers of Australia. Here, as elsewhere, some effort was made for the girls, but with little success. Schools for domestic training are established, parallel with the boys' camps, but it is hard to attract girls to them. Indeed, all the boys' camps are at the moment a bit under their complements. A good deal of care is taken to get a favourable public opinion, but the interest is not so widespread as in USA, I think.

The Heads of a GTC and of an IC are selected public servants. The cost of an IC building and equipment is about £15,000. ICs are non-residential and technical, six months training, and are in and near industrial cities. GTCs are in the country, forests, roads, drains, etc, the period here is 12 weeks only, reconditioning and recuperation. Rag and bag places. The best lads are sent on for IC courses. The improvement noted in health, weight, and physique in these camps is remarkable, I am told. There is a wide range of age, 18–45 years. There is a selection panel, intelligence, adaptability and medical fitness, but this is easy to pass. There are 3,200 in ICs, 1,170 in summer camps, and 400 in non-residential (farm and local) centres = 5,000 total. About the same in the GTCs.

28 July 1937, London

Government Training Centre, Gorst Road, and Mr Smeddle, the Principal. Spent the whole morning and until well after lunch going about the shops, and having lunch in the school (they have a waiters' school as part of the outfit). Mr Smeddle was one of the divisional officers in charge of the Returned Soldiers' Vocational Training here. In England this scheme was called 'the Ministry of Labour's Independent Training School for disabled ex-soldiers and soldiers'. This soldier's scheme did not fade right out in England, as it did in Australia. The incidence of the Depression was different. There was the coal strike of 1925, and the General Strike, and in 1931 the Depression was worst. So it came about that while there were still some ex-soldiers training centres (for disabled men) in being, a new movement was commenced for fit men who were unskilled and unemployed, or whose'skill had lost its marketable value'. So the peculiar term, the Training of Fit Men (TFM), still sticks.

Mr Smeddle has been continuously on this job right through. It is, in effect, just like our own Returned Soldiers Vocational Scheme, with non-residential places, trade workshops, weekly allowances, six months (26 weeks) courses, and so on. The place is generously housed and equipped for 650 men. Brick and steel with factory lighting. All the trades to be seen in one sweep of the eyes, except the sheet metal workers and panel beaters, who are noisy fellows and must be kept to themselves, and the waiters who are attached to the dining rooms. Work is from 8 am to 4.30 pm. Discipline excellent. Every man, 438 here now, gets one good hot free meal at midday. The Instructors are all selected tradesmen, and are enthusiastic and successful. The head gets £800 per annum. They are under strength, can't get enough men wanting training, and can't supply all the demands for men that they receive. They place their men easily, 95 this month, and most of them do well, charge hands, foremen, managers, etc. There is a Selection Panel, with one man from the department representing the employment section and one representing the training section, which interviews the applicants, who have often been sponsored and directed by the officers of the employment bureau. Each lad has one hour refresher work in arithmetic etc every day. I saw them all working steadily at this as if welcoming it.

The classes I saw were: fitting, turning, tool making, cabinet making, upholstery, motor trimming, coach painting, electric welding, oxy-welding, sheet metal working, metal spinning, motor mechanics, hairdressing and hotel waiting. This is the nearest yet to the kind of thing we shall have, but there is still much we could learn from the vastly different USA and German schemes.

25 August 1937. London, Broadcasting

Went to the BBC. Had a very entertaining and profitable time. Saw all the ins and outs of the editing that is considered by them to be the chief factor in broadcasts. The principle is 'Give nothing that the teacher can give.' This means experts, explorers, adventurers, famous men, etc., folk not used to teaching nor trained in broadcasting, but people with first hand original information, not something stewed up from a book. All are required months ahead (the Spring addresses are now being galley-proofed nearly nine months ahead!).

They are first told the type of thing that will interest. Then they submit script. This is edited, especially the opening to arouse interest. It is colloquialised, abbreviated, simplified. Saw some scripts and their notes and editings, much tact required. One of Sir Hubert Wilkins' (submarine), one of Dr Mackay's (excavation of ancient cities) and so on. I realise the importance of this. Means a long range selection of topics, and a staff for editing, selection and printing of pictures. Seems to me we try too much, too many topics, too many broadcasters. Schools will only listen to certain topics or a certain number of lessons. 'Atmosphere' (effects) used in almost everything, but with restraint, easily overdone, but necessary where possible. Most school listeners assumed to be 12-14 years old. Another good idea for criticism is the appointment of a special panel of headmasters or teachers, for a month, to frankly criticise all the broadcasts of a certain topic. This is good. Most reports favorable and friendly critical, some outspokenly severe—all helpful. Must get frankness here. Mr Williams assumes that frank criticism would be difficult to get in South Australia, as the Department controls the broadcasts. Maybe.

Comment by FF: As mentioned in the Introduction to this chapter, the purpose of the 1931 trip was to attend the Centenary meeting of the British Association for the Advancement of Science (BAAS) and Father spent most of his time in the Geology and Geography Sections. By chance, he was in England at the time of the 1937 meeting of the BAAS in Nottingham, and used the opportunity to spend most of his time there in the Education Section.

2–6 September 1937, Nottingham, BAAS

H. G. Wells was President, Section L, Education. His Presidential Lecture was great. It was largely to hear and see him that I came here, and stayed long enough to do so. And it was worth it. He gave a slashing address of one hour and ten minutes. He is just like his photos, and in the life he has a whimsical humorous expression, as if that was his characteristic way of looking at things. Confined himself to the information about the world that a child should have, and the development and growth and presentation of that information. Great stuff. Stimulating and provocative. Said that looking around on the people he concluded 'that what they think and know and what they are ready to believe is very poor stuff'. Spoke of the difficulty of getting teachers, most of them should be superannuated or re-conditioned. Said that most of our teachers, like our average doctors and lawyers, were a mediocre lot. And so on. It was good to have heard H. G. Wells at his best.

At Section L there was a discussion on the function of the University in Education, followed by one on the Relations between Technical and General Education. H. G. Wells was chairman, and sat there with downcast head most of the time, looking somewhat tired, and either bored or quizzical according as the speaker interested him or not.

Saw and talked with Dr Grafe, whom I had met in Geneva, and whose address on the aims of German education had aroused a great deal of interest and some opposition here. There is, by the way, a curious and often expressed friendliness with Germany in these north central counties. Doubtless there is some underlying reason or tradition.

To go back to where I was in these notes, I want to record something of the interesting discussion on Technical and General Education. Knowing the title I had thought up a few things that I should like to say, such as that Technical Education was the best kind of general education and so on. But all the things I wanted to say, and many more that I should like to have said, were said ever so much more cleverly and forcefully by the three speakers. I could not help thinking how our remoteness affects these things. Having now heard these men I am armed and reinforced to say the things I was already thinking. They in their turn said them well, that is, with special ideas and turns of speech that they had got from hearing a score of other men talking on the same topic. I am satisfied that any brilliant speech that one man makes is based on 100 others that he has heard and discussed. In Australia, each man is as a voice crying in the wilderness. He never hears any other speech on his topic but his own.

9 September 1937, London, Cinema in Schools

Last day in England. I had an appointment with Mr Oliver Bell, Head of the British Film Institute, Great Russell Street. We had a long talk about the film business. It is going well, the Educational Film he said is 'a rising market'. He gave me much informative literature, but what is said is always more meaty than printed reports. The Board of Education has become definitely sympathetic. Whitehall now believes there is a definite place for the cinema in schools. For instance: in their now famous *Handbook* there are paragraphs on the use of the film for Geography, History etc., and they will give 50 per cent grant towards the cost of projectors, local authority provides the other 50 per cent. The authorities do not specify projectors, but 16 mm is the preferred size. The majority of schools, on the basis of cost, prefer silent projectors, in one census 197 were silent, 72 sound.

The Film Institute advises to get a sound machine capable of taking silent films. The cost in London is about £100, £94.5 cheapest. The best machines are American. Very good ones for £120–140. The best on the market costs £300 English, and is being used by wealthy schools, museums (Bristol for example), Universities, and local news theatres.

The idea of 'Film Appreciation' is stressed. Here and there are clubs and societies founded for that purpose. In all Adult Education circles they work towards film appreciation, something resembling dramatic criticism, plus movie technique. Schools are held to show teachers what can be done with films, also the making of films. Re subjects of most value, all depends on the enthusiasm of the teacher. Sound films help in the teaching of music, languages, and history. It is insisted that with all films there shall be included teaching notes, but the full technique of teaching is admittedly yet to be developed.

25 September 1937, Cape Town

Mr W. H. Hemer, Principal, Cape Technical College, came for us. I went through the printing department and the mechanical engineering department. It works under the Union (Federal) government, and is governed by a Council, the chairman is a medico. The principal gets £1,000 per annum. Senior heads of departments get £750. There are 5,000 of all sorts of students. The upkeep is £34,000 from the Government. Several thousand in fees, and that leaves three or more thousands still to be made up from 'subscriptions', and Hemer complains that he is expected to go around cadging for money to that extent. The building is an excellent one. Quite good equipment, but could be much better considering the population and status of the place. They even asked my advice re their equipment, and I feel sure from their eagerness to say how that appealed to them, will follow it.

The story of the rise, progress, difficulties and tendencies of Technical Education is very much (but not quite) the same here as elsewhere. A late growth, not coming from the Education Department, forced upon the Government by pressure of private bodies and practical men, antagonistically regarded by both university folk and secondary school authorities, both of whom see in it work they think should be theirs or which they think they could do better. In some cases they are right. Coloured people not admitted, but (because this is the Cape, not Johannesburg) there is a separate school for them. But the Principal admits that it is often difficult or impossible to draw the line. He then refuses admission and if they protest, the onus is on them to prove that they are 'European'.

This is a land of upside down and topsy turvy. Here a Kaffir is not a 'coloured', and the third or later generation of white born (if without tarry admixture) is by law 'European' but neither he nor his parents may ever have seen Europe. In this hotel there is a notice 'This bar for Europeans only'. But the people meant are white native Africans. The hatred and loathing and scorn for the 'coloured' people here (who are a darkish and yellowish bastard mixture of English, Portuguese, Malay, Hottentot, Kaffir) is beyond belief. And they tell me that they are treated much better here than are the Kaffirs of the Transvaal. I am sure these people and the black people get something very far short of justice. But the gentlest women I have spoken to are just as keen and hard against them as the old hands from the Rand. It is an attitude that has got to be learnt. I expect its origin lies in two things. The only way to keep the blacks and the coloureds in their place is by hard definite fear, and it is essential to keep them in their place because of the fear of what would otherwise happen to the whites.

References

Fenner, C. 1931 MS178/6/3A5 (1931 diary) Basser Library, Australian Academy of Science. Also lodged in PRG 372: papers of Dr C. A. E. Fenner, held in the Mortlock Library of South Australiana.

Fenner, C. 1937 MS178/6/5B (1937 diary) Basser Library, Australian Academy of Science. Also lodged in PRG 372: papers of Dr C. A. E. Fenner, held in the Mortlock Library of South Australiana.

Chapter 16. The Scientist and Science Communicator

Introduction

As Bernard Hyams observed (see Chapter 14), Charles Fenner was a 'scientist who would be an [educational] administrator'. In fact, he was more than a very competent academic scientist, he was also what is now known as an excellent science communicator. He wrote over 30 papers on various aspects of geography and geomorphology and lectured in geography at the University of Adelaide from 1927 until 1939. He also wrote textbooks on geography for both university and secondary school students and fortnightly articles on science for a Melbourne weekly magazine for 23 years, from which he developed three books of essays on science for a popular audience. And his interests went beyond science; he was also very interested in the history of early explorers of Australia. I cover these aspects of his life in this chapter, and as in Chapter 14, which deals with his career in education, I have added as Chapter 17 an abbreviated and edited version of the parts of his diaries that deal with science, especially physiography and human geography, on the overseas trips in 1931 and 1937. I have included a classified bibliography of his scientific papers and books at the end of this chapter; this does not include items that are listed in the references.

Studies of Physiography in Victoria, 1913 to 1916

As related in Chapter 13, in 1913 Charles Fenner was appointed Headmaster of the Mansfield Agricultural High School. Although there for only a little over a year, he used the opportunity to produce his first scientific paper, on the physiography of the Mansfield District. He had become an Associate Member of the Royal Society of Victoria in 1913 and it was published in their *Proceedings*. When he moved to the Ballarat School of Mines and Industries in 1914, he immediately undertook field studies in nearby Werribee Gorge, Bacchus Marsh and the Glenelg River, and produced substantial papers on these, for which he was awarded the degree of Doctor of Science (DSc) by the University of Melbourne in 1917.

Science Notes, by Tellurian

Early in 1916, while he was still in Ballarat, he came to an arrangement with a prominent Melbourne weekly magazine, *The Australasian*, to contribute an essay on some scientific topic each fortnight. These were called *Science Notes*, for which he adopted the pseudonym *Tellurian*. He continued writing these, even while on his overseas trips in 1931 and 1937, until he became Director of Education in 1939. During the 1937 trip, he wrote 18 essays for *The Australasian*, two for *The Advertiser* and one for *The Argus*. In all he wrote some 620 essays, on a very wide range of topics. They provided the basic material for his three books of essays on popular science, *Bunyips and Billabongs, Mostly Australian* and *Gathered Moss*, each of which received excellent reviews.

Books of Essays

Bunyips and Billabongs

Published in 1933 by Angus and Robertson and illustrated with sketches by his eldest son, Lyell, this book contains 71 chapters, each based on one of his fortnightly articles in *The Australasian*. The foreword was written by the distinguished Professor of Anatomy in the University of Melbourne, Frederic Wood Jones, FRS, extracts of which are published in several of the 58 reviews I have been able to trace. They are worth quoting here:

> Charles Fenner has laid us all under a debt, for he has, with an art that conceals art, made easy the entry into the great world of wonderment in Nature in which we are all free to wander and to speculate and to make intellectual and spiritual gain...Australia is fortunate that within her borders there is a man capable of leading youth towards a wonderment that makes for reverence in a sunset, a gum-tree, or a bandicoot; and who does this out of the fullness of his knowledge and not of mere uninstructed sentimentality...There will come a day when young Australians will be jealous of any interference with a fauna and flora so full of interest as is that of their native land...when that day is fully accomplished then it will be admitted that the writings of Charles Fenner took a large part in its advent.

Mostly Australian

This book is also based on the *Tellurian* articles; it was published by Georgian House in 1944 and contains 12 charming drawings by John Goodchild. It received excellent reviews and a World-Wide broadcast by Radio Australia in October 1947. Of the 61 chapters, most are on Australian topics, but 16 draw on experiences during his overseas trips in 1931 and 1937.

Gathered Moss

To explain the title and the contents of this book of essays, Fenner wrote in the Foreword:

> Proverbs are usually true. But there are exceptions. It is not impossible for a rolling stone to gather some moss. Rolling stones have, indeed, the opportunity for gathering a wide variety of mosses. During a long life the author has been a rolling stone in two ways: first, in the realms of science and literature; and second among the peoples and places of the

> world. There are two sections: Travel notes from an Australian viewpoint, and Science notes, with two or three chapters which are of neither travel nor science.

Like *Mostly Australian*, this book, published in 1946 by Georgian House, is a collection of 28 essays, and is illustrated by 24 John Goodchild drawings.

Royal Society of South Australia

Until the establishment of the Australian Academy of Science in 1954, the Royal Societies in each State were the principal homes of science in that State; national conferences were convened periodically by the Australasian (later Australian and New Zealand) Association for the Advancement of Science. Already an Associate Member of the Royal Society of Victoria, in 1917, just after he had arrived in South Australia, Fenner became a Fellow of the Royal Society of South Australia. He regularly attended its meetings, published in its *Proceedings* and took an active part in the deliberations of its Council. He was appointed Secretary in 1924, was a member of Council, 1925–27, Vice-President, 1928–29, President, 1930–31, Treasurer in 1932 and Honorary Editor of the *Proceedings*, 1933–37.

Field Naturalists' Society

Fenner joined the Field Naturalists' Society (FNS), initially the Field Naturalists' Section of the Royal Society of South Australia, in June 1917. He was elected Vice-President in August 1917 and was President from 1919 to 1921. He was active on the Committee and at evening meetings of the Society, and conducted many of their field trips. He was a member of the group which in 1921 designed the badge of the 'Field Nats', with bright red Sturt's peas *(Clianthus dampieri)* as the emblem. He also served as the representative of the FNS on the Flora and Fauna Protection Society for many years.

Royal Geographical Society of Australasia (South Australian Branch)

Besides the Royal Society of South Australia, Charles Fenner was very active was the Royal Geographical Society of Australasia (South Australian Branch), since 1996 called the Royal Geographical Society of South Australia. He joined the Society in 1921 and was made a Life Member in 1950. During this period he was active on the Council of the Society, being Honorary Secretary, 1925–33, Honorary Treasurer, 1930–31, and President 1931–32, but resigning on 14 February 1932 because of ill health. In 1933 the Society set up a Library and Publications Committee, which, under Fenner as Chairman, was extremely active, 'at times assuming the role of virtual executive of the Society' (Peake-Jones, 1985). This Committee soon spawned an Editorial Board. F. L. Parker was the first to bear the title of Editor, followed by Fenner from 1933–41.

Historical Memorials Committee

The Historical Memorials Committee, since 1982 known as the Geographical Heritage Committee, was established by the Royal Geographical Society in 1927. The initiative for its establishment had come from the Historical Memorials Committee of Victoria, which had written to the Director of Education in South Australia about the need to mark the routes of Australian explorers, notably Charles Sturt. Fenner was a member of this Committee, 1927–53, and Chairman, 1938–41. Besides articles on Nuyts and the Burke and Wills expedition, he took a particular interest in William Light, who planned Adelaide, and Charles Sturt, one of Australia's greatest explorers.

Peter Nuyts

Because of their early occupation of the Dutch East Indies (now Indonesia), most of the early explorers of Australia were Dutch sailors. Peter Nuyts, about whom Fenner wrote an article in 1927, was the first person to explore the south coast of Australia. A memorial obelisk commemorating the tercentenary of Nuyts' voyage was erected at Streaky Bay, on the west coat of Eyre Peninsula, in 1927; a plaque was added and unveiled in 1938.

Colonel William Light

In 1926, a year before the Historical Memorials Committee was formed, the home of Colonel William Light, who designed the city of Adelaide, was demolished to make room for a factory. Fenner wrote a short paper about it (see Bibliography, Historical). Several years later, in 1935, he arranged for the publication of Light's last diary and wrote the introductory notes accompanying that volume. In the 1980s, Thebarton Corporation initiated a proposal the rebuild Light's cottage. A newspaper report about this proposal (*The Advertiser*, 1987) prompted Dr John Tregenza, a professional historian, to produce another paper about the cottage (Tregenza, 1989), which was highly critical of Fenner's paper—'It is a remarkably confused paper'. In 1997, to make these papers more accessible to local people, the Thebarton Historical Society republished both articles (Ralph, 1997).

Captain Charles Sturt

Charles Sturt is widely regarded as 'the father of Australian Exploration'. Fenner worked for many years and in many ways to ensure that his contributions were not forgotten. In 1929, the centenary of Sturt's expedition down the Murrumbidgee and Murray River to the Murray mouth, he was President of the Historical Memorials Committee, and with the cooperation of officials from South Australia, Victoria and New South Wales, he organized a centennial memorial trip of that expedition. Excellent maps and accounts of the voyage were published and distributed, and granite or concrete memorials with appropriate bronze plaques were erected near every community of any size along the route.

In 1931, when he and Peggy visited England for the first time, he met Captain Anthony Sturt (Charles Sturt's great-grandson) in London and later travelled by train to the home of Mrs Beartrix Sturt, daughter-in-law and biographer of Charles Sturt, at Bewdley, overlooking the valley of the River Severn. They went there again on their second trip in 1937, and subsequently corresponded for many years (see Chapter 17). In 1937, he persuaded Ivor Hele to paint a portrait of Sturt—on horseback in the central Australian desert in 1844—that was entitled 'Sturt's Reluctant Decision to Return' and is now in the State Art Gallery of South Australia.

Mr W. J. Adey, the Director of Education, was in London in July 1936 and in his absence Fenner was Acting Director. On a motion by him, the Royal Society urged the government to preserve Charles Sturt's home *The Grange* as a National Historical Memorial and offered to undertake and supervise the layout and planting of the surrounding grounds. On 28 July 1938 the government set up a committee to consider making the home a 'Charles Sturt National Memorial', but it was some 20 years before the idea came to fruition. In 1944, the end of the war was in sight and the Society's thoughts turned again to Sturt. The initiative came from Fenner, now Director of Education, and was successfully taken up by the Historical Memorials Committee, with the purchase of *The Grange,* built by Sturt in 1840 and his residence between explorations and a trip to England in 1847–49, until he returned to England in 1853. In 1982, *The Grange*, fully restored, was finally opened as an historic memorial of an early South Australian home and a famous Australian explorer and South Australian pioneer.

Application for Chair in Geography, University of Sydney

In the 1920s, there was only one chair in geography in an Australian university, occupied since 1921 by Thomas Griffith Taylor, as an Associate Professor. In 1928, Taylor, frustrated by the failure of the University Senate to promote him to a full professorship, moved to a full professorship in Geography at the University of Chicago. Fenner applied for the advertised position of Associate Professor in Sydney, with strong support from Sir Edgeworth David, the father of Australian geology, Mr W. T. McCoy, the then Director of Education in South Australia, Sir Douglas Mawson and others, but he was not selected. David wrote to McCoy on 14 May, 1929, as follows:

> My dear McCoy,
>
> As you were so kind as to give Dr Charles Fenner a strong testimonial for the Associate-Professorship at this University [Sydney], I regret that I cannot congratulate those of us like Sir Douglas Mawson, Dr Ward, Professors Skeats and Richards and yourself who supported Dr Fenner's

candidature, on the success of our advocacy, for the Committee of advice in London comprising the leading geographers in the British Empire, placed Dr Fenner second on the list, stating officially that he was well qualified to succeed Prof Griffith Taylor, but that on the whole they preferred to recommend to our Senate a Mr Holmes, now in charge of Geography at Durham University. A reason that Prof J. W. Gregory of Glasgow in a private letter to me mentioned in favour of Holmes as compared with Dr Fenner was that he was 33 years of age, compared with Fenner's age of 45, and moreover Holmes had for 8 years been a phenomenally good teacher of Geography on the most up-to-date lines at British Universities. Nevertheless I feel strongly that the Committee of Advice did not fully appreciate the brilliance of Dr Fenner's work, and his great efficiency as a lecturer and organizer as we do. The report however of the Committee being unanimous it was hard to gainsay it and the Senate adopted it. I may add that in placing Dr Fenner second it placed him above two full Professors of Geography at other Universities [in the UK], which shows what a high opinion they have of Dr Fenner.

Yours sincerely,
T. W. Edgeworth David.

On 17 May, Mr McCoy sent a copy of David's letter to Sir William Mitchell, then Chancellor of the University of Adelaide, suggesting that 'The information may be of use if the Council wishes to establish a Chair in Geography.'

In my opinion (FF), to some extent this choice represents 'cultural cringe' and also the fact that at the time Charles Fenner had never been to the United Kingdom. The situation might have been different if the position had been advertised after his visit in 1931 (see Chapter 17).

Lecturer in Geography, University of Adelaide

The year 2004 saw the publication of a Centennial Review of academic geography in South Australia (Harvey and Gale, 2004). Most of the information in this section is drawn from Gale's essay in that volume. The first teacher of geography in the University of Adelaide was John Miller Clucas, who died in 1930. He had been appointed as Librarian at the University in 1900, but undertook teaching geography in 1904, initially for the Diploma of Commerce course. By 1921, geography appeared to be established as a mainstream course, as Economic Geography I and II, still taught by Clucas. In 1928, however, it looked as though it was going to be scrapped. When Sir Grenfell Price heard of geography's possible demise he blew along to the Director of Education, who said 'You can

Figure 16.1. Charles Fenner with Sir Edgeworth David, examining a piece of fossil wood

tell the Vice-Chancellor that if he abolishes geography he will lose the whole £5,000 from the Government for his night classes' (Kerr, 1983). As a result, the Vice-Chancellor, Sir William Mitchell, sent for the distinguished geographer Charles Fenner and put him in charge of a one-year scientific course. Helped by his friends in the Royal Geographical Society, Fenner did much to restore geography and set it on a permanent course. He remained in charge of geography at the University until the latter part of 1951, but was initially assisted, while he was abroad in 1931 and 1937, by F. C. (Clarrie) Martin, who was a teacher at Thebarton Technical High School. Martin was appointed lecturer in geography on a part-time basis when Fenner became Director of Education in 1939, with Fenner nominally in charge until 1951, when Graham Lawton was appointed reader in geography.

In 1960, a Charles Fenner Prize was established in the Department of Geography of the University of Adelaide, for the student with top marks in second year geography. It was first awarded in 1961, and has been maintained annually since then. I donated $5,000 to the University of Adelaide in 2003 to ensure that it would provide a reasonable value in perpetuity.

Books on Geography

Fenner's first book was published in 1931, shortly after he had been appointed Lecturer in Geography at the University of Adelaide. It was *South Australia—A Geographical Study, Structural, Regional and Human,* a book of 352 pages designed for university students and published by Whitcombe and Tombs Ltd, Melbourne. This book is still listed on the internet as a 'book worth reading'.

Three years later, in 1934, he used much of the material in this book to produce a second book, *An Intermediate Geography of South Australia,* of 163 pages, by the same publishers but designed for secondary school students. This proved very popular and went through eight editions. In 1956 and 1958, it was considerably revised and extended by members of the staff of the Geography Department of the University of Adelaide and produced as the ninth and 10th (revised) editions, still with Charles Fenner cited as the sole author, but titled *A Geography of South Australia and the Northern Territory.* The 10th edition was reprinted in 1960.

Australites

The first mention of these objects occurred in 1857, after the explorer Major Thomas Mitchell handed Charles Darwin a small and beautifully formed object of black glass, which Darwin thought was of volcanic origin. Fenner notes in one of his papers that he first saw an australite in 1907, just after he had enrolled at the Melbourne Teachers' College, but his serious interest in them began after conversations he had with Dr L. J. Spencer, of the British Museum, in 1931. He then became very interested in australites (see Australites in Bibliography, below) which are the Australian forms of tektites, small objects thought to originate either as debris arising from major meteor impacts on Planet Earth, or possibly of meteoric origin. He started serious work on these in 1933 and continued after his retirement in 1946, when he took up volunteer work in the South Australian Museum. He thought that they were glass meteorites, principally because of their very extensive distribution, found on the surface over an area of some five million square kilometres of the Australian continent (Figure 16.2), and the fact that their their front surface had melted and solidified again before hitting the earth (Figure 16.3).

Comment by FF: Now, 57 years after his last paper, there is clear evidence that australites, like all other tektites, resulted from a major meteor impact. Studies of australites and micro-australites — about 1 mm in diameter — which have been found in marine sediments to the north and northwest of Australi, reveal that all australites are the same age (770,000 years) and are of a similar composition. The Australasian strewnfield, as it is now called, extends into the land mass of southeast Asia and the consensus is that the meteor impact was probably in Cambodia (Koeberl, 1994). 'They are produced by non-equilibrium shock melting of surficial rocks...and are then lofted into the atmosphere...quenched, and distributed over a geographically extended area, the strewnfield. Some tektites [notably australites] solidify in near-vacuum, re-enter the atmosphere and melt again, to form the ablation-shaped tektites' illustrated here in Fig. 16.3. It is interesting to note when Fenner was looking at the tektites in Professor Lacroix's collection in Paris (see p. 311) he commented on the Indo-Chinites, which had 'a wonderful regularity of form', although they were much larger than australites.

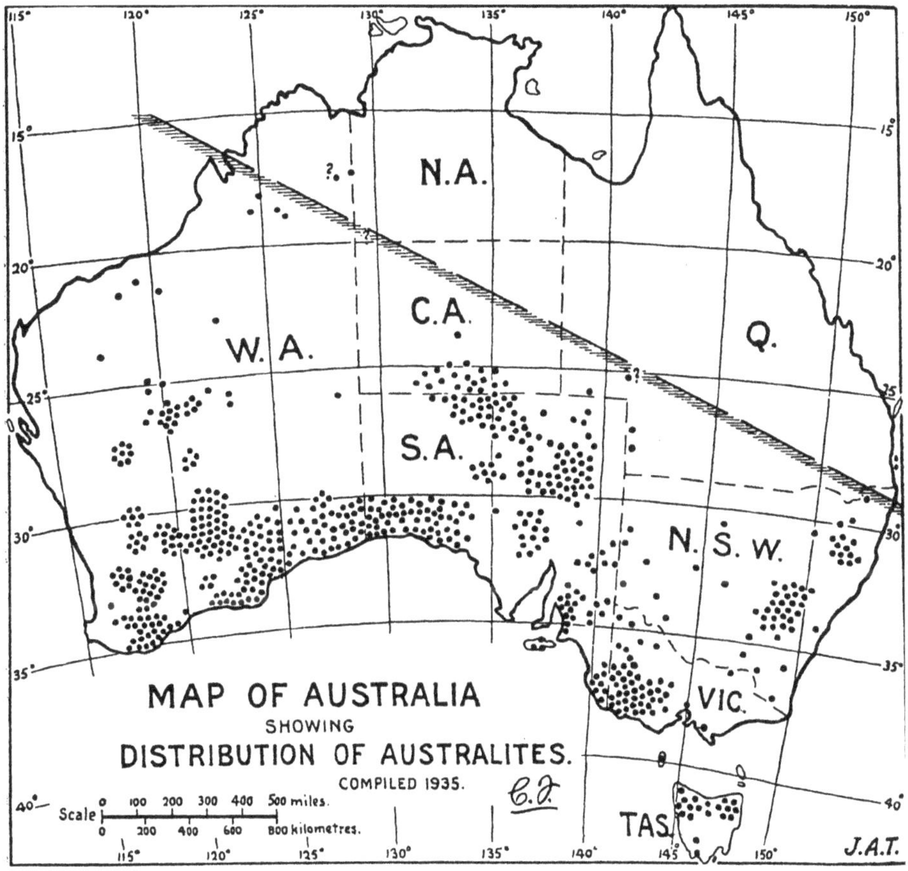

Source: From 'Australites', Part 2. Subsequent studies have shown that the strewnfield was much larger than this.

Figure 16.2. Map of Australia showing the 'strewnfield' of australites

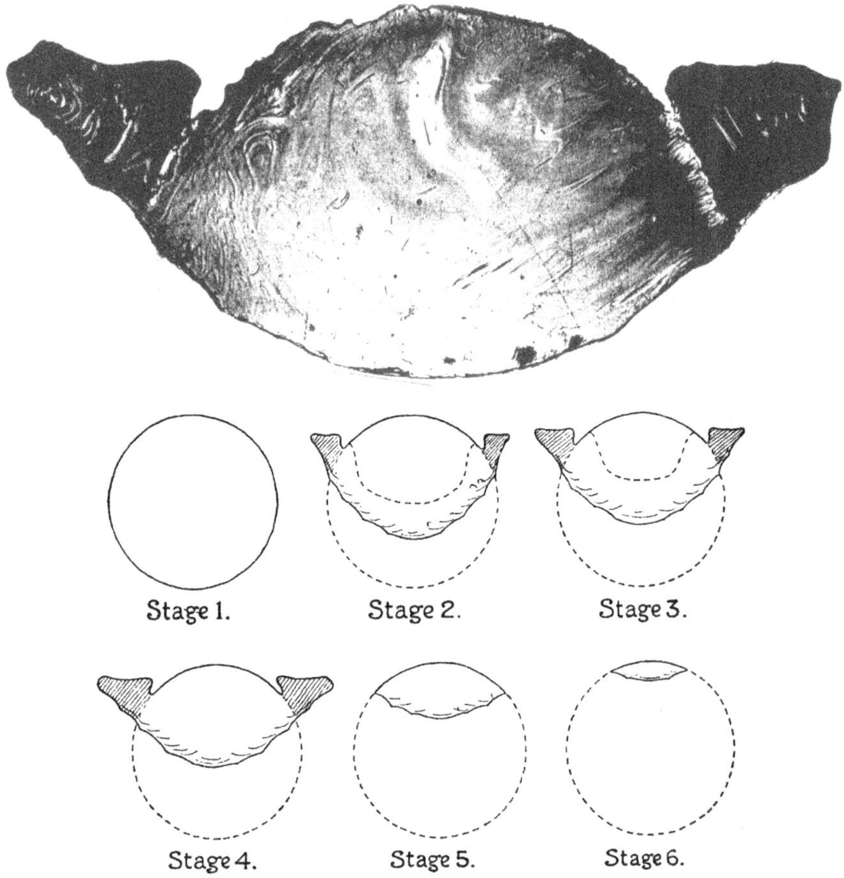

Source: From 'Australites', Part 3

Figure 16.3. Australites

Figure 16.3a. Cross section of an australite showing the original spherical shape and the flow-lines of the original sphere and the rim.
Figure 16.3b. Diagram of cross sections, illustrating the development of different forms of australite. Stage 1 is hypothetical, stages 2 and 3 are rarely found with the flange attached, stages 4, 5, and 6 are common forms.

Prizes for Scientific Work

At a time when prestigious awards were much less common than they are now, Charles Fenner received three awards for his geological and geographical work. In 1918 he received the first award of the Sachse Gold Medal of the Royal Geographical Society (Victorian Branch) for his 137-page paper on the physiography of the Werribee River area. It was handed to him by the Federal Minister for Home and Territories (Hon. A. Poynton) when he visited Adelaide on 1 August, 1920. There is a detailed account of the life of Mr Sachse in Gill (1962).

In 1929, he was awarded the David Syme Research Prize of the University of Melbourne for his 64-page paper *Adelaide, South Australia: a Study in Human Geography*. This was the basis for his subsequent books on the geography of South Australia. He (and I) were delighted when I was awarded the Syme Prize 20 years later, in 1949.

In 1947, he was presented with the first award of the most prestigious medal of the Royal Geographical Society of South Australia, the John Lewis Gold Medal, for his literary work on geography. Subsequent recipients include Sir Douglas Mawson and Sir Edmund Hilary. At the request of the President, C. M. Hambidge, the presentation was made by Dr J. Brooke Lewis. Fenner was made a Life Member of the Society the same year.

Figure 16.4. Medals

Figure 16.4a. Sachse Gold Medal of the Royal Geographical Society (Victorian Branch), awarded 1918.
Figure 16.4b. David Syme Research Prize Medal of the University of Melbourne, awarded 1929.
Figure 16.4c. John Lewis Gold Medal of the Royal Geographical Society of South Australia, awarded 1947.

The Wongana Circle

With his interests in local Royal and Royal Geographical Societies, the Historical Memorials Committee and the Field Naturalists' Society, Fenner had a busy life outside of his duties in the Education Department. He and his wife were also members of the Adelaide University Theatre Guild and the Dual Club. However, the social activity he enjoyed most was the Wongana Circle ('wonga' being the Aboriginal word for 'to speak'). It was a social group comprising Adelaide intellectuals of different backgrounds (men only, still, in 2006; the wives prepare supper). It first met in 1934 and Fenner was a foundation member; he attended every meeting that he could. They used to meet for supper and conversation (talk, talk, talk) in each others' homes, and they still meet. Members at various times during father's membership included Professor Jack Wilkinson, who chaired the first meeting, Archibald Grenfell Price, Ray Hone, Hans Heysen, Henry Basten, Peter Karmel, Leonard Huxley, John Horner and many others.

Activities after Retirement

Some months after he had retired, when he felt well enough, Fenner undertook volunteer work in the South Australian Museum, helping with the displays of geological specimens and especially the very extensive collection of australites and other tektites held by the Museum. He carried on with this until he had a stroke in 1954, from which he partially recovered, but died on 9 June, 1955.

References

Gale, F., The development of university geography, 1904–1960, in Harvey, N. and Gale, F. (eds), 2004, *The Centenary of Academic Geography in South Australia. South Australian Geographical Journal*, vol. 102, pp. 26–40.

Gill, F. 1962, *Proceedings of the Royal Geographical Society of Australasia, South Australian Branch,* vol. 63, pp. 73–8.

Koeberl, C. 1994. Tektite origin by hypervelocity asteroidal or cometary impact: Target rocks, source craters, and mechanisms. In Dressler, B.O., Grieve, R.A.F., and Sharpton, V.L. (eds), *Large Meteorite Impacts and Planetary Evolution*, Boulder, Colorado, Geological Society of America Special Paper 293.

Kerr, C. 1983, *Archie: The biography of Sir Archibald Grenfell Price,* Macmillan, Melbourne.

Ralph, G. (ed) 1997, *Thebarton cottage: The home of Colonel William Light and the great controversy surrounding it.* Wilmar Library, Lockleys, South Australia.

The Advertiser, 14 August 1987, Colonel Light's cottage may rise again.

Tregenza, J. 1989, Colonel Light's 'Theberton Cottage' and his legacy to Maria Gandy: a reconsideration of the evidence, *Historical Society of South Australia*, vol. 17, pp. 5–24.

Bibliography of Charles Fenner's Publications on Science

Geology, Geography and Physiography

Physiography of the Mansfield district, 1913–14, *Proceedings of the Royal Society of Victoria,* vol. 26, pp. 386–402.

Notes on the occurrence of Quartz in Basalt, 1915, *Proceedings of the Royal Society of Victoria,* vol. 27, pp. 124–32.

Physiography of the Glenelg River, 1918, *Proceedings of the Royal Society of Victoria,* vol. 30, pp. 99–120.

Physiography of the Werribee River Area, 1918, *Proceedings of the Royal Society of Victoria,* vol. 31, pp. 176–313.

The craters and lakes of Mount Gambier, 1921, *Transactions and Proceedings of the Royal Society of South Australia*, vol. 40, pp. 169–205.

Notes on the advance of physiographical knowledge of Victoria since January 1913 (with Frederick Chapman), *Australasian Association for the Advancement of Science Report 15,* January 1921, pp. 314–18.

Notes on the advance of physiographical knowledge of South Australia since January 1913 (with L. K. Ward), *Australian Association for the Advancement of Science Report 15,* January 1921, pp. 323–26.

Physiography of Victoria, 1923, *Pan-Pacific Science Congress Proceedings,* vol.1, pp. 719–21.

The Bacchus Marsh Basin, Victoria, 1925, *Proceedings of the Royal Society of Victoria,* vol. 37, pp. 144–69.

The physiography of the Adelaide region, 1924, *Australasian Association for the Advancement of Science, Adelaide, Handbook,* pp. 12–14.

Adelaide, South Australia: a study in human geography, 1927, *Transactions of the Royal Society of South Australia,* vol. 51, pp. 193–256.

A geographical enquiry into the growth, distribution and movement of population in South Australia 1836–1927, 1929, *Transactions and Proceedings of the Royal Society of South Australia,* vol. 53, pp. 79–145.

Major structural and physiographic features of South Australia, 1930, *Transactions and Proceedings of the Royal Society of South Australia.* vol. 54, pp. 1–36.

The natural regions of South Australia, 1930, *Australian and New Zealand Association for the Advancement of Science Report 20,* pp. 509–45.

The structural and human geography of South Australia, 1931, *British Association for the Advancement of Science Report,* pp. 413–14.

The Bacchus Marsh Basin, Victoria, 1933, *Progress in Australia,* vol. IV, no. 7, 4 pages.

The Murray River basin, 1934, *Geographical Review, American Geographical Society of New York,* vol. 24, no. 1, pp. 79–91.

Report of the Research Committee on the Structural and Land Forms of Australia and New Zealand, 1935, *Australian and New Zealand Association for the Advancement of Science Report 22,* pp.463–74.

A sketch of the geology, physiography and botanical features of the coast between Outer Harbour and Sellicks Hill (with J. B. Cleland), 1935), *Field Naturalists Section of the Royal Society of South Australia,* Publication No. 3, 35 pages.

Geology and physiography of the National Parks near Adelaide, 1936, *South Australian Naturalist,* vol. 17, pp. 16–25.

The growth and development of South Australia, 1934–35, *Proceedings of the Royal Geographical Society of Australasia, S.A. Branch,* vol. 36, pp. 65–89.

The significance of the topography of Anstey Hill, South Australia, 1939, *Transactions of the Royal Society of South Australia,* vol. 63, pp. 69–87.

The value of geography to the community, 1937–38, *Proceedings of the Royal Geographical Society of Australasia, S.A. Branch,* vol. 39, pp. 61–8.

The Kybunga daylight meteor (with G. F. Dodwell), 1942–3, *Proceedings of the Royal Geographical Society of Australasia, S.A. Branch,* vol. 44, pp. 6–19.

Aboriginal records near Broken Hill (with A. B. Black) 1945, *Records of the South Australian Museum,* vol. 8, no. 2, pp. 289–92.

Australites

The origin of tektites, 1933, *Nature,* vol. 132, p. 571.

Australites, Part 1, Classification of the W. H. C. Shaw collection, 1934, *Transactions of the Royal Society of South Australia,* vol. 58, pp. 62–79.

Australites, Part 2, Numbers, forms, distribution and origin, 1935, *Transactions of the Royal Society of South Australia,* vol. 58, pp. 62–79.

Australites, Part 3, A contribution to the problem of the origin of tektites, 1938, *Transactions of the Royal Society of South Australia,* vol. 62, pp. 62–79.

Australites, Part 4, The John Kennett collection, with notes on Darwin glass, bediasites, etc., 1940, *Transactions of the Royal Society of South Australia,* vol. 58, pp. 62–79.

Australites, Part 5, Tektites in the South Australian Museum, with some notes on theories of origin, 1949, *Transactions of the Royal Society of South Australia,* vol. 73, pp, 7–21.

Australites: A unique shower of glass meteorites, 1938, *Mineralogical Magazine, London,* vol. 25, pp. 82–5.

Sandtube fulgurites and their bearing on the tektite problem, 1949, *Records of the South Australian Museum,* vol. 9, no. 2, pp. 127–42.

Historical Papers

The first discoverers of South Australia; the tercentenary of Nuyts, 1925–6, *Proceedings of the Royal Geographical Society of Australasia, S.A. Branch,* vol. 27, pp. 23–8.

Thebarton Cottage—the old home of Colonel William Light, 1926–7, *Proceedings of the Royal Geographical Society of Australasia, S.A. Branch*, vol. 28, pp. 25–45.

Two historic gumtrees associated with the Burke and Wills expedition of 1861, 1927–8, *Proceedings of the Royal Geographical Society of Australasia, S.A. Branch*, vol. 29, pp. 58–78.

Colonel Light's last diary, with introductory notes by Charles Fenner, 1933–4, *Proceedings of the Royal Geographical Society of Australasia, S.A. Branch*, vol. 35, pp. 93–129.

Chapter 2, Foothills, plains and streams, 1956, pp. 7–10, in *The First Hundred Years: A History of Burnside in South Australia,* Corporation of the City of Burnside.

Books

South Australia—A Geographical Study, Structural, Regional and Human, 1931, Whitcombe and Tombs Ltd. (For university students).

An Intermediate Geography of South Australia, 1934, Whitcombe and Tombs Ltd. (For secondary school students, eight editions published).

A Geography of South Australia and the Northern Territory, 1958, Whitcombe and Tombs Ltd. (Revised by staff of the Department of Geography at the University of Adelaide, but published as the ninth (1958) and tenth (1960) editions of *An Intermediate Geography*, with Fenner as the author.)

Fenner, C., Parker, F. L., Portus, G. V., Price, A. G., Richardson, A. E. V. and Roach, B. S., (eds), 1935, *The Centenary History of South Australia.* Royal Geographical Society of Australasia, S.A. Branch, 420 pages. (Fenner wrote the first and last chapters: Chapter 1, The geographical background, pp. 1–15, and Chapter 25, Retrospect and prospect, pp. 378–92.)

Bunyips and Billabongs, 1933, Angus and Robertson, Sydney.

Mostly Australian, 1944, Georgian House, Melbourne.

Gathered Moss, 1946, Georgian House, Melbourne.

Popular Articles

The study and love of Nature; An appeal to youth, 1934, *The South Australian Naturalist,* vol. XV, no. 4.

Kangaroo Island, 1934, *S.A. Teachers' Journal,* 27 August 1934.

Our South Australian Climate, 1934, *S.A. Teachers' Journal,* 27 October 1934.

Myths and superstitions, 1935, *Progress in Australia,* 7 January 1935.

Blackfellows buttons, the remarkable glass meteorites of Australia, 1939, *The Sky—Magazine of Cosmic News,* New York, pp. 16, 17, 27.

Australites and other tektites, 1953, *South Australian Naturalist,* vol. 27, no. 4, pp. 2–8.

Chapter 17. Overseas Trips, Diary Extracts on Science

Introduction

The background to these extracts from the diaries my father kept on his trips overseas, to Europe in 1931 and to North America and Europe in 1937, is set out in the Introduction. This chapter contains some of his comments on scientific matters, particularly the geology and geography of places he and my mother visited, discussions about these fields of science and his other great interest, australites. Only a small proportion of the entries dealing with science have been selected and these have been severely edited. Each entry is distinguished by the accompanying date and place.

1931 Trip

Travelling over in the *Balranald*

27 August 1931, Aden

Hartung and I, by courtesy of the skipper, went up on the bridge, and I had a good look at the charts. It is really marvellous what a collection of valuable information they represent. It is equally marvellous that they have several errors, such as the geology of the peninsula of Aden, a high rugged land joined to the mainland of Arabia by a low sandy swampy isthmus. The pilot book asserted that it was limestone. On the other hand the chart showed a beautiful, and unmistakable, example of a symmetrical crater, breached to the east, 1,800 feet to the rim, and heavily eroded by valleys.

Aden is really beautiful as seen on a moonlight night and it was one day from full moon. Captain Duncan, who was stationed there in 1924, told us a number of very interesting things about it, and one of them was that it was basalt. And so it proved, or rather it is a basic igneous rock—perhaps middle Tertiary, by the erosion, but it may even be Pliocene. The arid land erosion is absolutely beautiful.

The second peninsula was said to be granite. There may be some granite there, but a great part of it is certainly a basic igneous rock, much rotted, stratified; brown, green and ochreous colours. As a compact lesson in the erosion forms of arid lands I had one of the most interesting few hours of my life. In the magic of dawn, and before sunrise, the hills were flat, like black cardboard cut out and placed against the sky.

Coming out from Aden the sea was chocolate brown; indeed in some cases it was a real blood red. I thought it was due to red mud but an old friend got his bucket out and examined some of the water. Great glee. It was due to billions of minute animals. Mrs Dodwell thinks they are a form of peridyne; it was a thrilling experience to be able to trace such a large-scale phenomenon to so small a cause.

6 September 1931, the Coast of Spain

Now we have had three hours of excellent observations of Spain—of Granada, of Moorish fortresses and towers, of white clusters of what are probably monastery villages high up on the mountains. The mountains are the Sierra Nevadas—the original Sierra Nevadas. We see also smelting works and difficult coastal roads. The towns are most numerous and very distinct. The air must be very clear, for we can see trees and lines of plantations, though the coast is seven miles away and the plantations much more. In spite of all this, the general aspect is utterly bleak and barren. The first town we saw was Almira, noted for its wines. Now we come to Malaga, noted for wines and raisins. The strongly ferruginous slopes are planted, and here may well be grown the vines that make port wine. It suggests to me that we might plant more vines in our less well-watered and less fertile areas, on hills, to get wine to the European taste.

One thinks of the Moors and the Arabs—who were also the Moors, and of the Cid and of Ferdinand and Isabella, and of the mighty struggles and conquests that went on here for centuries. Here science, in the form of alchemy, entered Europe with the Moors. The windowless flat-roofed houses are suggestive of Arabs and the names of some places show this influence also. Gibraltar, for instance, is such an Arab name—a corruption of Jeb-el-Tar or something of that kind. It is just behind these hills, which are of ancient rocks, of which some geological structure can be made out in the colours, and in the positions of the villages, etc.—just behind is the land where lies the Alhambra, greatest work of all Moorish architecture, so near and yet so far. It is most extraordinary that so barren a country should have such a population.

There is a village at each place where a larger valley emerges onto the coast. The mountains must be well over 5,000 feet high, unbelievably rugged, red, cream, brown, with just a few tufted trees. I fancied I could see a eucalypt grove or two along the deltas further back. The river valleys are wide. A lady from Burma told me they had similar 'rock-waste' valleys in Burma, dry most of the time, torrential at flood time, called *chaungs*. The same thing as the *wadys* of Egypt and Arabia. We have them in northern South Australia but have no specific name for them.

Evening. This coast of Spain is altogether entrancing. Glorious mountains, bare, reds and browns and creams, thousands of feet high, scarred and gullied with most gorgeous stream sculpture. They look as if, in Australia, they would carry about 10,000 sheep and support one squatter. Here it 'runs' scores of villages, and some quite large towns. Up the mountain side there are glittering white villages. Mostly at sea level, but some at other levels one or two thousand feet up and barren looking red hills planted with vines—all this on a stupendous scale.

The scenery grew on one. I did not like it at first, too grim and bare, but long before we got to Gibraltar, at Almira, Salabrena, Motril, Malaga, and at least one hundred other towns and villages, one fell in love with the exquisite scenery, not lessened by the white buildings, by the great forts, by the greenness where great rivers came out of these mighty mountains. It was all very beautiful, very interesting, a wonderful lesson on the geography of Spain.

At about five o'clock we approached Gibraltar. The south point, where the lighthouse is, is Europa Point. The Rock is a mere fragment of limestone, huge and precipitous, very steeply dipping, caves with rough stalactites at sea level. Very steep on the east where they have built a huge concrete rain catchment. Some scrubby growth on the western side. It is a maze of fortifications and the outline shows guns and guns. It seems a most extraordinary thing that this place should have had such strategic importance. Algeciras, of historical fame, was opposite on the north side. Genta, the great Spanish fortress stood up on the south side. Then following some fine geology and physiography came Point Tarifa, strongly fortified Spanish, a thriving well-built and beautiful city, where early customs dues were levied, and whence all the countries of the world get the word 'tariff'.

Great Britain

10 September 1931, England

The coast looks still and silent and green and watchful, and the white cliffs look forbidding, and I recall that from those cliffs came England's name of Albion (Alb = white, same as Alps)—to the French, perfidious Albion, and that these chalk cliffs are the home of the Cretaceous rocks (creta = chalk). Yesterday as we neared Plymouth and saw the fairly steeply dipping stratified rocks there, I guessed they might be Mesozoic. If I hadn't been such a goat I might have known that since the land was Devon, there was a 90 per cent chance that the rocks were Devonian! So I have possibly seen two of the source names of the geological series.

21 September, London, Faraday Centenary

It has been a gorgeous day. Instead of coming to London as strangers, we are received on all hands as guests, and given opportunities for meeting people and seeing places in a way that does not occur, for mortals like us, more than once in a century. This morning Chapman and I went to the Royal Institution. Received tickets for a beautiful full-day trip tomorrow to Windsor Castle, 'by special permission of his Majesty the King' to view state apartments and chapels and what not. Then to the National History Museum. Saw the Director, Dr Herbert Smith, and signed my name, as a delegate, in a lovely parchment volume. Also got an invitation for Peg and me to a Conversazione on Thursday 29th—by H. M. Govt.—guests to be received by Ramsey McDonald and his daughter. Then to the Royal Institution Reception of Delegates—a most wonderful function. I should consider it to be the most inspiring function I have ever or shall ever attend. To celebrate the memory of Michael Faraday, in the beautiful and sumptuous hall where he lectured, all the great men, mostly physicists, from all lands of the world.

Australia came near the beginning, and when my name was thrown on the screen, I rose and duly bowed to the President, Lord Eustace Percy. As each country was mentioned, pictures of that country were thrown on the screen. Two of my slides of Adelaide were proudly prominent. After that I met, among others, Dr Fritz Paneth, who has determined the age of iron meteorites; 30 determinations, 20 of them about 3,000M years, none older.

Home, dined, dressed for the evening treat at Queen's Hall. A tremendous and wonderful gathering. The most exquisite music by the BBC Symphony Orchestra (100 instruments under Sir Henry Wood). A great gathering. Speeches by Lord Eustace Percy, the Prime Minister, Lord Rutherford, Duc de Broglie, Marconi, Zeeman, Debye, Thomson and Sir William Bragg. It was broadcast to millions here and relayed to remote Australia and elsewhere. A great night.

23–30 September 1931, London, BAAS Centenary Meeting

23 September 1931

The twentythird! The day that I have looked forward to for months as the opening of the BAAS Centenary Meeting. Now we are here. We are shifting over to the Tudor Court Hotel, Cromwell Road, right next to the places of meeting, this morning. It is a great pity one hasn't time to set down impressions in some detail; for instance to tell of the Gainsboroughs and Reynolds and Romseys at Kenwood, great masters in abundance, and of the Rubens and Vandykes and Holbeins at Windsor Castle. It is overwhelming—(this is the correct word)—to those from our little remote homeland where a Corot and a Van Eyck in the Melbourne Gallery have in the past caused our hearts to swell and our minds to think that we really had something of the real art world of the Old Masters.

Sir William Bragg, last evening, in Faraday's own lecture hall, and with the same or duplicate apparatus, performed all Faraday's century-old experiments as he felt his way towards the 'first electric motor', the 'first transformer', and the 'first dynamo'. There are functions all today and well into tonight, so we must go slowly and not waste nervous energy. I must set aside a half day to go more thoroughly into the preparation of my own paper, which is still one week away.

Went to the Installation of Jan Smuts as President in the Albert Hall. A vast show. Seats 10,000. Sat with an elect crowd of some 200 or more. Had to walk up and be received as delegate for the Royal Society of South Australia and shake hands with Smuts. Then a Royal Message, etc. and out. Went again to reception rooms, etc. Then to this hotel, very convenient. Dressed for dinner—in the glory of white vest and all. Peg with her new green silk dress, in which she looks very nice. Then, with Hartung, to Westminster Hall (Wesleyan, a fine building, with a glorious view from a great window of the Abbey) and heard the Presidential Address. Enjoyed it very much. A learned philosophical dissertation by the soldier, lawyer, statesman, philosopher, Smuts. A most brilliant gathering. It was curious to note that while Smuts eulogised the imperishable glory of certain men in unveiling the mysteries of the atom, etc., these very men (the Marchese Marconi, old Sir Oliver Lodge, and Lord Rutherford) shivered and put on their overcoats because there was a draught!

24 September, 1931

At 10 am I was in Jehangion Hall, a good crowd, several hundred geologists, to hear Professor J. W. Gregory's presidential address on 100 years of geology in Britain. Saw Skeats and Bronwen (of Delft). At 11 left and walked the half mile to Section E (geography). I see they have put me on the Sectional Committee. I shall attend tomorrow morning. It is very hard, as it always is to me at AAAS meetings, being torn between three sections; Education, Geography and Geology.

Heard the last of Vaughan Cornish's address. A very good speaker. Author of the *Great Capitals*, and of several other books I have studied and quoted from. Sir Halford Mackinder was in the chair, pal and co-student of Sir Baldwin Spencer. Quite different from what I had pictured but a very pleasant gentleman and also a delightful speaker.

Then listened with much joy for an hour to H. R. Mill, geographer, on his reminiscences and readings of 100 years of geography in the British Association. Most delightful. One of the readiest, most humorous, and most fluent speakers I have heard. And he is 71!

The Royal Society's *conversazione* was a most brilliant affair. There were hundreds and hundreds of people there—the people, Lords and Duchesses and Counts and Earls. Mere knights and their ladies came in droves, while professors and doctors from all the Universities of Europe and America were in profusion. Some beautiful women and magnificent dresses. And decorations! And the supper and wines! The sherry was very good. But the outstanding things there were the wonderful exhibits. Manuscripts and apparatus, historic, of Charles II, Faraday, Cook, Hooke, Priestly, Newton, Rainford, and many others. Most entrancing things.

25 September 1931

Went to the Sectional Committee meeting in Geography. Then heard the Presidential Address by Sir Halford Mackinder, which was very good. Then followed Griffith Taylor, on geography and nation planning. I brought him home to lunch and we three (Peg and he and I) had a pleasant talk. The third lecture, apart from an interlude by Baron de Geer, the Norwegian who has done such wonderful work in measuring post-glacial time, was a paper by one Professor Verinskiöld, of Oslo, on the recession and advance of glaciers. I have a most dreadful headache, but must now set about dressing for dinner, and then we shall go to the reception by His Majesty's government and the Imperial Institute, at South Kensington. 11.30 pm. A very pleasant affair in many ways, but like all these receptions, most boring in others.

Went into the galleries of the Imperial Institute and had quite an interesting hour looking at the wonderful exhibits of the economic resources of the Empire so beautifully and attractively displayed. Australia's exhibit is good, and parts of it made me feel quite homesick. But there is a lack of the dioramas that are so interesting of other lands, even of little places like Trinidad and Fiji.

26 September 1931

Great Hall of the London University this morning. A crowded audience, 8 or 10 loud-speakers in the hall. In a discussion on Population Problems; heard six of the leading men, whose names I have heard and whose writings I have read for years. Julian Huxley (who is young, slight and an excellent speaker), Hogben (small, witty, delightful) his quips re soap, electric light, modern hot baths and the population problem were most delightful, E. W. MacBride, a great old Battler, have admired him from afar for years, F. A. E. Crew, who has done all the genetic work on fowls, sex changes, etc., J. B. S Haldane who wrote that great book that Birkenhead copied, Carr-Saunders, leading British author on population problems.

27 September, 1931

Excursion to the Chiltern Hills. The thing I should like to write about, had I the pen and the time, would be the English countryside. What a land! What a country to be born in, and to love! It is wonderfully beautiful. Stable. Complete. Rich. Peaceful. Fertile. The trees! The people! The crops! The villages! In the copses where we wandered over the uplands pheasants rose in coveys. The people told us there was other game. I saw a dead hare, shot and not found, and many rabbit burrows, and 5 young dead moles. There were foxes also, but these are of course not shot. The glorious, tree-clothed, hedge-rimmed emerald-green flats were great places for fox hunting in the season. This is taken as a matter of course. No one sees the immense amount of class privilege wrapped up in this idea.

Thousands of other impressions. Curious houses, and curious signs, and strange thatched or tiled or shingled roofs, and touches of autumn, crimson and gold, and reeking fertility of unwanted (at any rate unused) grass. It is always green, though one man explained to me that, while it didn't alter much, a little browner in winter, and a little greener in summer, it was never really green! The driest part of it was much greener than our Botanical Garden lawns!

28 September 1931

Up early and off to the Royal Geographical rooms. At 9.15 the Sectional Committee met. Quite a big crowd of geographers on it, including many people whose books I have read for years. I may note here that after I retire to rest each day, mostly some time after midnight, there come to my mind scores of interesting things of which I made no note, interesting people I have met, or new ideas obtained, or curious things noted. But, from a talk at the University—with Rudmore Brown, whose charming book I have loved for so long, down to the episode of the 'small beer' and cheese at the strange little inn at Royal Oak, at Chapelford, Hertfordshire, I don't think of them. One's mind gets full to overflowing of new impressions. Then the time comes when one's mind is so tired that no more can be received.

At the Geographical Section there was a great discussion on the Earth's Crust, its composition, temperature, movement, etc. J. W. Gregory, Sir Halford Mackinder, A. R. Hinks (Sec. Royal Geographical) A. Holmes,

H. Jeffreys (who is an FRS, and one wonders why—though I believe he's a marvel at mathematics), G. C. Simpson (who is an out and out supporter of Wegener) etc. At 1 pm we adjourned to the de Vere Hotel for a luncheon. Vaughan Cornish presided. I walked across with two Americans (Brogg and Huntingdon) who were introduced to me by Griffith Taylor. Met also Slöss from Heidelberg, and talked of Geisler etc. Sat at table with Ogilvie (Glasgow), Verenskiöld (Oslo), Wellington (Pretoria). At 5 pm set off for the Albert and Victoria Museum to hear the lecture on *Sinanthropus pekinensis* by Professor Elliott Smith. It was a great lecture. I would not have missed it. I expect I should have said that also of a hundred others that I have been compelled to miss.

Dressed after dinner for the Reception at the University of London. It was down for academic dress, but I loathe the conspicuousness of that scarlet and green gown, so went without it. When I got there, and had parked my coat and scarf, I saw that the placed reeked, literally, with scarlet gowns so my desire to be quite ordinary asserted itself and I hurried, bareheaded and in my dress suit, back through these unquiet streets to my hotel. Got my gown and hood, back to the reception and so appeared in full plumage.

Tuesday, 29 September 1931

Cut Geography and Education today and went to the Central Hall, Westminster, via underground, to hear the discussion on 'The Evolution of the Universe'. Very fine. Unique. Impressive. All the masters of the question from both the material and the spiritual side. Consider them, each at his best, and reacting to the thoughts of his most notable colleagues and opponents, and stimulated by the vast audience in that great and capacious hall. Sir James Jeans, FRS (introduced it) Professor E. A. Milne FRS, Head of Cambridge Astronomy (Peg says Oxford), Professor de Sitter (Leyden) a most delightful speaker. Sir Arthur Eddington, very decisive and original, Professor R. A. Millikan, chief American, cosmic rays, The Lord Bishop of Birmingham, Bishop Barnes, very good, popular and pleasing, General Smuts, A1 as he always is, Monsieur L'Abbé Le Maître (whom I have quoted so often, quite clear, for a foreigner) and lastly old Sir Oliver Lodge, as precise and clear and definite and loud-voiced as one could wish. He brought in spiritualism a little, but not too much, did not do so except by implication.

Wednesday, 30 September 1931, My Lecture

This has been a great day, a very wonderful day, a day never to be equalled. It is marked by three things, and though it is now after midnight, and I must be up early and off to Cambridge, still, I must set down some notes of them.

The three things are:

1. I delivered my first, and perhaps last, address to a London audience, a big audience for a Sectional meeting, at least 250, all geographers, and representative of all countries, and it went off 'beautifully' (This is Peggy's description; she was there).
2. I went, by invitation, to Downe House, where Darwin lived and worked; and enjoyed and absorbed the inspiration of that very wonderful shrine.
3. I (we) in all cases, nearly, I, of course, means 'we', went to the Guildhall Reception, and were duly announced to and received by his Worship the Lord Mayor of London, with hundreds of others, in the most wonderful building that we have yet seen, surrounded by some of the most wonderful treasures of time and art and history – all displayed so well that one could spend months in looking at them. And I ate *paté de foie gras* and drank champagne (here I does not mean we).

Felt very nervous about my paper this morning, for after all it was really the most important thing of the whole meeting for me that I should do this thing well. There were three papers and good audience of English, American and foreign geographers. Sir Halford MacKinder in the Chair. The fine geographical hall all in order. The first paper was 'Land Utilization in South Africa' by Professor Wellington (Pretoria), the second by me, and the third by Professor Burfee of Canada, a real Imperial affair. Wellington got through his in good time. Very interesting and much along the same lines as me, with spot maps and so on. Then me. Was ghastly nervous till I got a start, and then felt perfectly at home and went on and told them, with maps and pictures, all about this unimportant little country of South Australia. I got a very good reception at the conclusion and as I sat down, someone behind me caught me by the shoulders and congratulated me. It was S. H. Smith, of Sydney, and very kind it was of him to come around to hear me. Then another man came up and sat in front and was complimentary, and Dudley Stamp and Huntingdon of Columbus, Ohio, and Kissing of Birmingham, and Verenskiöld of Norway, and Griffith Taylor who had come in late. The chairman made quite a long complimentary speech, and had evidently seen my book [*South Australia—A Geographical Study, Structural, Regional and Human*] and told them that if they wanted to know more of my work they should buy the book. So Peg and I walked home on air, and I felt less tired than I had done for the whole week.

After lunch we went by coach through the miles and mazes of South London into the glorious fields of Kent, and out to Downe House. It was really wonderful to walk along the 'sand walk' that Darwin called his thinking path, to see the lawns where he did his earthworm experiments, to see his study exactly as he used it, with his pens and papers and opened envelopes, and a few fossils and minerals (selenite and pyrites, etc.) lying about, and so on.

1–3 October 1931, Cambridge, Clerk Maxwell Centenary

On Thursday morning Hartung and I set off for Liverpool Street by Tube and then on to Cambridge for the Clerk Maxwell Centenary celebrations. A delightful place. I had thought to arrive at a town in which there was a University. Quite different. It is a town which *is* a university. Everything is so different from what I expected, despite talks and photographs and descriptions. It was so quiet and restful after London, and all the places one wanted to go to were so near.

The taxi dropped Hartung at Pembroke College and took me on to King's College. There I was, alone almost in those great grey stone buildings and wide squares and age-old traditions. No one met me. It is not done that way. The porter at the Lodge conducted me to Guest Room A, a fine large room with marble fireplace and a nice coal fire and a huge Gothic low window overlooking King's Parade. Monarch of all I surveyed. There were several letters and papers and invitations awaiting me. It was obvious that the Cambridge affair, the Clerk Maxwell celebrations, were to be more generous and much better organised even than the London ones.

In the official number of the *Cambridge University Reporter*, one sees oneself listed with what is perhaps the mightiest and most brilliant array of chemists and physicists that have ever come together. One [C.F.] is set down as the chief representative from Australia, and President of the Royal Society of South Australia. Towards one o'clock the scarlet gown is donned, and off we go. I was diffident at first about even carrying that red gown on my arm in the street, but before I left, having lived and walked and dined in it, and walked all over the place in it, and seen the streets filled with gowned men of all the colours of the spectrum, well, I got as used to it as I am to my normal overcoat.

At 1 pm we arrived at Corpus Christi College. A fine banquet in the Great Hall. Sat with and talked to Professor Wasastjerna (Finland) opposite, Perrier of Lausanne and R. H. Fowler, of Cambridge, son-in-law of Lord Rutherford. All very pleasant; so was the food. This country (at banqueting times) is a wonderful place for food. The caviar was fine, and the champagne and other wines most excellent. I dined and wined well. Then we formed a procession, as set out in the printed syllabus, and up the main street we marched to the Senate House, the Marchese Marconi, in his blue and silver and ostrich feathers, and gowns and uniforms and tassels of scarlet and green and blue from all the countries of the world. My partner in the procession was Professor Schonland, of Cape Town University, very agreeable.

At the Senate House we sat in serried rows, and when our name and standing were announced we approached and were welcomed by the Vice-Chancellor (Baldwin is the Chancellor, but couldn't leave London owing to government difficulties). Then J. J. (Sir Joseph) Thomson delivered the Maxwell oration. Very fine. I have it, and the eight other great addresses I heard at Cambridge, all in print in a book so that you may read them.

The young man who makes the beds and who 'dresses' one had promised to call me at 8 am. He did so. He also took all my clothes and boots and brushed them, and then set them out in order as one dresses, socks half turned, etc. Most curious and interesting. And he took my flung-off evening clothes and folded them away in the most professional way. It was allright as an experience. Still, I prefer to dress myself, and I forestalled him the next morning, by being dressed and brushed up when he came to call me at 8! He was equally surprised, and his unspoken query caused me to say 'I was restless, and got up early.' 'Then you are ill, sir?' 'Oh, no,' I replied. 'Ah, sir, you have been overworking.' 'Yes', I said. These people are well trained.

At 10 am was with the gowned mob at the Arts School. Sat with Kerr Grant and Hartung, an island of Australia in a sea of foreigners. Lord Rutherford in the chair. Had three of the most delightful addresses by Max Planck, of Berlin, the discoverer of the quantum, possibly the most brilliant man in all that brilliant crowd, and some most charming reminiscences by Sir Joseph Larmor, who knew Maxwell, and by Neils Bohr, of Denmark, unraveller of atomic structure. His address reads better than it sounded.

Dr William Garnett told us that when 'J. J.' (Sir Joseph Thomson) came as a young man to study under Maxwell that Maxwell summed him up after a couple of weeks. He said 'Garnett, Thomson will never be able to tell the things that are hard from the things that are easy.' Now, well over 80 years of age, Master of Trinity, and one of the most loved of these great physicists, Thomson is still full of life and energy. Back to Kings and lunch in the Hall. In the afternoon at the Arts School, and then a wonderful series of lectures. Sir Richard Glazebrook in the chair (whose textbooks I was partly reared on), Sir Arthur Schuster present as guest. Addresses—brief, direct, and full of meat by Sir James Jeans, Dr William Garnett, Sir J. A. Fleming, and Sir Oliver Lodge. They may say what they please about the old man, but Lodge is still one of the giants, mentally as well as physically, even among all these mental giants, and that in spite of his spiritualism.

At 8 pm we were due at Trinity for the chief Banquet. A magnificent affair. It was indeed beyond description. The conversation and speeches were fine. The caviar was excellent. The wines, including 91 year-old Madeira (Old Burl 1840) were fine, and I did quite well on all. The Latin graces, chanted, the singing from aloft by a choir of boys, the age-old candle lighting, all these were wonderful. There were 206 guests at the tables, and I supposed I was the least among them. However, I did enjoy it all.

Slept well. Breakfast in the Combination Room. Then Dr Rothenberg ran us out to Ely to see Ely Cathedral. It was a beautiful run. His car was fitted with a drive that needed no awkward gear-change; I think it is called the 'Wilson' gear. Most admirable. Ely Cathedral, with its traces of the story of Hereward the Wake, and its tombs of ancient men of every age, and its vandalism—where beautiful stonework was smashed by the Danes when they massacred the inhabitants and plundered the cathedral in AD 870. Cambridge is young compared with this; just as this is young compared with the pyramids; but we in Aussie are spell-bound with them all, for all we have is so new, so superficial, so lacking in tradition. What we must do is to preserve all we have, and put some tradition into it.

Was glad to have a good long run over the Fens. They are wonderful and remarkably fertile. Here in the marshes, according to A. G. Fenner, lived the bloodthirsty robbers, who were the ancestors of the modern [English] Fenners, and who waylaid travellers across the marshes. It must have been a great life. To think that one must fall from that to the dull routine of a Superintendent of Technical Education. God help us all.

Australian History, Memories of Charles Sturt

19 September 1931, London

At one o'clock, arrived at the Empire Club, 69 Grosvenor Street—the club of Mrs G. R. N. Sturt. Mr and Mrs Sturt were there waiting for us. He is the grandson of Captain Charles Sturt. He was most interested in all we have done and what we think about Charles Sturt. Mr Sturt mentioned the Crossland portrait of Sturt, in the Adelaide Gallery, saying that Mrs Sturt didn't like the colour of his waistcoat in that picture and had another painted, with a yellow waistcoat, which they have inherited. Also that a popular picture of Sturt on the Murray, showing him with a large pipe in his mouth, offended Mrs Sturt, as Charles Sturt did not smoke. This Mr Sturt is a most cautious gentleman and in stature and appearance seems to me to resemble Charles Sturt. I hope I may be able to run up to Bewdley, Worcestershire, to see Mrs Beatrix Sturt.

10 and 11 October 1931, Bewdley

Winterdyne, Bewdley—the home of Mrs Beatrix M. Sturt, daughter-in-law and biographer of Charles Sturt. In musings on this great trip to England the desire has always been paramount to me, and the mental picture most clear, of a trip into the English countryside—the real England. I never dreamed, nor did Peggy, that our wish would be so amply and so bountifully fulfilled, as in this weekend we are spending here with a most delightful and most mentally vigorous lady, Mrs Napier Sturt, and her daughter Catherine. Here we are, in the heart of this fine old English home, receiving the warmest entertainment, among the oak and beech woods and the copses on a high cliff overlooking the River Severn and its perfectly beautiful Worcestershire Valley.

We left Paddington Station at 9.45 am. Thence along the Thames Valley to Reading, Oxford, Everham, Worcester, where there is a beautiful cathedral and where poor forlorn old King John is buried. Then we got to Hartlebury, and got out and found the chauffeur and Mrs Sturt, and had a lovely purring drive along the close well-grown roads of this part of England. Altogether beautiful, and fertile and bounteous, and quite indescribable. What a land to love and to fight for. What a land of comfort and culture men like Sturt gave up to come out to the hardships of Australia! Winterdyne, with its three storeys overlooking the Severn, and its farms and coppices and lush green fields, is a lovely old home, and Mrs Sturt is a wonderful hostess, and we are enjoying our stay most keenly.

When I think of Frank and Tom coming to dinner without a collar, or even in a rough sweater, and remember Mrs Sturt's apologising for the fact that they always dress to teach her grandsons 'to be decent and tidy', I blush for them and for ourselves. Factory smoke can be seen from Winterdyne. Three or four separate huge power transmission lines run across the country. So the tail of the dragon is here and still, there is much of 'beechen green and shadows numberless' that nothing can touch, and the softness of it all, the glory of the landscape, and its richness in history and mystery.

This is quite a big establishment for Mrs Sturt and her daughter, unusually so to our Australian eyes. Some seven or eight maid servants, and seven men. I should really have made full notes of the manuscripts of Sturt's overland journey with cattle, and of the pictures and medals and souvenirs, and of the incidents that Mrs Sturt has told us about, but it is quite impossible.

1937 Trip

United States of America

Note by FF: *The 1937 trip, made with Mother, Draper and Elizabeth Campbell as companions, involved crossing the Pacific Ocean and arrival in San Francisco and then visits by train to various places in western United States before going to the east coast and then on to Europe by ship. There follow selected entries containing comments of scientific interest.*

28 March 1937, Yosemite

By train from Oakland, and on through the foothills, like the Berkeley Hills, but not so much built on, then at last we got through them, and there before us lay the immense, rich and wonderful San Joaquin Valley, a natural wonder of the world. 250 miles widest place east to west, 150 miles north to south, really runs north and south, between the Coast Range and the Sierra Range, 'the dim Sierras far beyond uplifting, Their minarets of snow'.

Well, this San Joaquin Valley surpasses description, its wealth and fertility and the teeming thousands of its population. We have little patches somewhat like it at Berri and Tanunda in South Australia. But here there are untold thousands of acres, and all richly fertile. The density of the houses, and the fertility and the belts of trees remind me of the Po Basin in Italy. And of course it is like Italy and South Australia—a Mediterranean climate. One sees sheep and wheat and figs and olives. But such vast fig orchards and such immense rich stretches of lucerne (alfalfa), grasses, orchards, and vineyards I have not seen. All the valley floor appears to be good. No bad or salty patches.

We rose from the flat irrigable land quite suddenly. It became grazing, cattle, dairy, and later (higher and more rugged) beef land. These were the foothills of the Sierra Nevada (Snowy Mountains). Trees mainly live oaks. As we gradually climbed higher, other trees occurred, digger pines and cypress pines, and later giant redwoods, all rather young.

From Merced to this very comfortable and beautifully appointed Awahnee Hotel, from which we travelled next day by car. Bang up against the Falls and the Precipices and the majestic halfdome—which impresses me vastly and which no words can describe. A light or two has appeared through the pine forest down the valley towards Yosemite Lodge. The threatening clouds have cleared up a bit. It is absolutely calm and peaceful. Elizabeth peeps in to enquire whether she should change her dress for dinner. There is a very bright planet to the West. It must be our own old Venus.

Have had dinner. As usual it is too abundant and too rich. Prodigious helpings are given. Some of the foods with the same names as ours are different—richer and greater in quantity than one expects. Today's drive has been a glorious one to me. It is the complete realization of satisfaction to travel through new country such as this. If only a Western geographer, a geologist or both could have been on the trip to tell me some of the things they have found out about the areas, it would have been the Summit. Still, as it was, with a very good driver-guide to answer questions it was excellent. The human geography of the vast San Joaquin Plains was obvious for the most part. The gorgeous sculpture of these great Sierra Nevada Mountains was magnificent to see. Trees, hills, gorges, canyons, Merced River, abundant falls, the few houses and people. Then the magnificent Yosemite itself.

29 March 1937, Monterey

At Del Monte, Monterey, old capital of California, home of the Monterey Pine (*Pinus insignis, P. radiata*). That beautiful waterfall that I saw and admired from our window in Yosemite at the Awahnee yesterday and last night was not the Bridal Veil, but the upper part of the Yosemite Fall. It is very like the Bridal Veil, but in these seasons of melting snow, there are twelve or more beautiful major falls and an infinite number of minor falls pouring into the Yosemite Canyon.

Decided that Peg and I would go to the Mariposa grove of big trees. We went 36 miles up, up, up the sides of the canyon to the tops of the snow-covered pine forests of the High Sierras. A great experience. The country is composed of massive schists and granites. The story is one of prodigious and repeated earth movements, uplift of the Sierras and Coast ranges, down-throw of the San Joaquin and Santa Clara areas, and also of the Pacific.

The car was fully loaded. We were told to put on cloaks and gloves. We were given blankets and rugs. We hired snowboots and were glad we did. Went through the remarkable tunnel that leads upward out of Yosemite – nearly half a mile, with three ventilating tunnels. Special fans, electric, and carbon monoxide indicators and semaphores. Electric lit. Then on, winding around valley sides through the cold pine forest area, a fine geographical and climatic experience. Saw no bird life. Two bears were about but another 180 were still hibernating. Saw a skunk, and a physician on board told how one had spit in his eye one night at his rubbish bin at home and how he had smelt for 10 days despite all efforts. The pines, cedar pines, sugar pines, yellow pines, cedars, reduvad (sequoia), black oaks, live oaks, maples, all in their order. No lesser trees, no undergrowth, and no grass, only a carpet of pine needles. Not sure as there was six feet of snow everywhere. In the end, past Wawoona, to the Giant Grizzly, said to be the biggest and oldest living thing in the world, there was a bank of snow 12 feet high. Walked across the snow to the Giant Grizzly 3,800 years old, 90 feet to first branch.

All very wonderful. Rock slides everywhere. Gangs at work clearing them all the time. Made first contact with the US CCC—the Youth Unemployment scheme of America. Saw lads up in the forest planting trees, $30 per month, and keep, have to send $20–25 home. Speak warmly in praise of it. Costs money. But it has great popular support. In this democratic land there is a real regard for the welfare of the unemployed youth.

Got back, fresh, cold, crisp, and rosy. Just in time to pack up, and feed up, in the luxurious Awahnee. To Mariposa, poor little village, but quite a beautiful new school. Stone or brick, low spreading well lit, with coloured glazed panels like Mosaics. School just out. Six large buses there to take the lads and lasses home. All very happy and cheerful. All transport is free, paid for by the State.

Down and down, to the cattle ranch country, and to the dairy ranches, and to the level land and to Merced. Saw again at Tuttle the Fancha monument. Very fine and ambitious. Erected by Mr Fancha to himself during his own life. Others do it too, but not quite so openly and unequivocally. At Merced railway station—Jack de Witt, smiling, awaiting us. Got into our Parlour Car de luxe. All to ourselves and off to Monterey, on the trail of the early Spaniards. Miles and miles and miles, maybe 100 or more of the marvellous San Joaquin Valley. In some ways equally remarkable a physiographic feature as Yosemite, and financially and socially immensely more important. A Bacchus Marsh and Tanunda mixed, but much more fertile, and extending for hundreds of miles as far as one can see—to the Sierras on the east and the Coast Range on the west. And abounding water for irrigation. Channels everywhere, and lush high green crops of lucerne and cereals. Some cotton. On through Las Barras, into the defiles of the Coast Range, up and down the striking Pacheco pass, through Hollister—a fine progressive town, Mexican.

Through San Juan Bautiste where the legend is that the swallows return every March 8 (Peg says 17th). They grow very good wine here—more mellow and mature than ours. And they sell it cheap! At Merced here was the only place in all my travels, and in all my experiences with the wine of the country, where I have got a glass of wine (very good sherry in this case) with my meal with no additional cost. Indeed wine (no matter what kind) is 10¢. Tea is 15¢! Peggy's drinks cost more than mine. The tiniest roadside place, abundant ordinary shops, any eating shop, most chemists, *all* sell alcohol. The Marne shop pays $80 per annum for a license.

Then, in the dark, through what appeared to be attractive and mysterious country to Salinas—a fine energetic progressive town – full of light and comforts and bustle. About as big as Gawler or Port Pirie, but with ten times the kick and energy and enterprise. Australia (I love thee still) but it's a dead old place.

30 March 1937, Monterey to Los Angeles

Having the huge luxurious parlour car to ourselves, and Jack de Witt being so obliging, there was no rush. We left about 8.30 am. Did the world-famous '17 mile drive' of Monterey, now 26 miles. Hills, plains, beaches, cliffs, forest, golf links, a most gorgeous panorama. Saw several places that have been used by the film companies for pictures, one for the coast of France (*Les Miserables*) and another place that was palm fringed and was used for tropical pictures in *The Mutiny of the Bounty*. Monterey, apart from the 21 mission centres of Fray Junipero Serra (buried here), is by far the most historic centre in western USA. Was the old capital under Spanish rule. Then it became the capital when this was part of the Mexican Republic, but was displaced by Sacramento when the US Government took over.

Around Salinas many thousands of acres of beautiful land, the Salinas Valley. In one place vast never-ending fields of lettuce—supplies New York etc. At another the biggest sugar beet factory (Mrs Spreckels) in the world. At another, all walnuts, another small pink garden peas, another artichokes which grow along our roads, and so on.

In each centre, whoever one is talking to, the High School is always shown as an important building. It is well to the front, well cared for, with beautiful grounds, and everyone is interested in it and proud of it and expect others to be. Quite different from our country towns and high schools.

At Barkly it was again more hilly, uplifted alluvial country, rounded hills, and here the chief export crop was almonds. Seems a good idea, this specialization of one area in one crop. The river terraces of the upper Salinas River were most interesting. The stream is wide, heavily loaded with silt, and braided. Many long bridges. From dead level in the north, about 100 miles south they are uplifted. Naturally these alluvials are profoundly eroded. Very well developed river terraces, affecting crops, roads, railways, bridges, everything. This was a long study, the erosion, old and recent of the uplifted alluvials. Most of them look Pleistocene to me. But they may in places be Middle or earlier Tertiary. Saw four or five oil fields. Small, some in valleys, some extending into the sea, some on the very hilltops, some on the slopes.

In the irrigated fields, vast things they are, often unfenced along the road, Filipinos are the chief cheap labourers. One 'section' we passed through was Mexican, peons, poorish but varied houses, low standard of living.

The signs along the road that are commonest are first EAT or EATS, and after that GAS, REST and so on. These then are the prime needs of men. Everywhere there were 'Auto Courts'—with rows of cabins for travellers. Excellent ideas of comfort, service, business, enterprise. We are dead and stiff by comparison. Americans appear to feed *very* well. We have been with Draper and Elizabeth Campbell for four weeks now, feeding American style. We have fed under all sorts of conditions. We have never ceased to marvel at the amount, the richness, and the variety presented.

Through El Paso de Robles. Pass of the Oaks. Had lunch there. Very good. A narrow pass. Still traces of snow on the highest peaks of the Coast Range. But it is fast disappearing before the coming spring. Past St. Simeon Ranch, where William Randolph Hearst lived till he was taxed out of this State. St. Simeon Castle. Vast ranches. Warehouses of antiques and treasures of art.

At the risk of repetition I will say again how geographically blessed this country is. Vast series of level fertile plains, with abundant water for irrigation at all seasons, zone of western winds, mild winters, and summers not so very hot. We are seeing things at their best, clothed in the green of early spring. In summer the hills and the non-irrigated parts must be dry and brown.

After lunch we left for Los Angeles. Glorious drive, an absolute geological treat. Most of the road cut along the sea edge of a steep gullied coastline. Been a recent uplift, and the road is partly on raised beach, with wave-cut truncated ridges inland. Tertiary, sands and heavy alluvials, and some hard rocks (? Cretaceous or Mesolithic, generally) of the Santa Monica Range (4,000 feet high)—bang against the coastline.

5 to 7 April 1937, Grand Canyon

5 April, 1937

In a Pullman car, Grand Canyon Ltd, on the way to El Tovar. This is the day of our leaving the lovely land of California. When we were kids we sang, in a game, 'Over the Rocky Mountains, over the hills so cold, We'll line our pockets with the golden dust of the Californian gold.' Which reminds me how very much we young goldfields youngsters in Australia were nourished on American things. Absolutely geographic and historic. It means more to me now that I am familiar with the every day appearance of numberless canyons and caverns. In a cavern, in a canyon, excavating for mine placer material here is quite different from ours, darker and deeper, different, more vigorous erosion.

So we set off. The train journey was excellent. Views for miles of the glorious Sierra Madres in snow and forest. Vast plains of San Bernadino and numberless others. 'Little' places here and there of vineyards and orange groves that would leave Renmark in the middle and not notice it. Up a high pass, getting dryer. Arrowhead Mountains. Torrential streams. Unoccupied land, though always a few homes of men even if some very poor.

6 April 1937

Grand Canyon is a miracle. Quite beyond description. So I will not describe it, nor attempt to. It is the greatest thing of joy and enjoyment and scientific and scenic pleasure that I have seen. Its vastness and depth and variety and purples and crimsons are vivid in my mind as I write. I hope they may ever remain there, for no pictures, even the most exquisite photos and the coloured ones, can give any true idea. Then there was the geology, and the Indians, and the lectures, rocks, fossils, watch tower, ever-present snow and forest, the Painted Desert, the Kaibab Forest, and Bright Angel trail. Sir Victor Wilson and his wife and daughters were with us all day, so that our car was dubbed by an American 'the Australian car'.

7 April 1937

Spent the day crossing the wide desert of Arizona and New Mexico. Americans are quite interested as they pass from one State into another. Trainmen even, who must become used to these things, remark: 'Now we're in Ari-zorna' and so on. If we could have got away from the track the desert might have been even more interesting. For the influence of the Little Colorado River was everywhere. And there was some swampy area, and there was a little grazing, and ranches. Vastly more occupation than our desert. But such ranches, some of them, must have been Mexicans. The most lonely, tiny, miserable places. Numberless. And also there were some fine places like 'Station homes', but not often.

The braided heavily-burdened Colorado was not many miles from the rail track anywhere, sometimes crossing beneath a broad bridge, sometimes washing against the side. Here and there were remarkable evidences of river terraces, here and there no cutting at all. But always red to salmon mud with fine silt. The water, of course, comes from the melting snows, the dust from the friable fretting ferric rocks of the Painted Desert. Grays, creams, and other pinks very abundant. As we got further on, the Vermillion Sandstones (Jurassic) stood up like vertical red walls, but carved into most curious shapes. Great cathedrals, streets, warehouses. I conceived a project in one case for carving out the inside of one great red-rock formation on the inside and forming a rose-red cathedral, like Petra, 'a rose-red city half as old as time'.

Passed close by Meteor Crater and necessarily by Canyon Diablo. Great meteorite lumps as hotel doorstops in one case. Have got to know the look of meteor crater stuff. Petrified forest close by too, but hope to see some of that later. Will get some petrified wood souvenirs. Permian? Gallup is a large coal mining town. Desert. Cretaceous. Largely Mexican, by the look of the living areas. Here and there also real Wild West towns. Girls and boys in groups and in bright coloured dresses. This I think was Buffalo Bill's country, too, along the old Santa Fé trail, though I believe he is buried at Denver.

Shall never forget the cliff panorama as we slipped down into the valley of the Rio Grande in New Mexico. Permian, Triassic, Jurassic sediments. Granites, also. And I must not forget the basalts. I twigged the first tiny outliers of the basalt flows yesterday. Great bare black raw rock masses, cracked and broken as they cooled. No change since. Then there were more and more, then dense basalt flows, then gorges, not deep. Flows mostly very mobile, about 20 feet deep. Later thicker. Mesas topped by basalt Roman wall columnar, scoria, dense as *pahoehoe* etc, a vast and very fine lava field—Recent or Late Pleistocene.

How suitable that Spain should have peopled these areas first, people who knew a land of desert plateaus, with snow mountains. No-one else, maybe, could have done it, put the missions across successfully. Near Zuna, which we saw, is the only remaining pueblo of the historic tragedy of the Seven Cities of Cibola. Though parts of these deserts look like ours, there is a variety, colour, and some quality due to elevation that is quite distinct. Another thing is the highways—road and rail.

Magnificent roads and railways. Incredibly long freight and passenger trains.

8 April 1937, Santa Fé

La Fonda Hotel. Santa Fé. Forgot to mention yesterday that when we got to the lowland there was a vast grey dreary sandy waste—all desert. Deep broad level-bedded alluvium. Through a thick pall of dust, yet not too thick, we could still detect a ghostly glimmering snowcapped mountain. Of the formations seen, there was the Mesa Verde of Gallup, with its coals, Upper Cretaceous. Navajo sandstones, Jurassic. Sauceage: the vermillion cliffs, of which the Rainbow Bridge and other things hereabouts are made; the Shinarump formations, Triassic, in which the petrified forests occur, and the Moenkopi shale of Cedar Mesa (May-sa) seen from the Watch Tower (Grand Canyon), Triassic.

The American Indians here are smiling, homely sort of folk, with impassive faces that *can* laugh. Kids real American. Dirty and spoilt. The Mongolian epicanthic fold very well marked, particularly in kiddies. This afternoon went on a 40-mile drive up the valley of the Rio Grande. Magical geography and geology. Vast terraces and plateaus and fan deltas of heavy alluvial drift. Incredible. Must have been a prodigious period of river erosion and aggradation. Positively incredible. Read Darwin's descriptions of something similar off the Andes eastward. He was shaken by the phenomenon.

10 April 1937, Santa Fé to Kansas City

The closing of my notes didn't close the day. I had just put down my pen when Draper rang me in my room at La Fonda to say that Mr Ernest Thompson Seton, the naturalist whose books and illustrations we have loved so long and so much, was in the lounge, with his wife. Went down. He is a fine figure of a man, nearer 80 than 70 I believe, but looks like a Red Indian, and clear, definite and vigorous in manner. His wife, who is most devoted (Peg says she is not Mrs Seton) is as tiny as he is big. But she too is clever and chatty. We all talked together easily and freely for an hour and a half. He was just back from a lecture tour in Bonn and elsewhere in Germany and full of enthusiasm for that country. Told us much of his work and his recent writings, his 'Indian Bible' about to be published, about Australia, in which both are keenly interested, and also about the character of the Indians. He appears to have quite dropped all his old naturalist work and taken up the cause of the Indians. Very enthusiastic. Told several stories re Buffalo Bill, whom he knew, and many other Indian chiefs and Wild West leaders. Recall one story about

Sitting Bull and the gentleman who was made Chief Scout and Sitting Bull's comment. He was most entertaining.

As we went east the land grew better and better, the rainfall was higher and higher, the fields greener, the houses more and better. Saw scores of Middle West towns, large like Emporia, Garden City, Dodge City, and multitudes of small ones. The layout and appearance of these towns is unlike anything in Australia. No fences, yards or gardens, you just see clean through the town at a glance. River terraces, weirs, oilfields, mules and darkies, wheat, corn, pigs, truck, bustle and business. Utter flatness till we entered the Missouri Valley.

28 April 1937 to 4 May, New York

28 April 1937, New York

Set out for the Natural History Museum. Reached Dr Clyde Fisher, Curator of the Planetarium and of Geology. A charming white-haired gentleman. He was interested in me (because of Professor Johnson's introduction) and in Australites, which I had taken. I promised to send him a set for the Planetarium's exhibit, which I must do as soon as I get home. He called in Dr Chester Reeds, Geology, and the meteorite assistant, Mr Arthur Draper. I showed them the Australites and the other representative tektites. They, or rather Dr Chester, was definitely antagonistic to the idea of glass *meteorites*. He thinks they are volcanic. However, each one of the four of us chipped in and said his piece. I advocated their meteoritic origin very strongly. But if I could see or hear any reasonable volcanic explanation (which I do not think probable) I would forsake the meteorite readily. Today I determined to write that paper Dr Leonard asked for, for the American Association for the Advancement of Science, and I shall call it 'Tektites (Australites)—Are they Glass Meteorites?'

30 April 1937, New York

To Columbia University, where Professor Douglas Johnson had a group of geologists to meet me. Several not there owing to other engagements, but I heard one remark to another that he'd not seen such a showing of University and Museum geologists for a long time. There were eleven professors, including Directors of Museum Departments. Professor Johnson, Professor Sharp, his chief man, Professors Krieger and Law, his department, Professor W. K. Gregory, dinosaurs and vertebrates, Professor Lobeck, petrology, Professor Chester Reeds, Museum, Dr Clyde Fisher, planetarium, and three other geological professors. Eleven. Most pleasant. Dinner in the Faculty Club at Columbia. Yarning. Australites.

Everything was very pleasant. Professor Johnson is a most remarkable host in putting everyone at ease.

4 May 1937, New York

Gave the lecture. 70 to a hundred there, six professors. The slides went well. I think they enjoyed it, and understood something about Australia. It lasted over an hour and though I was told that some of them had lectures, no one left. Then Professor Johnson handed me a cheque from someone's funds for $50, which was very fine, and quite undeserved. Professor D. W. Johnson has made this New York visit memorable.

Canada

20 April 1937, Ottawa

Passed over an exposed part of the Great Laurentian Shield. Enormous *roches moutonnées*, for miles and miles. Granites and gneisses smoothed and planed by continental glaciers. Very level in between. A great sight. Some of the great smooth rocks so large that forests were all over their surfaces. Farms between. Glad to have discussed and been reinforced by Griffith Taylor that the structure and geology of a country are the fundamental facts of geography. I feel I have a better understanding of the land of the beaver and the maple. Not seen the great north and north-west, but have talked with men who know it intimately.

Getting to know the trees a bit. Vast forests of evergreens in places today; larch, spruce, fir, hemlock and cedar. Others are maples, oak, birch, elm, osier, etc. Steep-roofed houses, set solitary, no gardens or verandahs, unadorned wooden, lonely houses. Draper took a photo or two. Hope they turn out well. When not glaciated, then moraine and boulder clay. Boundless erratics or boulders. Here and there, Kingston, Brockville, some level-bedded dark coloured ?Cretaceous.

All these houses have central heating. Saw the ordinary plant in Taylor's basement. Coal up the basement at opening of winter. Janitor comes in and stokes it at morning and evening. Otherwise no attention. Always hot water circulation. There appear to be far fewer cars in Canada than in the United States (except at Ontario). No negroes seen. And no negro quarters in the towns and villages that I could see. Shingles everywhere. Two galvanized iron roofed barns. Stook and log fences. Real pioneer fringe. Fighting cold and water—not heat and drought. Just as hard? Evidence of abundance of wood everywhere. Arrived at Chateau Laurier, the finest hotel we have yet stayed at.

Continental Europe

10 June 1937, Danube Valley

It has been a busy day for me. From 9 am to 6 pm I have been going, with maps, panorama, dictionary, Baedecker, and worrying the anxious couriers. Talked to folk now and then. Used the field glasses on the quaint and ancient castles, fortresses, and strongholds, and also on the abbeys, stifts, churches, monasteries, and on the fields and the houses and the people.

The Danube as I have seen it for over 100 miles is comparable in beauty and interest with the Rhine, of which I have seen much more and shall soon see still more. There is, it seems to me, no saying that one is better or richer in interest than the other. Both are wonderful, uplifting, thought-provoking, inspiring. So too, in richness of interest and beauty is the Thames. And, in quite another way, the Mississippi, where I have spent only a couple of days, or the Hudson, of which I saw much more, or the Nile of which I saw Cairo to Zapazig. Indeed, all rivers, as rivers, and as the homes of men, fascinate me, even more than mountains and plains, of which they are the sculptors and builders. The Murray has its interest, and the Yarra and the Torrens, and even McCallum's Creek, at Dunach.

12 June 1937, Vienna, Natural History Museum

I saw the meteorites, a gorgeous collection. They said it was the finest in the world. And it may be. But I have now seen many fine collections, especially in London and New York, that look as if they were the largest. However, the variety here was remarkable. And the tektites were there in full recognition, labelled Tektiten oder Glas Meteoriten; Moldanites, Billitonites, and Australites. The Director of the museum is Dr H. Michel who has written several papers, and we had quite an interesting talk on the ways and methods of meteorites, especially glass ones.

Saw one of the most perfect geological and physiographic maps of the high Alps that one could imagine—the sort of map that many geologists and geomorphologists have dreamt about or imagined. It was really wonderful. A German Doctor of Science, Haecker did it. What a monument to be remembered by! There it is, forever, every detail of strata and erosion, and forest and talus, the most delicate detailed perfect thing imaginable. It is an education to have seen it. I wish our University would buy a copy of it. Went through the animals, birds—very good collection of moas here, also moa's feathers, abundant insects in amber, coal forests, models of Vienna. One can see that lack of money presses.

Nothing new seems to have been done for 30 years—doubtless since 1914!

14 June 1937, Kecskemét, Hungary

An unforgettable day. Up early. We knew that we were going somewhere, but had no idea of the great treat and novelty that had been prepared. For myself, looking at everything from the point of view of human geography; houses, vehicles, dress, crops, occupations, associated with physiographic factors, it was an immense privilege to get well out on to the great Plains of Hungary, the vast level stretches of the Lower Danubian Plains.

Driving out on our 80 kilometre motor-bus trip, along the main highway from Calais to Constantinople (excellent road), there was much of interest in the suburbs, the city buildings, the great Danube and the way its shores are intensely used here. Everywhere in the ornaments is the eagle, or rather a mythical bird the Turul which guided the first Hungarians here in AD 896. As we got out into the country there was much the same. The same fertility (though there has been a few weeks drought and rain is needed) the villages and paths. The avenues of fruit trees. The woods. But the villages were differently spaced, the houses of a somewhat different pattern, there were more ponds and geese; there were many more cattle and, for the first time, a good few horses. We were coming to the plains, the home of pastoral occupations. Cherries persisted but apricots outdid them. The cows of the Puszta, like those of the Northern Swiss, are curiously of one breed, the Puszta cattle are big horned, creamy coloured. One old bull had silver balls tipping his horns.

The light four wheeled wagon of the Tyrol persisted irregularly and gave way to a heavier typical wagon, nothing like so heavy as ours, with bent wood and iron pieces going over the wheels to the axles. These carts have straw in the bottom and a couple of seats set in, and they form the family conveyance. We drove for many miles across the plains in such a vehicle, with an old Hongroise as driver. The wells and their gear thrilled me for they are in every geography and story book of these great grasslands. The lever is weighted and the water lifter handles the rope or chain that goes into the well. One must assume that all the wells everywhere are at the same level. A great place to develop arguments against the dowser, the omnipresence of water is so obvious.

The crops were largely maize and spuds (both the gift of America) but there were also poppies, sunflowers, wheat, rye, and many others. Much lucerne. Many women worked in the fields with the men, and almost always all were barefoot. There are no stones on these plains. Everywhere as I saw it a blue-gray fine sand. Is it loess? I strongly suspect that it is. Too even for anything else. The land where cultivated is under strip culture, but everywhere fenceless. We saw cattle, sheep, pigs, and geese out in herds and with their herdsmen (sometimes women or children). We saw also goats and cows tethered and hobbled. Reaping is done with the scythe, and this is men's and youths' work.

The houses are often prettily red-tiled, but many are thatched. There are no building stones. The houses are not of wood, I suspect mud brick, always neatly painted white. Each house has its well. Towards Kecskemét the houses were more like ours in their spacing, but in no other way, except that they have rooms, doors, and windows. But often one saw houses set in their own gardens, on their own property, not clustered in villages so much. Often too one saw a sort of communal place for the grass hay stacks, 20 or 30 or more, all belonging to different people obviously and all in one space.

On the plain we saw three great black and white storks, and a storks' nest in a tree. Also spur-wing plovers, mole hills (many and new), and a little grey animal that may have been a weasel or stoat. The herds of horses and cattle on the *puszta* (prairie) we saw well. The cattle and horses have tinkling bells. The herdsmen are nomadic for the summer grazing season, and have movable houses.

When we got to Kecskemét, the mayor welcomed us in English, German and French, at some length in a very charming and well-decorated town hall. Then, escorted by soldiers, we took the train. Rattled along, on board seats in ordinary trucks, with a calico covering overhead. Uninterrupted view. Waved to everyone, in the fields or by their houses. There are few 'great open spaces' here where you can't wave to someone every minute or two. At Bugac, met by retinue of cowboys. Got into the village waggons, two or three plus a driver, and away we went over the *puszta*. Jolted and all that. But great fun. Then we got to the place where we fed. Great preparations. Abundance of apricot liqueur and apricot brandy, very cheap. On the tables an abundance of wine. Announced that all the wine was free, but that coffee cost extra! A rush for the wine. It was very good wine, the liqueur was exquisite, and the apricot brandy very good, *tres bon, sehr gut*. All the dishes of the country, cooked their way: Soup, Chicken, Sweets, Goulash, Wine, Wine, Wine.

17 to 20 June 1937, Italy

17 June 1937

Again we passed through some hundreds of kilometres of fascinating country, rich, green, fertile, well tilled and cropped. No fallow. Grass hay in abundance. Finally stored in high places, may be 20 feet high. We were not far from the Italian border all day, and at least three times we passed considerable numbers of armed mounted soldiers, with batteries and machine guns. These regions are troublous places to live in. The boundaries can never be fixed. England is fortunate, she has about the same mixture of races, but there are no boundary lines where one race may merge into another, or where doubt may exist. 'Around us indivisible the sea', as in Australia.

Noted the flat floors of the valleys and decided to write a paper on them. Lake formed. Saw two places, particularly one at Lake Dobiacco, where the whole process could be seen in progress. Shrines of all types very numerous. At times three close together, within 20 yards. Houses a bit different. Larger, better painted than Hungary. Brighter. They look very comfortable. I felt that I should like to live in one. Here in these Alpine fastnesses, where everything seems to be done as it has been done for ages, it is curious to see the power pylons striding across the land! The New and the Old. There was a vast amount of timberworking. Wood distillation. Sawmills in abundance. Water power. Valley of the Drave (Drau). Rushing waters everywhere, and it needs only a little channel to divert some to a water wheel and there you are. Village follows upon the heels of village. Churches, shrines, wells, and other villages lie higher on the hills where the woods have been cleared, or away across the fields where the valley becomes a little wider. An anti-erosion scheme on the cutting faces was made of plaited willows. These sprout and hold the soil. The fences are interesting also, no nails, strips of planks, two uprights tied with reeds.

18 June 1937

Forest, snow, and this type of house have been with us for a long time in Switzerland, Austria, Germany and Hungary, and now in Italy. But there are differences. I should love to get a book on these houses, and their differences from place to place. There is a general type of the mountain and valley peasant home, but the Swiss are far the most trim, the German and Austrian neat, the French the least clean and tidy and orderly, were it not for the Italians.

On the one side, what a contrast between the houses and villages and fields of France and Switzerland. And again now, what a contrast between those of the mountain villages of Austria and Italy. Then there are the churches, more numerous than the villages. With their different typical towers, and the extraordinarily numberless and varied shrines. Both these villages are Austrian, towards the Tyrolean (Versailles) border. As we got lower and lower the forests gave way to fruit trees and other warmer trees, for we were coming to the Mediterranean lands. And at Treviso it was a garden. The long flat fertile plains from there south to Merazo and Venice are a miracle of fruitfulness. They are astounding; there is a riot of growth, but orderly, with rich soil, abundant water, and abounding sun.

The crops, three at a time—fruit (hard pruned) vines and roots or cereals or lucerne all at once. And the glorious long avenues of great-boled planes (sycamores)! It was truly a revelation, though I had seen already (though at another season) the wonderful flat lands of the Valley of the Po River farther west. So on we came to Venice itself. The approach is much more attractive and romantic than it was by rail [in 1931], for you can see the whole magic city there before you as you approach, towers and campanile and all.

20 June 1937, Milan

There is evidence of very heavy erosion, mostly kept in check by man, in the rushing streams from the Dolomites. The abundance of limestones in all these great mountains is a surprise to me. Not only in those made known (in analysis) by Monsieur Dolomieu. It was Sunday, but in many places peasants were in the fields, turning the grass hay that had got wet. Men and women and girls.

This evening I have written an article (for *The Australasian*) on the trip around the lake, so I shall not put very much here. White walls, blue water, vivid green trees, red roofs. It is most beautiful and romantic. The geology at Torbole is very good, also the obvious geomorphology. The abundance of limestone has much to do with the good roads and buildings of Italy, an important geological contribution. The western walls are steeper and are tunnelled for the roads in a wonderful way. We tore around some of these bends and through the tunnels in a hair-raising way. The circuit of the lake is 100 miles or more. All the roads are in limestone or dolomite. The upper road is newer and untunnelled. The cost must have been considerable and the engineering skill very great. All the tunnels are pierced every few yards to give both ventilation and a view of the lake.

22 June 1937, Grande Corniche

Just as I always hoped to travel once upon the Appian Way in Rome (which I did) so also have I always wished to travel on the Grand Corniche. Now I have done the whole thing from Genoa to Nice, under gorgeous weather conditions and very pleasantly. The mountains, good and high, come right to the coast. Try as I will, and I do try everywhere, I cannot get views of mountains, villages, roads, crops, etc. Plenty of castles, and churches, and pleasure grounds, of man and his urban works but little of man and the country and his work there. Steep terraced slopes, facing the hot Southern sun. Vines abundant, with olives, figs, Mimosa (some blossoms, some dry pods). The rivers are sometimes dry, heavy with shingle always, but the majority had an outlet to the sea. Everywhere women washing clothes on the shingles of these aggrading braided streams, and it was quite common to see the women carrying home the washing on their heads. Here and there the patient ox in cart or plough, here and there a donkey in cart or with packs, here and there an ox and horse harnessed together.

First heavy limestones, I am astonished at the amount of limestones in these mountains, a most important geographical fact. Later heavy mica schists and schists generally. Then, very interesting, extraordinarily heavy conglomerates; Flysch, glacial, or heavy pluvial and coastal action due to overdeepening and rapid uplift of land? Often silicified. Here and there at a sunny small port such as Alassio there were holiday makers, speed-boats, ships, sun-bathers, regular rows and rows of them lying on their bellies on the sand. Savona seems to be an important coal port, small good harbour, heavy cranes, etc.

In addition to other crops there now were fields of trombones, asparagus, tomatoes, artichokes, potatoes, vines, figs, carnations. Plenty of Australian acacias planted at the roadsides, some kurrajongs, mesembryanthemum (like ours but maybe native), olives, figs, prickly pears, agaves, all the things that grow well with us in South Australia. Also albergos, ristoranti, pensioni, castles on peninsulas and islands, relics of the 'good old days', when 'they might take who have the power, and they might hold who can'. Must not forget the fields of castor oil plants. There were the usual strange old churches looking as old as Time, thick stone walls, close green shutters, iron gratings on all first (ground) floor windows. Some gum trees, some tamarisks, sheoaks.

24 to 30 June 1937, France

24 June 1937, Nice to Grenoble

We were on Napoleon's route of his return from Elba in 1815 all day. We came to most astounding country. Was it Jurassic? Valleys in synclines. Two formations, one capping the hills, and resulting in some of the most striking and bizarre forms I have ever seen, on a huge and magnificent scale. Castellane with its great rock, musketeer legend, church, vermouth and cognacs were very good. Digne, good also. At Digne Napoleon reformed his troops. Met Louis. Poorer valleys, smaller villages, dingier chateaux. Sisteron, very old, with the great braided

Durance. Saw wild geology along the old French-Italian boundary. Got out, picked Salvation Jane, wild chicory, and lavender. Magnificent gorges, narrow defiles, huge rocks, briar roses, dwarf junipers, like the scrub of Santa Fé. Then came snow peaks to the east and high needles and crenellated tops to the west and broad fertile villaged valleys between. Always perilous clinging roads. Huge river terraces, mighty ones. And so, with interest and conversation and speculation we came on to Grenoble.

25 June 1937, Dijon

A remarkably enjoyable journey and much memorable geology, physiography, villages, statues, crops, trees, and famous towns and cities. I was sorry to leave unmentioned all we saw at the charming and historic old town of Grenoble, with its Roman wall and its Revolution declaration of 1788, and all that. Left Grenoble early, cloudy and dull, but cleared up to a fine sunny day, except when we climbed above the clouds on the High Jura Mountains, north of Geneva, and got into cold and fog and wind once more. The scenery and human geography of Savoie and Haute Savoie were very fine. High mountains, dense forests, serrated snow mountains, Belledon Range, to the east, deep gorges and defiles, with miraculously high bridges. Wide valleys and pleasant slopes with the characteristic French culture of villages, vines, crops, people; grapes, wheat, beet predominating.

Then borders, *douanes,* passports, etc and we were in Switzerland. Distinctly unlike German Switzerland and Italian Switzerland. This remarkable country is just the most mountainous valleys of three separate countries, and the peoples keep their own languages and customs with the laws, peace and prosperity of the one. Then Geneva, very beautiful, and the statuary and the League of Nations buildings.

Then out and away after a lovely lunch high above Lake Lucerne with Mont Blanc dim in the far background. Across a fertile plain and rising upland into the barrier mass of the mighty Juras. 'Scythe and hoe culture'—the phrase I coined for it. Up. Up. Up, till we crossed the Pass des Fantilles at 1323 metres, about 4000 feet, which we had climbed in half an hour. Down and down—different people and occupations, but much the same. The charming old towns of Champagnolles and Poligny, where we strolled a bit, then Dole, where Pasteur was born, and now we were out on the wide fertile plains of the Cote d'Or, where we had five years ago travelled across in the train by night. And so through all these miles of beauty and interest we came to Dijon, the chief city here, with its churches, monuments, arch, soldiers, fortresses, mustard, cassis and chocolates.

30 June 1937, Paris

Professor A. Lacroix is old, tall, impressive, with a cast in one eye, and a long white beard. He speaks some English and with a word or two of French we had a long and fascinating conversation. I have now known or met all the most famous workers on tektites. I gave him a flanged button. He gave me a beautiful collection of Indo-Chinites, from five separate localities and types. And he gave me his last paper on Indo-China and the Ivory Coast, and he asked me to write him a few lines, with my signature to add to his collection of signatures of 'famous mineralogists'! His assistant, an excellent man, with good English, showed me through the collections. Saw where he had melted Darwin Glass and Indo-Chinites and made new tear drops out of them by gravity. The Indo-Chinites are wonderful. Very large and with wonderful regularity of form compared with what I had thought. I was permitted to handle the largest ever found, four kilos, and was given a fragment. I handled also the treasured two specimens from Columbia, South America, which Professor Lacroix thinks are not tektites. But I do not feel convinced he is right. They are so like them. Water-clear, nearly, when looked into. Sub-spherical and pitted. Saw also the astonishing Ivory Coast tektites, undoubted!

Among the meteorites was a whole case of 'seen to falls'. No rusting. Many chondrites, some moulages, beautiful widmaustalten figures, one the best I have seen. Their largest iron meteorite is from Touat, Sahara, 500 kilograms. Saw and handled poires, larmes, sphères, ellipsoïdes, disque baguettes, plaques, and bulles fazeuses. They show much of piezoglypts (thumb prints) and Monsieur told me that these were to be seen also microscopically on exploded grains of gunpowder. And so I left Professor Lacroix with much understanding and goodwill.

11-12 July 1937, Prague, Czechoslovakia

I was puzzled about the river. Peg suggested the River Moldau. Inspiration. Of course. Moldau. Moldavia—Moldavites!—the first of the Tektites. Next day I went too see the collection of Moldavites at the National Museum here; it was glorious. Two huge fine cases with seven long rows each, 35 large, well selected specimens in each row. Seeing them in the mass gives one a better and a different idea of their story. I have the characteristic types: but all these were fine large specimens. Although they are very far from being as regular in form as the Australites, yet there is a very definite series of form types. Those I have are quite characteristic. The amount of etching and pitting and erosion generally is notable and all the kinds of pits, such as one gets in other tektites, are here abundantly. There were three Australite specimens also, very poor; three of Darwin Glass, one Billitonite, four from Cambodia and three from Annam. The Australites were from Lake Eyre. Among the Moldavites were the cusp shaped pieces of burst bubbles, irregular as in the Indo-Chinites, and much rarer than with the Indo-Chinites. I had a good eyeful of these, and it was very good, indeed.

31 July to 20 August 1937, Holiday Trip to Spitzbergen

Comment by FF: So far their travel had been primarily to get from one place to another to work. The Spitzbergen trip was a well-earned holiday, on a ship called the Stella Polaris. It was also Father's first real experience of glaciation, of which he had, of course, learned, in theory, during his geology courses in Melbourne years before.

2 August 1937

It is now 10 am, and we have just entered the Norwegian coast. It is a miracle of newness to me. The landscape is different from anything I have ever seen. A low peaceful island, with its barren grey granite rocks, its white houses and churches, and red roofs, and small level green patches here and there where food is grown. There are no trees, not even a shrub.

It is the island of Karmöy, on our port side, near the north of the Boknfiord. On the starboard side are other islands and peninsulas of the mainland. These rocks are a very beautiful grey. In this broad morning sunlight a purple-silver-gray. The rock is bare, mossed, jointed, and worn, by aeons of sun and wind, and sea, and before that by ice and snow. It looks very old. Gives me the impression, purely psychological, perhaps also a little bit physiographic, of being the oldest landscape I have looked at. But it is, of course, new also, still in the making, but in such hard and enduring rocks, hard bare granites, the wearing down process must be very slow.

Bays, with deep water, natural harbours, large and small, are countless. These, as I see it, are the *viks*. From these cosy *viks* as hiding places and homes, the old Vikings emerged in the years gone by. I had known and read something of them, but I have now to learn how far afield they went, ruling over Ireland, parts of England, parts of France, and surprise of all to me, entering the Mediterranean, going up the Rhone, seizing Pisa, and besieging Constantinople (865 AD).

4 August 1937

Have watched the glacial physiography of the islands for a while. It grows more and more marked, more and more severe, naturally enough, as we are at the moment within half an hour of the foot of a retreating glacier. It cannot indeed be so many thousands of years since these islands were covered. The sculpture is unmistakably different, all these bare granite faces (schists too, and dykes, all contorted) have been so evenly levelled off. The ice mass, with its grinding rocks, is no selector, or not much so, and there is less, much less, of that intricate selection of the softer or more erodable rocks, less differential erosion, less of the etching effect that comes with slow subaerial decomposition plus the work of running water, or the equally selective, but quite distinct, etching of hot sun, dry wind, and sand blast. Here and there, there are green grasses and small patches of soil, and a few trees, but they are something of the later effects of rain, sun, and air, and running water. Still, not much impression has yet been made on the original job done by the glaciers.

From the point of view of human geography, it still impresses one that houses and homes should exist on such bare and stony coasts. Most of the harvest must, in all cases, come from the sea. The big event today was the visit to the Svartisen glacier. It looks to be no distance from the water's edge to the glacier, but it is a long and rough and rocky path, through dwarf birch woods and over acres of glacial pavements, and huge moraine boulders and gravels. The crevasses and hollows in the ice revealed the most deep and wonderful blue. I climbed out upon the glacier surface, huge ice crystals, one inch cube average, and if I had not bought a spiked stick I should have slid all over the place, as others did.

A big river runs from the glacier to the fjord. The walls are steep, and apparently it is stronger in winter. Also, as at the Rhone Gletscher, one gets the impression of small misfits. The glacier is just a remnant, left in a valley built to fit a huge glacier. This one is of course huge by comparison with the Rhone glacier. But here beyond the Arctic Circle, where we have now encountered perpetual snow at sea level, one expects to see glaciers at their best, and they will doubtless grow bigger and better yet. For there is a long way to go.

Among the thousands of thousands of boulders there were only a small percent striated. I guess many boulders are carried and never get ground. The variety of folded schists, gneisses, marble, actinolite schists, Kyanites, etc was exhilarating. Got a better idea than ever I have had of the general scheme of glacial erosion, the narrow sharp ridges, the place where the descent starts and where erosion is most rapid. Above is the old peneplain, upon which the glacial period commenced. One pictures it covering all, and going on away across the North Sea to Scotland and England.

5 August 1937

There was beautiful evidence on the island beaches, for miles, of an old coast line 100 feet above present sea level, with wave-cut cliffs, etc. Practically all the settlement was below that line. The houses do not cluster, but are lonely and separate as a rule. Though they are wood, and look like English houses (or Canadian, etc) they are really thick-walled, of hewn wood, covered with pine boards. After lunch we came to Tromsø, on an island, quite a busy town, a town of fish and furs, fishers, hunters, etc. Salmon drying arrangements all about. Quite a lot of shipping. One sees more big ships at once than on most sea roads.

Lyngseidel has been a treat. Not so much for the Lapps' Camp, though that was remarkable and interesting. But because of the walk home, through green woods and pleasant fields, crops of oats and potatoes, nice homes, perfect blue fiord seas, blue mountains, indescribably beautiful, jagged and snow capped, not exactly capped, the peaks are free from snow because it is summer. But there are numberless glacial patches. The air, chill now we are out on the fiord again, with the ship bowling along, was mild and beautiful. Which shows what a breath of warm air can do—the warm air that comes from over the North Atlantic Drift. It is a remarkable climatic experience.

Saw also at very close quarters 200 of those small invaluable timid beasts, the reindeer. Very timid. High proportion with fine velvet antlers, many calves. As they run their feet click in a very curious way, click, click; don't know whether it is the joints or the hooves. Saw the papooses, with the epicanthic mongol fold more obvious than in the old folk. The broad cheekbones were marked, square heads, etc. Mongolian, Scandinavian a bit too. But living at a distinctly lower culture level than the Norwegians who have fine homes here, with electric light, abundant heat, gardens, radio, and all the comforts of the present day, except nearness to big centres.

6 August 1937, Hammerfest, Norway

Hammerfest, three to four thousand people, whose harbour never freezes—the most northerly town in the world. Naturally, it is a strange and fascinating place, with its narrow streets of wooden houses, sheltering along the narrow strip of low land at the foot of the cliff of metamorphics; fine schists and gneisses. The vast racks of drying fish, the houses, three storeys, filled with dried fish, a literal fact. Neat trim houses, sturdy fishermen, good shops, abundant souvenir and postcard places, a fine old wooden church with a good organ, good stones, an altar-piece of Christ helping Peter the Fisherman and other pictures of hills and ice and heaven that were quite geographically appropriate. It is here that we are to have our week without night.

At about six o'clock we came to one of our chief goals, the great grim North Cape, Nord Kap. To the west there is a low peninsula of reddish granite, with a beautiful margin of intrusion into the schist, which is black. Beyond, North Cape stood up black and steep. The nearer peak is called Knieskjollen. We sailed beyond it and around to a sheltered bay on the lee side. Then the ascent of the Cape. In a steep chine or 'valley', a winding rocky path had been cut, very steep, with ropes here and there to aid one, and with wooden ladders over the worst bits. But the ladders were steeper to climb than the rocks. The whole climb was about 1,000 feet, 972 I think. And if you think that's not much, just think what it would mean to climb the Eiffel Tower, step by step, or the Empire State Building.

We landed. Peg came up far enough to get a good view. I pegged on. And by sticking at it, despite my age, indolence and obesity, I beat it. But when one got to the top the North Cape light (and restaurant) were far away. So away I went across the moor, real dinkum tundra, wouldn't have missed it for anything. An old mature land, almost a peneplain. But such a 'soil', such vegetation! It was a geographical treat. The soil was full of boulders, most peculiar and remarkable. Formed I think by freeze and thaw, freeze and thaw. Low vegetation, blueberries, grasses, mosses, peats, springy soil, but everywhere full of stones, an utterly new physiographic phenomenon to me.

16 August 1937. Gerainger Fiord, Norway.

We have had a wonderful journey into a strange wild country—high bleak desolate glaciered Norway. Five of us, of a party of 150, in motor cars, with excellent drivers, travelled for the whole day in these uplands. First through the wild, narrow, and beautiful Norangsdal Valley, past the lake caused by a rock avalanche, and the ruins of the buried houses still visible, up past poorer and poorer homesteads, in country where one cannot imagine that man could make a living. Peat bogs and peat digging. Glaciated rises and massive moraines. Small pine forests, birch and beech. Stunted woods, then mosses, bogs, peat, and morasses, but always berries and wild-flowers.

We had three stops on the journey, all at good viewpoints; Visnes (morning lunch), Videsaeter (lunch), and Grotli, high up in the snow country for afternoon coffee. Finally of course, beautiful Merok, one of the loveliest imaginable of places, source of half the drop scenes of the theatres. We must have been up 5000 feet or more, in the country where the treeline goes only a few hundred feet above sea level. Roads that wound up like the Furka and Grimsel Pass Roads in the Alps, but Peg and I thought them higher and steeper. The whole scenery was alpine, and yet totally different. For it was all in these massive schists, gneisses and granites. One huge scale impression is of the barriers several hundreds of feet high that the old glaciers rode over, smoothing and glaciating rock surfaces at all angles.

I have acquired a wonderful lot of knowledge, that can only be got by seeing, about glaciers, first in the Rockies, then in the Alps, then in Spitzbergen, and now again in Norway; about the way glaciers do their work, the long low valleys and then the line, somewhat marginal, of abrupt descent, and the economic impact of that. Glorious rock basins with alpine lakes, glacial-gouged, we saw in plenty. Bare bleak rock faces where not even mosses and lichens grew. Vast moraines. Upland moors, etc.

17 August 1937

Just before dinner a young American came and asked me if I would give a lecture on the 'Geology of Norway and Spitzbergen' to the young folk on board, in the forward lounge, starting at 10 o'clock, at which time the ship left Balholm. I agreed. So they provided a blackboard and chalk, and I took my glacial boulder. At 10, when the usual three cannon shots announced our departure, I gave my first sea lecture. Quite a good audience, and a good number of questions at the conclusion. So that I was answering questions to a late hour.

It has been fine seeing this naked and unashamed physiography of Norway, with the unique chance, as we went north to Spitzbergen and the polar ice, of having the same experience as if one stayed here in the south and went back in time, for tens of thousands of years, to the various successive stages of freezing. Spitzbergen is almost completely glaciated, but not so much so as Greenland and Antarctica; and Norway must have been as deep under ice as central Antarctica at one time. The naked schists and granites have huge crevices, master joints, faults, and thrust planes. These are *always* occupied by streams. Most interesting. One guesses this in soil covered places like the Mt Lofty hills. Here it is clear and obvious. One sees the elements of physiographic development in a wonderful way.

The other point that is fascinating in these valleys is the rhythm. Fundamentally, maybe, this rhythm depends upon the regularity of the occurrence of master fissures. Therefrom comes the rhythm of streams, of talus slopes, of bastions, of peaks, of fiords themselves, and there is a rhythm of glacial lakes and (coming to the realm of geography) there is a regular recurrence of fertile flats, and therefore of villages, of schools, and so on. One gets an idea too of the whole great scheme of continental glacial erosion, the 'push' of the vast snowfields, the point or line where fiord making starts, the influence of 'plucking', which is most important, the headward erosion of glaciers, the concentration on the valley, and the consequent profound deepening there while the upland is not so greatly changed. In many ways the river cycle is paralleled, yet different.

2 to 7 September 1937, Great Britain

2 to 3 September 1937, Nottingham, BAAS Meeting

Got a note that I was on the Section C (Geology) committee, and I believe I am now on Section E (Geography) committee also. Met Dudley Stamp, who had very kindly had me elected a Vice President of Section E (Geography) and asked me to their Section Dinner on Monday evening as their guest. Since I last saw him, he has visited every country in South America, and (at invitation of USA) inspected every state of that country, and (for British Government?) has inspected Nigeria re soil conservation, and who has now been invited to visit and report on the mapping of India and China. He is a smart fellow, and an indefatigable traveller. But I fancy it's a fine thing for an ambitious man if he has the good fortune to live in London and has the further good fortune to have so influential an uncle as Sir Josiah. We talked of world geography and of Australia's north and soils, and soil utilization, and culture patterns in England. He stresses (as I have lately in these notes) that the land pattern of England is something that has evolved through 2,000 years of struggle between man and nature, roads, rivers, products, traditions, and cannot be lightly set aside by the whim of thoughtless 'town planners' who have never allowed these things to enter their heads. And he spoke of opening up the waste lands of the world, and of us in Australia having to give Australia to Indians. And I said *Cui Bono?* If lands are productive, population increases, and the pressure becomes as great as ever. Why should we try to fill up the earth with struggling people. Look at Fiji. To whom the good? Why indeed should we try to increase the world's population at all?

Professor Fawcett (London) gave his Presidential address on World Movements of Population. He stressed the fact, and supported it with unanswerable figures, that the Southern Hemisphere could never be of any great importance in world history. Have often stressed to my students the same thing, based on the same facts, but presented differently. He brought out many other remarkable facts about world tendencies in population. It was a very fine effort, indeed, one of the best I've heard at any time.

I am just back from the Geography dinner. A most excellent and pleasant meeting. I was seated by the President as an honoured guest, with Mrs Haile, wife of the Trent waterworks engineer, who has made all those wonderful flood models, on my right. And Mrs Roach, sister of Brigadier General Winterbottom, who has had charge of surveys all over the world, on my left. Both were very pleasant and we talked freely so that the hours went by without noticing. Next on the left was Dr Bews, of Natal, South Africa, who has written a geographical book that I must get hold of. Next on the right was, of course, Professor Fawcett.

Fawcett, London, is a very fine man, and I have told you of his excellent Presidential address. Opposite was Dudley Stamp, who has written geographies of all the world, and all its countries. We talked of possible Land Utilization mapping in Australia. He it was, I suspect, who gave the President material for the remark in his speech of the quiet-spoken man from Australia whose books they knew and who had lately travelled 35,000 miles at the rate of 150 miles per day. A bit of a stretch of my figures. Just before I left a young man of the London University came up and said he must meet me, as he gave two lectures every year from my books! And, in general, everyone was very friendly and nice. Saw Stevens and Maclaren to say goodbye. General Winterbottom is the most charming fellow, fine after-dinner speaker.

3 September 1937

I knew the day and date to-day. It was imprinted on memory as August 5 was in 1931. This was the day set down for the delivery of my paper on Australites to the British Association for the Advancement of Science. How far that little hobby of mine has led me! With no special knowledge or ability to deal with the problem at all, but just because I was so intensely attracted by the puzzle of them that I wanted to interest other people in the problem and so try to find out all about them. And now I have presented papers on them to four different scientific societies, in South Australia, Australia, America, and England! And it all dates back to one evening at our little Science Club in Ballarat when I gave a popular talk on them!

The Presidential address in Geology was 'drooled on' (as Skeats put it) far beyond its time. Then my paper came. Late. He asked me to be as short as I could and I promised to stick to my slides alone. Which I did. But everyone seemed very interested and appreciative and so I didn't think any more of time. Took my half hour and a bit more. So that the two succeeding papers were very cramped. Still, they were not of such general interest as mine, which should have been given more time. Felt a bit self conscious about my pronunciation, for the first time in my life, and especially tried to avoid the word 'Austrylia'. All my jokes went off very well. I feel from what was said that they are prepared to accept cosmic theory. One commentator drew my attention to the fact that Professor Pickering, in the *Astronomical Journal* a few years ago, had advanced a theory that the tektites were formed when the moon was torn out from the Pacific Ocean. Must look that up, and include it in future publication.

Then Dr Fritz Paneth, a very charming and brilliant chemist, works on meteorites, etc, helium content, and the age of the solar system, etc. He is now in London, an exiled Jew, from Königsberg. Seems a rotten shame to turn him away from Germany. Well, he had seen my name and paper, I had met him in London in 1931, and he came over from his Section. Asked me to dine with him at his hotel, which I did. And we had a long and valuable talk on the possibilities and probabilities of the arrival of the Australites.

In case there might be someone on the trip to-day (4 September) who was interested in tektites, I took a pocketful of choice specimens with me. At the first stop a gentleman came to me, asked a host of questions about them, wanted to see them, and had some interesting suggestions to make. He sat with me then and continued the talk. Turned out to be Dr E. B. Bailey, FRS, Chief of the Geological Survey of Britain.

Next stop another gentleman tackled me on the same thing, even more interested, and with more suggestions to make. His wife also was very charming, and knows quite a lot of geology. He was Sir Lewis Fermor, OBE, FRS, late Chief of the Geological Survey of India. Referred me to two papers of his, of which he has no reprints left, one on the 'Origin of Meteorites', *Journal of the Asiatic Society of Bengal*, and another on 'Garnets as a geological barometer', *Records of the Geological Survey of India*, both about 1911–12. He has a theory that explosive (deep) earthquakes, and isostatic balance, and basalt production can be explained by the alternate solidification and fusion of garnets. Most attractive. Remember most of it, but won't record it, as I hope to find it in the journals he speaks of.

Next came the President of the Section, Professor L. J. Wills, of Birmingham University, who is interested and wants some for his museum. Promised to send paper No. 2 to Professor Tilley, and Professor Paneth, Cambridge College of Science. Various other people also were interested including a man from Leicester University Museum who wants some specimens. Oh, and also Dr Rudolf Richter, the *Universität*, Frankfurt am Main. He was very keen, and we had long yarns. He asked if I should mind him publishing a popular article in one of his journals on them, and including some of the illustrations from my papers. Since the object of my reading this paper was to try to convince English geologists of the interest that lay in Australites, and more so of the fact that they must be accepted as of meteoritic origin, it was pleasant to have several of the folk refer to them as 'those meteorites'. Mr I. S. Double, of Birmingham, was also keen on them.

7 September 1937, London

Decided to go down to the Museum at South Kensington, and see whether I could get a copy of Darwin's Australite. Saw Dr Key, temporarily in charge. Nice fellow. Glad over a new mineral from Cornwall, a Bismuth Template!—fancy a new mineral from Cornwall at this late date. Then we saw the new tektite collection. Most beautifully set out, with my Shaw collection stuff duly honoured. Colombian tektites also, and I saw all the original Wabar silica glass blebs etc. Had a general good inspection, under Dr Key's direction, of all the tektites and meteorites. Went down to the crypt, and was given two casts of Charles Darwin's Australite. More important still, I saw and handled that classic object, the first Australite recorded. Around the inner margin of the flange is some of that red stuff that has not yet been analyzed, but which may yet help to solve the problem of origin, it may have excess of the original metallic content of the speculative parent meteorite! Somewhere in these notes is recorded the first brain wave I had to hunt up Darwin's Australite when I got to London, and to interest folk in them from that focus. This has been successful beyond my anticipations.

South Africa

24 September 1937, Cape Town

Of course, one always leaves out the really important things, the striking moments, and records the trivial. One BIG thing was the first sight I had of that great grim grey-purple precipice that is Table Mountain. It was just a glimpse caught over the top of some railway carriages, but that sight and memory will probably still be with me when all the other things I write of are forgotten.

Mr Sisson Cooper, the Manager of the *Cape Argus*, asked me to lunch, and he placed a beautiful new Buick car at our disposal for the afternoon. We did most of the journey through magnificent mountain, coastal, and fertile valley scenery. And so one gets comparisons and contrasts and we had a good South African (English-born) chauffeur who knew the country, and one learns (with prior reading) something of local ideas and local values. For a good part of this trip the road (convict-built) was cut in an incredibly steep cliff face, possibly thousands of feet high in places. Geologically a treat also, the ancient peneplain of grey granite as clean cut as possible and the red sandstones laid down on those. We saw the Cape itself very well and the marvel of Table Mountain, and the miracle of the Table Cloth, which is not just a dead mass of ordinary level cloud lying on top, but is an ever moving fleecy cascade coming over the top and flowing down and disappearing at a certain level.

25 September 1937, Cape Town

Several times I have thought of all the things Dad told me of 'the Cape', which he visited on his way to Australia 70 or more years ago, and of which he often spoke—particularly of the black men who had such 'long arms'. Noted today the prevalence of things we have from here in Australia, such as Cape Weed, Cape Broom and South African daisies, etc. On the other hand one might go for miles here through wild wattles and gumtrees and imagine one was in Australia. But there is more variety on the Cape Peninsula than anywhere I know in Australia, because of the tremendous contrasts of mountain, cliff, coast, plain, and valley.

26 September 1937

Have had another magnificent drive, in quite a different direction and much interesting experience of South Africa. Mr and Mrs Hemer, with Cynthia at the wheel, called for us at 2 o'clock. We set out north this time through Milnerton, then turned sharp east toward Paarl and Stellenbosch. The whole way across the sandy flats and the low rolling sandy hills that separate this part from the Hottentot Holland mountains was covered by wattles in bloom. They call it Port Jackson mimosa. And abundant eucalypts, largely sugar gums *Eucalyptus corynocalyx* and blue gums *Eucalyptus globulus,* and pines.

Went first to Stellenbosch, a most interesting town, with its *sluits* and valves along the streets, and old Dutch appearance—the home of the Stellenbosch University—the Dutch (Africaans) University of South Africa. In a rich moist valley, lying among high crenellated mountains. Then we drove further on into the mountains, the Drakenstein mountains, and Simons Berg, to Helshoogte. Had tea there, coloured girls, at a Dutch farmhouse tea rooms on a beautiful *nek*—or col, with a view down into deep and beautiful valleys to both east and west. Then we went further on, on until as a matter of fact we couldn't go any further. We were at the home in the V of the mountains, that rose like purple precipices all around, the farm of Mr Buller. Lovely Dutch home 150 years old. Friends of Hemers. Saw his vineyards, and hot-houses and proteas and bulb acres, and heard of the baboons and leopards that he shoots on the farm, 16 in one morning in the oaks at the house—baboons. With the bodies they trap leopards. Mr Buller is a fine old gentleman, and it was interesting to see the inside of South African farm conditions here. 49 inches rainfall. It was so interesting that it was nearly dark before we set off for home, and some miles of bush track.

27 September 1937, Cape Town to Johannesburg

On the 'Blue Train' between Cape Town and Johannesburg.

The level-bedded rocks, with their characteristic arid erosion, look like the ruins of forts and castles. Even here there is an occasional small lonely cemetery, testimony of some battle of the past. Near the line, or not so far away, there is usually some river or stream. About here, in the gathering gloom, the soil is everywhere dominantly yellow, such soil as there is. And in the *kopjes* one sees the remains of an earlier peneplain. There are some remarkable erosional effects to be seen.

Mud huts, goats, and donkeys, very rare. Like the half-breeds' huts on the Colorado desert. Here in the stony desert is a most remarkable thing: a Botanic Garden! The Whitehill National Botanical Gardens—the Karroo Gardens. All the Karroo plants. So that now we are really out upon the Great Karroo. I watch interestedly for Dwyka, and recall the indomitable E. J. Dunn, who did so much to discover and explore the famous Dwyka conglomerates! In a cutting I saw rocks very like varve shales. Also thought I saw glacial boulders along the base of the netting wire fence.

28 September 1937

Looked out upon a vast, rolling, dead khaki coloured landscape. My God. South Africa. Passed close by to the Modder River and Paardeberg! One sees more of the little forlorn and neglected cemeteries. How these all bring up memories of 37 or so years ago. And the 'purple patches' of the Boer War and Kipling's South African war poetry, and my 17-year-old mad enthusiasm for the flag-flapping jingoism of those days. How well I was led, with my fellows, by the well arranged public opinion propaganda, sending off our fine Australian boys to help England smash this little handful of Boers, in their own country, in what everyone now recognizes as a thoroughly unjust war.

29 September 1937, Johannesburg

Went to the Mines Department and saw Dr L. T. Nel, assistant director of the Survey, most obliging. Long and interesting talk, but not enough of the geology and physiography of South Africa. He gave me a full series of maps and books covering the geology of all South Africa, for me to read on the ship. Have had a go at the maps already. He is Dr Malherbe's brother-in-law. Here also I saw some of the most beautiful aerial photographs, and the way they make them 'come solid' with mirrors, and the way they interpret them on to maps for geological base maps. A miracle. It is almost as good as the American Geological Society's work on Alaska, but with a cheaper plant. The photos are about one foot square, are immensely superior to our Australian military aerial photos, and are on sale by the Aircraft Operating Company, 23 Rogers Street off Booysen's Road, Johannesburg. Might get the University to buy a few for me when I get back.

1 October 1937, Johannesburg

I shall be for ever grateful to Fate that I decided to stop off on our homeward journey to see South Africa. I am specially glad that, having done so, I was able to come to this marvellous city of the champagne air, the City of Gold of the world, Johannesburg. And, having got here I thank the Fates for having handed out to me such a day as today, for these reasons:

First. I have walked the streets of the wonder city where East meets West, and where the Magic Magnet of Gold has attracted the brains and energy of the most adventurous citizens of the world. Here on the High African Plateau, more than a mile above sea level.

Second. This morning I have been down the richest of the mines of this richest treasure house of nature that has ever been known, and I wandered in the stopes, east and west along the Reef, 850 feet below sea level.

Third. This afternoon I have flown over the city, over 2,000 feet in the air, seeing the reef from east to west, more than a mile and a half above sea level. What a day!

I will elaborate on the second and third adventures. I went down the shaft of the Crown Mines for 3,500 feet, and then down another shaft for a farther 3,500 feet, total 7,000 feet, which is 850 feet below sea level. I saw and had explained to me the whole great business. This mine has an area of over 40 square miles, which means a solid 60 or so square miles of payable continuous reef, dipping gradually and consistently south at an angle of 30 degrees. The manager's job is to take this out and put it through the stampers, the mills, and the cyanide plant. They do this, losing only .02 dwts [pennyweights] of gold per ton which goes to the dumps, and helps to pave the streets of Johannesburg. They crush

350,000 tons per month, pay 100 per cent dividends to their shareholders, give work to 25,000 natives and several thousand whites, and treat their manager like a prince and their staff like lords. I wish I had the time and pen to tell of the impressions of going down these depths, the casual men in charge, the winding gear, the ubiquitous natives, the marvellous ventilation system and its efficiency, the beautifully organized haulage system, the pressure on the ear drums, the fact that at top the descent is 3000 feet per minute, the feeling that you are moving up again when the skip slows down, and so on. I forgot to say that before I went down I was introduced to the General Manager, Mr A. V. Lange, and we talked for some time, looking at the coloured plans and sections on his walls, about the geological structure of the reef, etc.

Practically the whole of the mining is done within the reef itself, so there is no waste effort mining and hauling country rock. The whole of this is a most stupendous geological miracle, nothing comparable to it anywhere in the wide world. I got a full idea of the mining methods, the hauling and the ventilation. At that great depth I saw the natives drilling in the stopes, angle holes 42 inches long, 18 inches deep from the surface. The gold is not in the pebbles of the conglomerate but in the pyrites and silica of the matrix. At 7,000 feet the air was as fresh as at the surface. In places it was warm, but not so much. 2,000 feet lower the rock temperature was 103 degrees but they are cooling it down in an ingenious way by ventilation so that when they come to work it, the temperature will be about 93 degrees.

On this mine they have I think 19 shafts, used for various purposes, and they haul from three. An interesting interlocking system of chutes and bins and inclined shafts and electric trains and eight-ton trucks provides for a haulage system comparable to the train system of a city. Then we came up, and went to the roaring stampers. Then another mile or two to the cyanide plants. No, I forgot. Then to the ball mills, where the 'balls' are larger pieces of the banket itself, a most ingenious idea. We also saw the picking belts and the crushing mills for the larger pieces.

Then to the *sanctum sanctorum* where the gold is poured. Luckily it was the pouring day, and I saw thousands of pounds worth of the red-green fluid gold poured out. And I saw 15 great ingots, valued at more than £15,000. And lifted one, which would not be easy even if one did not have rheumaticky hands.

Lastly, my flight over Johannesburg. It was to have been in a seven passenger cabined Junker, but the pilot was on leave. Would I go up in an open plane, a Moth? I was quite comfortable and immensely thrilled and astoundingly interested in all that was set out below. The chief thing was the mines, the cyanide plants, the dump heaps, the numberless gum-tree plantations, the winding streams, and I could detect those that flowed off to the south and the Vaal and the Atlantic, from those to the north and into the Limpopo and the Indian Ocean. I flew right over the Crown Mines shaft, over 2000 feet in the air, and saw the spot where I had been this morning down to 850 feet below sea level. Was that not an experience for one day!

Then there were the towns and gardens and race-tracks and kaffir 'locations', and grass fires, and some most beautifully sculptured hills and valleys, with their southward dipping Rand rocks. And from the air, in open fields, as you know, one sees all manner of curious things that date back to the long past. And I speculated on some round marks that I thought were maybe the sites of kaffir villages, and others that seemed to have been the homes of animals, for here not so long ago the baboons barked, the lions roared, and the wonderful animal life of Africa flourished. As I looked at one side and then the other, and was now and then delighted to recognize the true significance of some strange marking, I now and then also lifted my eyes to the far horizon, where the brown veldt stretched away to interminable distance in a hazy horizon. And maybe I thought that was the best part of the flight. The best physiographers and geologists (and engineers) of the future must be aviators as well. It is not enough just to go up in a flight. One must be able to fly often and wherever one wills. There was an extraordinary and unexpected number of dams and reservoirs, and always there was that impressive line of white heaps, the Rand, the mines where all the energy and science and skill of engineers and chemists and geologists, and all the effort of tens of thousands of black men and white is concentrated on bringing together into yellow bars the fine gold that is minutely disseminated for over a mile in depth through that vast series of rocks.

8 October 1937, Cape Town

'These are the last sad days.' Driving home with Professor Young, wise in his old age and tolerance and his many years of Africa, I asked him what was the future of the whites and coloureds here. (Coloureds are Hottentot–bushman–Malay–white crosses.) He said: 'In 300 years there will be no distinction between white and coloured.' I believe that this statement, made public here, would bring down the most profound public indignation on his head. He said he based his belief on the exceptional 'coloured' folk. Some of them were cleverer than their fellow whites, and of more value to the community. There was a coloured man in Transvaal. They wanted him to go for Parliament. There were objections. It was said he was the cleverest man they had with figures; they needed such a man. He stood, got in, and was the first Minister of Finance in the Union Parliament. (His son was put off a train in Natal for being a coloured!). Then there's Dr Abdarahman, who is the smartest man in the Cape Town Municipal Council today, an Indian 'coloured'. Then there was Brink, who became De Beers's chief diamond expert and valuer, no white could touch him. When he retired the Cape Government took him on as their diamond valuer, against immense social opposition. But he was the best man available. So there you are.

Chapter 18. Reflections, Frank Fenner

In this final chapter, I want to reflect on two matters, one that spanned my working life, namely special friendships, and the other an analysis of the relative importance of nature, nurture and chance, as they affected my father and myself.

Special Friends

Friendships are an important element in everyone's life. Lifelong friendship with one's wife is the most important and one that I enjoyed in full measure, but I have had many other friends. Here I want to acknowledge the debt I owe to several special friends who have been important to me at various periods of my life, many of who still are. I mention a number of them at some length in the 'boxes' in earlier chapters; those who were close personal friends were Ted Ford, Cecil Hackett, Francis Ratcliffe, Ian Marshall, Gwen Woodroofe, D. A. Henderson and Isao Arita. Two of my mentors, Macfarlane Burnet and René Dubos, were also close friends. Here I will list some other friends chronologically and alphabetically and provide a few comments on each of them.

School Days

The one person who I remember as a special friend from primary school days is John Dowie, a noted sculptor and painter who still lives in the same house as he did in the 1920s. I greatly enjoyed visiting him in his house and studio, with my sister Winn and my brother Bill, as recently as April 2005.

The one person I remember from Thebarton Technical High School is Alf Chittoch, who I met again when I visited Tennant Creek in 1968, where he has lived since the war and risen from being a 'battler' in the mines to Mayor. We have maintained annual correspondence ever since.

University Days

The friend that I remember best from Adelaide High School and university days is Denis Shortridge, who subsequently became an ophthalmologist in Sydney. Later he moved back to Adelaide and I used to visit him every time I went there until he died in the mid-1990s. In recent years, whenever I visit Adelaide I arrange to have lunch at the Army and Navy Club with all those of my year at medical school who are able to come. We had a special meal there in 1994 to which about a dozen came; now the 'regulars' have fallen to two, Alan Campbell and Malcolm Newland.

Army Days

My best friend in early Army days was Noel Bonnin, a surgeon who was six years older than me but who was initially in the same unit as I was (2/6 Field

Ambulance) and carried out some research with me when we were in the Woodside Camp. Later, when we were both in the same unit in Lebanon (2/1 Casualty Clearing Station) we went on two wonderful week-long trips, first to Petra and then to Luxor and Aswan. Later my close friends in the 2/2 Australian General Hospital were Orme Smith, Ian Wood and Rod Andrew, and in New Guinea, Ted Ford.

Melbourne Days

I lived in Melbourne from October 1945 to August 1948, and again from February 1950 to November 1952. During the second period, Bobbie and I lived in and she looked after the houses and children of Alan and Mavis Jackson (I knew Alan in New Guinea as a pathologist and we co-authored one scientific paper) and Macfarlane and Linda Burnet. Both Bobbie and I became close friends of the Burnets and the Jacksons.

Canberra Days

Bobbie was a much more sociable person than I was, and several of my current friends in Canberra derive from her friendships. In the ANU and especially in the JCSMR, I met a very large number of people, many of whom became close friends. I also met a number of scientists during my overseas travels who became, and remain, close friends. Co-authorship of books was important in developing friendships. I will list some of them as Local Friends and Overseas Friends, arranged alphabetically.

Local Friends

Gordon Ada

I knew Gordon from early Hall Institute days and he came to Canberra several times to give seminars here. In 1962, I tried to recruit him as Reader in Biochemical Microbiology, but the Hall Institute was still too attractive. Then, in 1968, he was appointed as my successor. I moved to CRES in 1973 and we became really close friends in 1991, when he came back to the School as a Visiting Fellow and occupied the office next to mine. Since then we have always had morning coffee together and discuss the woes of the modern world.

Rod Andrew

Rod was a senior physician in the 2/2 Australian General Hospital when I was there in 1942 and after I went to New Guinea he was deeply involved in looking after volunteers who had been given malaria at the LHQ Medical Research Unit. After the war he became Dean of the Monash Medical School, and in 1964 he was responsible for recommending the award of my first honorary degree (MD). I visited Monash University intermittently after that and, in April 2004, I gave the first Roderick Andrew Oration there.

Stephen Boyden

Stephen, an Englishman, worked with René Dubos a year after I did, and was subsequently appointed to the Department of Experimental Pathology in JCSMR. He was elected a Fellow of the Australian Academy of Science in 1966 for his work in immunology, but in 1965 he had decided that he wanted to get away from reductionist science and transferred to my department in that year, to work on human ecology. He subsequently transferred to the new Department of Human Biology and in 1975 to CRES, where he worked until retirement in 1990. In 1991, following the interest expressed by several people in the audience when he gave a lecture at Questacon, he set up a non-governmental community body, the Nature and Society Forum, of which I became an early member and later a patron. I regard him as perhaps the most creative younger scientist whom I have known well (thus excluding Burnet), and he and his wife Rosemary have been among my closest friends, especially during the last decade.

Ruth Conley

Ruth was the nursing sister at Canberra Boys' Grammar School and a good friend of Bobbie's. Since Bobbie's death and her retirement to a cottage near the Yarralumla shops she kindly invites me to have lunch or dinner with her and other friends.

Jack Crawford

Jack was a senior public servant and a member of the tennis group that I had joined as soon as I was settled in Canberra. In 1960 he was appointed to the ANU as Director of the Research School of Pacific Studies, and from 1968 to 1973, a period coinciding with my term as Director of JCSMR, he was Vice-Chancellor. From the mid-1960s, we were very close friends, and he and his wife Jess played bridge with Bobbie and me every Saturday night for over a decade; we visited Jess every weekend after Jack died in 1984.

Walter Crocker

Coming from a South Australian farming family, Walter had extensive experience in the British diplomatic corps in Africa during the war. In 1948, Copland persuaded him to join the Research School of Pacific Studies as Professor of International Relations. We got to know him well because he built a house almost opposite ours in Torres Street, Red Hill. He transferred to the Australian diplomatic corps in 1952 and initially filled important posts in India and Indonesia. In 1956, he invited me and two colleagues to Indonesia for three weeks to advise him on the best location for a medical school in Sumatra. Then, in 1960–61, he organized a six-weeks visit for me to virology laboratories all over India. After he retired in 1970, he went back to his farm, but in 1973 he was appointed Lieutenant-Governor of South Australia; I stayed with him once at a time when he was living in Government House. Eventually he retired to a house very near the beach at Grange, an Adelaide suburb on the coast. For many years I used to visit him there whenever I went to Adelaide; he always plied me with penetrating questions. He died in 2002 at the age of 100.

David Curtis

David had joined Jack Eccles as his first PhD student in 1954. He rose to be Professor of Pharmacology when that department was created in 1973 and was appointed Director of the JCSMR, 1989–92. He was elected a Fellow of the Academy of Science in 1965 and appointed President, 1986–90. He and his wife Laurie were always good friends and he hosted many lovely dinners which we enjoyed at his home in Campbell. In 1999–2001, we collaborated in producing a history of the first 50 years of the John Curtin School.

Max Day

Max Day was born on exactly the same day as I was, one year later. For many years, Max and his wife Barbara and Bobbie and I would celebrate our birthdays with dinners, alternately at our respective homes. In the early 1950s, he worked in CSIRO Entomology on the transmission of plant viruses and he was the first person with whom I collaborated in laboratory work in Canberra. I used to come up from Melbourne to work with him on the mechanical transmission of myxoma virus by mosquitoes. Later he became a member of the CSIRO Executive and then Chief of the CSIRO Division of Forestry. Appointed a Fellow of the Australian Academy of Science in 1956, he became very much involved in environmental problems in Kosciusko. He was an early member of our tennis group and we played on his court in Melbourne Avenue for over twenty years, until he sold the house in 2002.

Jack Eccles

Jack was older and more distinguished than the other early professors when he joined us in Canberra in 1954. I got to know him well when he was President of the Australian Academy of Science in 1957–61 and I was Secretary, Biological Sciences. In the early days, I also used to play tennis with him and some of his children in their home just down the street from ours and Bobbie was very friendly with his wife Rene.

Marcus Faunce

Marc Faunce came to see me shortly after he came to Canberra in 1957, enquiring about research work in the JCSMR, which at that time had no connections with Canberra hospitals. He was a first class physician, attending no less than five Governors-General and three Prime Ministers. He also examined me once a year and more often if necessary, essentially he was my GP until he retired. Then we used to visit each other on alternate week-ends, in Wanniassa and Red Hill. After his colon cancer became incurable in 2004, he stayed for several months with the former chief pharmacist of the Canberra Hospital, Enid Barnes, and I used to visit him there at least once a week. He was cheerful until the very last days. I still visit Enid and see two of Marc's other close female friends there every Monday afternoon.

Bryan Furnass

Born in Manchester in 1927, Bryan came to Australia in 1966 as physician to the ANU Health Service, in which he developed the concept of 'Wellness', established walking trails around the ANU campus, and served until 1993. He then became very active in the Nature and Society Forum, where I came into closer contact with him and his wife Anne. He was a skilful writer on popular science, which included books on nutrition and microbiology and letters to the *Canberra Times* and the *Guardian Weekly*.

Adrian Gibbs

I had met Adrian, a plant virologist, on a visit to Rothamstead Experimental Station in England in 1965. He joined the Department in 1966, went back to England in 1969 for a few years and then came back, with my support, to a senior position in the Research School of Biological Sciences, ending up as a Professor and a Fellow of the Academy of Science. Our common interest was viral taxonomy, and as an animal virologist I found his insights very valuable. One of his sons was an expert balloonist, and he and Adrian took me for a wonderful flight over Canberra.

Alfred Gottschalk

I first met Alfred, a refugee from Hitler's Germany, when I went to the Hall Institute in 1946, and I got to know him quite well then. After he retired in 1958, at age 65, he joined the Department of Microbiology as a Visiting Fellow, funded by an NH&MRC grant. He had separated from his wife and lived alone in a flat in Northbourne Avenue, and we used often have dinner together. He returned to Germany in October 1962, to a position in Tübingen; he died there in 1973. I was the executor of his will, which included a donation of $36,231 to establish the Gottschalk Medal of the Australian Academy of Science.

Alan Jackson

Alan had worked with Burnet from 1936 to 1939 and then spent six years working as a pathologist in the 2/9 Australian General Hospital. I saw quite a lot of him when we were both in Port Morseby and we collaborated in a long paper on infections with *Salmonella blegdam*. Bobbie and I really got to know and love the family when we looked after them while he and Mavis spent a

Keogh-organized year overseas just after we had come back from the Rockefeller Institute. Each of us always visited them whenever we went down to Melbourne, and we remained very good friends with Alan, Mavis and the children.

E.V. (Bill) Keogh

As I mention in Chapters 3 and 4, Bill Keogh played a major role in my career. We were friends, but he was a reserved man and much older than me, and I don't know if he had many 'close' friends, although he greatly influenced the careers of many army medical officers. However, he and fellow Adelaide University medical graduate of mine, John Funder, Sr., went on a couple of long drives with me through the countryside in Victoria.

Kevin Lafferty

Kevin joined the Department of Microbiology in 1957, as a PhD student with Stephen Fazekas. After postdoctoral studies in Canada he came back to the Department as a Research Fellow and later rose to be a Professorial Fellow in the Department of Immunology. After several years as a professor in the University of Colorado in Denver, where I visited him several times. In 1993, he returned to Canberra as Director of the John Curtin School and was very supportive of my work in the School.

Jim McCauley and Doris; Cam Webber and Joanna

My GPs after Marc Faunce retired, Jim and then Cam, have been good friends. Since Jim's retirement, I always seem to be running into him and Doris at various 'Friends' events, at the National Library and the National Portrait Gallery.

George Mackaness

George had spent a few years with Florey in Oxford before he came out to Canberra in 1954 as Head of a small Department of Experimental Pathology. He carried out some excellent work on cell-mediated immunity in intracellular bacterial infections, for which he was elected FRS in 1976. A couple of years after Colin Courtice had been appointed Head of Experimental Pathology in 1960 George accepted a chair in the Department of Microbiology at the University of Adelaide. We became friendly because we both had an interest in cellular immunity. Then, in 1965, René Dubos, with whom he had spent a study leave year, invited him to be Director of the Trudeau Institute for Medical Research in Saranac Lake. Later, in 1976, he became President of the Squibb Institute for Medical Research and went to live in an enormous converted barn in New Jersey. Bobbie and I visited him in Saranac Lake and also in his 'barn'.

Barrie Marmion

I first met Barrie when he came from England to the Walter and Eliza Hall Institute as a Rockefeller Foundation Visiting Fellow in 1951. He came to Australia again in 1962, to be the foundation Professor of Microbiology at Monash University. In 1968, he moved to Edinburgh for 10 years, then came back to the Institute of Medical and Veterinary Science in Adelaide. He officially retired in 1985, but continued to be very active in investigations of various aspects of Q fever. I have known him throughout this period and seen him most often during the last ten years. I see him and his wife Diane whenever I visit Adelaide.

Cedric Mims

In 1953, I visited the research institute in Entebbe, Uganda, primarily to see John Cairns, who joined the Department of Microbiology two years later. Cedric went to Entebbe in 1954, and joined my department as a Research Fellow in 1956. His work on the pathogenesis of infectious diseases was of special interest to me and he worked his way up the ladder to become a Professorial Fellow in 1968. In 1972, he was appointed Professor of Microbiology at Guys Hospital Medical School in London, and proceeded to write a book on the pathogenesis of infectious diseases. I visited him several times in London and also in his lovely home, a couple of hours by train south of London. All their four children were in Australia and he and his wife Vicki came out periodically to see them, but in 2002 they decided to come out here permanently. They used to come for gossip and a drink on alternate Sunday evenings; Vicki had to move to a nursing home a year ago but Cedric continues to come.

Geoff and Margaret Rossiter

Geoff was Executive Officer of the US Educational Foundation (Fulbright scholarships) in Canberra from 1950 and Bobbie became a close friend of his wife Margaret. In 1965, he became Warden of Burton Hall, at the ANU. After he retired and after Bobbie had died and I had moved into the extension of the house, I initiated the pattern of drinks and nibbles with them, which changed when Margaret had to move permanently to Morling Lodge, a nearby nursing home, because of severe arthritis. Gwen Woodroofe, Geoff and I would then have a drink at my place and then drive up to Morling Lodge and have another, and more gossip, with Margaret. Both Geoff and Margaret died in 2004.

Richard Smallwood

Professor of Medicine in the University of Melbourne, Richard resigned from that position in 1999 to come up to Canberra as Chief Medical Officer of the Commonwealth of Australia. I met him when he was developing procedures for handling a bioterrorist attack with smallpox virus. We lived close to each other and he and his wife Carol often invited me to dinner at their home. They also used to come to my weekend drinks and nibbles.

Michael Studdert

In the early 1980s, when I decided to publish *Veterinary Virology,* I enlisted the help of several overseas veterinarians, but I also wanted an Australian co-author, and Michael, the first Professor of Veterinary Virology to be appointed in Australia, was the man for the job, and we became close friends. As an expert in equine virology, he made valuable contributions to the original book and all subsequent editions.

Bob Walsh

Bob set up the New South Wales Blood Transfusion Centre during and after the War, and in 1973 became Dean of the Faculty of Medicine at the University of New South Wales. I first got to know him through the Australian Academy of Science and later the Medical Research Advisory Committee of Papua New Guinea. We became very close friends and served together on various Academy committees. I would always call and see him, and often stay overnight with him, if I went down to Sydney during the 1970s.

Hugh Ward

Hugh's brother, Dr Keith Ward, was Government Geologist in South Australia and I knew him in prewar years. I got to know Hugh when he was Professor of Bacteriology at the University of Sydney. During the 1950s, he used to ask me down to Sydney to lecture to his medical students and I would stay overnight with him, at a time when I was never asked to lecture to undergraduates in ANU. A very unassuming man, he had a remarkable record in World War I; as a medical officer he was awarded the Military Cross with two bars. He, Mac Burnet and Bill Keogh exerted considerable influence on medical research in Australia in the early post-war years.

Wendy Whatson

Wendy was another friend I inherited from Bobbie. As well as periodic dinners with her and John at their lovely home in Shackleton Circuit, in Mawson, since Bobbie's death in 1995 Wendy is the most generous friend anyone could dream of in terms of dropping in beautiful cooked meals for me every now and then.

David White

David's father, Harold White, the creator of the National Library of Australia, lived close to us and regularly invited Bobbie and me to his garden parties. His eldest son, David, a Sydney medical graduate, was one of my first PhD students and later became Professor of Microbiology in the University of Melbourne. He was an outstanding teacher, and when I decided to produce *Medical Virology* as a student's textbook, I persuaded him to be a co-author, and we collaborated in all later editions and in *Veterinary Virology*. Naturally, we became very close friends. He and his wife Marjorie came to Canberra at the time of my 80th birthday celebrations and David was one of the three speakers at the celebratory dinner.

Overseas Friends

Derrick Baxby

With poxviruses my early and continuing scientific interest, I soon found that Derrick, who had trained with Alan Downie at Liverpool University, knew more about the history of vaccinia virus that anyone else in the world, and we conducted a long and protracted correspondence. I also visited him whenever I was in England for more than a few days, and we became and remain good friends.

Joel Breman

Joel worked in the Smallpox Eradication Unit in Geneva for three years and in 1978 he accompanied me on a visit to China to check its eradication status. We have kept in touch ever since and I have occasionally collaborated with him to produce articles on smallpox. He is currently Director of the Fogarty International Center at US National Institutes of Health and we still exchange papers and emails on medical topics.

Bob and Beth Chanock

The Chanocks are my closest friends in the United States. Bobbie and I first got to know them well in 1972, on my second term as a Fogarty Scholar. He was a virologist and head of the infectious diseases unit in the NIH; they lived in a large house in Bethesda and apart from the periods when I was on a Fogarty scholarship I always stayed with them when in Washington. When I was there for prolonged periods as a Fogarty Scholar with Bobbie in 1973–74 and 1982–83, they also took us to concerts at the Kennedy Center and introduced us to the many exhibitions in the galleries in Washington, and were most loving and hospitable. They came out to Canberra at the time of my 80th birthday celebrations and Bob was one of the three speakers at the celebratory dinner.

Alan Downie

Alan, Professor of Microbiology in the University of Liverpool, was the leading British expert on smallpox in the 1940s. I first met Alan and Nancy when I visited Europe in 1949 and always went to see him on later visits. Keith Dumbell and Derrick Baxby were among his students, all were experts on poxviruses. Alan and Nancy both came out to Australia and stayed with us in Canberra.

Keith Dumbell

A PhD student with Alan Downie, Keith Dumbell became Professor of Virology at St Mary's Hospital Medical School and was the leading authority on variola virus during the smallpox eradication program. We met often at committee meetings in Geneva. After his wife died he married a South African woman and they went to Cape Town, where he was appointed Professor of Microbiology. I used to stay with them whenever I visited Cape Town and although he has now retired we still correspond regularly.

D. A. Henderson

I have outlined the career of DA (as he is always called) in a 'box' in Chapter 16. Naturally, I saw a lot of him in Geneva throughout the campaign, and much more when I was writing *Smallpox and its Eradication*, for which he was the author of almost all the 'operational' chapters. He did this while carrying the heavy task of Director of the Johns Hopkins School of Hygiene and Public Health. I stayed with him and Nana whenever I visited Baltimore, and I was delighted when he and Nana came out here for my 80th birthday celebrations; he was one of the three speakers at the celebratory dinner and awarded me a unique accolade, Grand Master of the Order of the Bifurcated Needle.

Zdeno Jezek

I saw a lot of Zdeno, who was on the staff of the Smallpox Eradication Unit, in 1980–88, when I visited Geneva several times each year while producing *Smallpox and its Eradication*. He was a great help in going through the WHO Archives. From 1980–86 he had also been in charge of a major WHO study of monkeypox in Zaire, and in 1987 we collaborated in producing a small book, *Human Monkeypox*. In 1995, I was able to visit Zdeno and his wife Eva in their home in Prague, where he took me on a wonderful tour through that beautiful city.

Edgar Mercer

I list him here, but Edgar was just a year ahead of me at Adelaide High School and the university, and I got to know him well when I spent three months in St Andrews College in Sydney in 1940 studying for the Diploma of Tropical Medicine, since he also boarded there. In 1963, he came to the JCSMR as electron microscopist. He was also a sculptor and made two sculptures for my garden. In 1967, he went to Phoenix, Arizona, and made a living as a sculptor, but left there a few years later to become Professor of Biology at the University of Hawai'i and lived on the Big Island in Hawai'i. I visited him there several times and he drove me everywhere. He retired to the mountains in Switzerland, where again I often used to visit him.

Fred Murphy

I had met Fred several times when he was at the Centers for Disease Control in Atlanta, and he was a Visiting Fellow at the JCSMR in 1970–71, but our close friendship dates from his involvement in the production of *Veterinary Virology,* the first edition of which was published in 1987. The chapters he wrote were excellent and his comments on those written by others very helpful, and after I had pulled out he supervised the production of the third edition in 1999. Our friendship was cemented by a week-long trip with him and his family through Yellowstone National Park in 1997, just after a meeting of the American Society of Virology in Bozeman, Montana. I have visited many interesting parts of the world, but none more exciting and unusual than Yellowstone, in such company. Very recently, he sent me a copy of his 'Memoir', a marvellous life story with superb colour photographs and a most illuminating text.

Nature, Nurture and Chance

I believe that the three words: nature, by which I mean the combination of genes that each of us inherit from our parents; nurture, by which I mean the way that our physical and our social environment, especially during childhood, influence our physical, mental and emotional characteristics; and chance, in its dictionary definition, the 'way things fall out, fortune', cover every other happening that influences our careers as human beings. Within these definitions, I would now like to compare my father's life and mine.

It is difficult to compare the genetic component. Neither my father nor I had any worrisome disease with a genetic basis, although possibly his addiction to smoking had some such components (I never smoked, so I don't know if I carry those hypothetical genes). In terms of the capacity to work hard and carry out scientific research effectively, we resembled each other more than either of us resembled our male siblings. I believe that my liking for writing books about science and science history, rather than exclusively in journals and review articles, which is the practice of most experimental biologists, must have been inherited, for Father was a very able science writer. The examples of two of my mentors, Burnet and Dubos, both of whom wrote many books as well as hundreds of papers on experimental biology, may also have had an influence.

In relation to 'nurture', father was born into a poor family and had to work as an 'apprentice printer' (before the days of formal apprenticeships) from the age of 11, although all the evidence that I can obtain is that the large Fenner family had very supportive parents. On the other hand, I was born as the second child of a middle-class family and had the best opportunities available in the country, at that time, in my school and university education. Father had to make his own way from printer's devil through pupil teacher to Teachers' College and university, whereas my childhood and adolescence were arranged by my family as I wanted them. There could hardly have been a greater difference.

The principal element in both our careers, given our innate abilities as shaped by nature and nurture, was chance. I presume that his marriage to Emma Louise Hirt contained an element of chance, in that they both happened to be boarders at the Melbourne Teachers' College at the same time, and she was a most supportive wife as well as a loving mother. Otherwise, it is difficult for me to evaluate that element in my father's career, except to express the view that the great disappointment of his life, I think, was the failure of the University of Sydney to select him to fill the place of Griffith Taylor as Associate Professor of Geography in 1929. Here, chance undoubtedly played a role, for if the opportunity had come after his trip to England in 1931 as one of the Australian representatives at the Centenary Meeting of the British Association for the Advancement of Science, I feel certain that he would have been selected, for he made a very good impression on the British geographical fraternity on that trip (see Chapter 17).

Although I have more empathy with his work as a scientist than with that as a bureaucrat, some comments on the latter are relevant here. At the time of his 'importation' from Victoria in 1916, he was regarded as an outsider by many South Australians, and he was finally appointed to the influential position of Director of Education just as World War II began, and resigned because of illness just after it finished, so that his visions for the future could not be realized. It is useful here to recall two tributes to his work in the Education Department, the entry in Hansard in 1947 by the Hon. S. W. Jeffries, long-time Minister for Education, and the comment by the Director-General of Education, J. S. Walker, in 1967 (see Chapter 14).

Chance elements in my own life were undoubtedly influenced by my father's standing and friendship with colleagues in the Royal Society of South Australia. Thus, I was given the opportunity to take part in an expedition organized by the Board of Anthropological Research when I was a second-year medical student; this led directly to my work on Aboriginal skulls which constituted the basis of my MD degree. When I began writing this autobiography, I thought that the event that played a major part in my military career and most of my subsequent career, the decision to go to Sydney and obtain a Diploma of Tropical Medicine (DTM) before I enlisted, was a matter of chance. None of my fellow-interns did that and my father did not know enough about medical training to have advised me. However, thinking over the early history of World War II, Australian troops were being directed to the Middle East in January 1940, and I may well have decided that in the Middle East a DTM would be likely to open up career possibilities not available to a raw graduate.

It was that step that led to the switch of my interests from physical anthropology (and in a university sense from a career in anatomy and cell biology) to infectious diseases and thus to my post-war career. On the way, it led to my close acquaintance with Neil Hamilton Fairley and Bill Keogh, and through them to my experience as a pathologist at the 2/2 Australian General Hospital, and, later, my appointment as a malariologist. Most importantly, if I had not been appointed as pathologist at the 2/2 AGH, I would never have met Bobbie Roberts, my future wife.

It was Bill Keogh who arranged that I should work with Burnet after the war, and later, in consultation with Burnet, to take a year's study leave with René Dubos. Three later career decisions were not matters of chance. In 1967, I decided to apply for the position of Director of the John Curtin School because I had spent so much time writing *The Biology of Animal Viruses*, and, without a training in biochemistry, felt so out of touch with molecular virology, that I thought it would be difficult to take up bench-work again. Then in 1973, approaching the end of that appointment as Director of the John Curtin School, I was influenced by what I saw as Hugh Ennor's mistake, namely the risk that I would think I knew all about the job if I took on a second term. Directorship of the newly established Centre for Resource and Environmental Studies offered the opportunity of indulging my interest in environmental matters, which had become prominent in my writings in the early 1970s, while still Director of the John Curtin School.

Finally, the fact that I have been able to continue writing books, book chapters and review articles, and giving occasional lectures, ever since I retired from paid employment in December 1979 depended on my continued good health, and the fact that I was working in the John Curtin School of The Australian National University, institutions that encourage such activities.

Index of Names

Numbers in bold indicates that the person is the subject of an illustration on that page, numbers in italics indicate a 'box'.

Abbott, C. L., 236
Abdarahman, Dr, 326
Abou-Gareeb, A. H., 149
Ada, G. L., 101, 180, 192, 328
Adey, W. J., 222, 225, 226, 233, 234, 269
Adler, S., 29
Agol, V., 76, 105, 127
Akimoto, Crown Prince, 178
Albert, A., 52, 55, 63, 64, 70, 98
Alibek, K., 170
Alpers, M., 28
Anderson, S. G., 86
Andrew, R. R., 328
Andrewes, C. H., 26, 65, 107
Aragão, H. B., 82
Arita, I., 142, *144,* 151, 152, 153, 154, 155, 160, 163, 164, 165, 166, 169, 170, 178, 179, 180, 192, 327
Atkinson, N., 38
Avery, O., 51

Bachman, P. A., 172, 173
Badger, G. M., 99
Bailey, E. B., 319
Bang, F., 36, 66
Barnes, Bishop, 288
Barsky, J., 131, 172
Basten, H., 275
Batman, J., 10,
Baxby, D., 335
Beale, J. G., 116
Beare, E. H., **28,** 29
Beaton, C., 173
Bedding, R., 179
Bell, O., 261
Bellett, A. J. D., 71
Belozersky, Academician, 76, 77
Bennett, J. H., 56
Bennett, J. H., 58
Bensusan-Butt, D., 98
Berns, K., 105
Best, J., 183, 184
Bews, Dr, 318
Birdsell, J. B. 24
Bishop, P. O., 101, 112
Blamey, T., 34, 35
Blanden, R. V., 104, 127
Blythe, M., 119
Bohr, N., 291
Bonnin, N. J., **28,** 31, 32, 50, 327
Bornemissza, G., 129
Boyd, J. S. K., 29
Boyd, R. 59, 61
Boyden, S. V., 71, 100, 101, 124, 129, 133, 191, 192, 329
Bradman, D., 19
Bragg, W., 284, 285
Brandl, M., 134
Breman, J. G., 131, 151, 153, 154, 177, 335
Brennan, K., 42
Briody, B. A., 50
Brogg, Prof., 288
Bronwen, Prof., 285
Brown, D. J., 98
Brown, R., 287
Bull, H., 127
Buller, Mr, 322
Burfee, Prof., 289
Burkitt, A. N., 25
Burnet, F. M., 26, 39, 40, 42, 43, 44, 47, *48,* 49, 50, 53, 55, 56, 58, 74, 77, 81, 83, 85, 86, 88, 92, 104, 115, 122, 139, 169, 182, 184, 187, 327, 328, 329, 337, 339
Burnet, I., 77
Burnet, L., 43, 58, 74, 328
Burston, S. R., 25
Butlin, N. G., 116
Butts, Miss, 257

Cairns, J., 63, **64,** 65, 66, **72**
Cameron, Mr, 250
Campbell, A. G., 327
Campbell, E., 239, 294, 295
Campbell, T. D., 24, 239, 294
Carmichael, N., 17
Carr-Saunders, A. M., 287

Casey, R. G., 86
Catchside, D. G., 70, 75, 98, 100, 102
Chanock, E. (Beth), 126, 129, 177, 335
Chanock, R., 126, 129, 131, 177, 192, 335
Chapman, R. W., 239, 284
Chapple, P. L., 88
Chase, M., 51, 65, 131
Cherry, T. M., 114
Chifley, J. B., 59
Chisholm, B., 141
Chittoch, A., 327
Chou, E. L., 69
Christie, A. B., 166
Clark, D., 219
Cleland, J. B., 23, 24
Clucas, J. M., 270
Clunies Ross, I., 85, 86, 89, 187
Cockcroft, J., 74
Cocks, A. W., **23**
Comben B., **64**
Condliffe, P., 177
Conley, R., 329
Cook, J., 285
Coombs, H. C., 124, 134
Cooper, P. D., 71
Cooper, S., 320
Copland, D., 329
Cornish, V., 285, 288
Costin, A. B., 124, 134
Courtice, F. C., 97, 100, 183
Cox, L., 67
Crawford, J. (Jess), 109
Crawford, J. G. (Jack), 97, 106, 107, 109, 111, 121, 127, 178, 329
Crew, F. A. E., 287
Crocker, C., 73
Crocker, W. R., 66, 67, 73, 74, 329
Crompton, R. W., 117, 174
Curtin, J., 33, 59
Curtis, D. R., 100, 160, 176, 330

Dales, S., 169
Darling, Dr, 126
Darwin, C., 183, 272, 290, 301, 303, 311, 320
Davern R., **72**
David, T. W. E., 269, 270, **271**
Davis, B., 52
Dawe, V., 17
Day, M. F. C., 87, 330
de Baca, C., 244, 245
de Beer, Dr, 151
de Broglie, Duc., 284
de Burgh, P., 34
de Crespigny, T., 25
de Geer, Baron, 286
de Kika, Mr, 50
De L'Isle, Viscount, 81
de Quadros, C., 192
de Sitter, Prof., 288
de Witt, J., 296, 297
Debye, P. J. W., 284
Deck, S., 165
Delille, P. F. A., 88
Dimmick, S., **67**
Dimmock, N. J., 71
Dods, L., 62
Dohi, Dr, 69
Dolman, C. E., 50
Double, I. S., 320
Douglas, G. W., 87
Dowdle, W., 131
Dowdy, N. 19
Dowie, J., 17, 327
Downie, A. W., 65, 166, 335
Draper, A., 302
Dubos, R. J., 49, 50, *51*, 53, 56, 65, 122, 127, 131, 177, 327, 329, 337, 339
Dumbell, K. R., 65, 147, 161, 162, 169, 168, 169, 336
Dun, R., 176
Dunn, E. J., 322
Dwyer, F. P., 70
Dyce, A. L., 87

Easterbrook, K. B., 71
Eccles, J. C., 59, 64, 65, 100, 101, 176, 330
Eddington, A., 288
Elton, C. S., 88
English, J. C., 34, 36
Ennor, A. H., 52, 55, 56, 59, 64, 72, 73, 92, 97, 101, 114, 339

Epstein, M. A., 105
Ermacora, E., **72**
Esposito, J. J., 177, 181
Estes, M., 181
Evans, J., 114
Evans, L. T., 117, 174
Everingham, D. N., 107
Ewart, A. J., 239

Fairley, N. H., 25, *29-30,* 33, 39, 40, 183, 339
Fancha, Mr, 296
Fantini, B., 160, 175
Faraday, M., 284
Faunce, M., 330
Fawcett, Prof., 317, 319
Fazekas de St Groth, S., 58, 66, **64, 72**
Fenner, A. C. (Catherina), 199, 200
Fenner, A. G. 197, 292
Fenner, C. (Christopher), 13,
Fenner, C. A. E. (Charles), 9, **12,** 16, 52, 197, 209, **210, 211,** 212, **213,** 214, 215, 217, 218-231, **225, 226,** 232, 233, 239, 265, **271,** 339
Fenner, C. L. (Lyell), 9, 10, **11, 12,** 12-14, 204, 215, 268
Fenner, D. (Daniel), 201, 202 , 205
Fenner, E. (Ekkhardt), 202, 206
Fenner, E. (Emilie), 202
Fenner, E. L. (Peggy), 9, **11, 12,** 200, 201, 203, 204, 205, 212, 227, 239, 285, 289, 296, 338
Fenner, E. M. (Bobbie, *née* Roberts), 15, 30, 34, **41,** 42, 43, 44, 47, 49, 51, 52, 53, 55, 57, 58, 61, **62,** 74, 75, 109, 128, 130, 131, 160, 177, 178, 181-182, 188, 189, 190, 191, 192, 205, 328, 339
Fenner, F. J. (Frank), 9, 10, **11, 12, 16, 23, 28, 35,** 38, 40, **62, 64, 67, 72, 91,** *90-92,* 108, 123, 154, **157,** 165, 169, 170, 173, 178, **187,** 188, 189, 190, 191, 192, 201, 204, 205, **207,** 216, 222, 239, 293
Fenner, J. (Jack), **210**
Fenner, J. (Johannes), 197, 209, **211**
Fenner, J. A., 200
Fenner, M. (Max), 13,
Fenner, M. (May), **210**
Fenner, M. (Murray), 15,
Fenner, M. A. (Marilyn), 58, **62,** 74, 75, 126, 182, 192
Fenner, O. (Otto), 126, 209
Fenner, P. (Patricia), 15,
Fenner, P. (Peter), 15,
Fenner, T. (Ted), 13, 217
Fenner, T. (Thea, *née* Kleinig, 13, 14,
Fenner, T. R. (Senior), 197, 201
Fenner, T. R. (Tom), 10, **11, 12,** 14-15, 58, 222, 293
Fenner, V. (Vicki), 15, 58, **62,** 63
Fenner, W. (Wolfgang), 253
Fenner, W. G. (Bill), 10, **11, 12,** 15-16, 209, 222, 327
Fenner, W. J. (Winn), 9, 10, **11, 12,** 14, 16, 204, 216, 327
Fennessy, B., 87
Fermor. L., 319
Finger, A., 22
Fisher, C., 302
Fisher, H. M., **28**
Fisher, R. A., 56
Fleming, A., 183
Fleming, J. A., 292
Fletcher, N. H., 117, 174
Florey, H. W., 52, *55-56,* 57, 63, 65, 70, 72, 111, 182, 183, 184
Florey, Lady, 183
Follett, R., 192
Ford, E., 27, 31, 34, *35-36,* 39, 43, 67, 68, 327, 328
Fowler, R.H., 290
Fraser, J. M., 115, 178
Freeman, J., 44
Freeman, M., 40, 44
French, E. L., 86
Freund, J., 52
Funder, J., 39
Furnass, B., 331

Gajdusek, C., 129
Gale, F., 270
Gallo, R. C., 170
Gandevia, B., 119

Garnett, W., 291, 292
Gasser, H., 53
Gay, P., **64**
Gear, J., 149, 151
Geisler, Prof., 288
Gibbs, A. J., 71, 125, 171, 331
Gibbs, E. P. J., 172, 173, 177, 192
Gibson, F. W. E., 56, 98, 99, 103
Giffen, G., 19
Glazebrook, R., 291
Goebels, ., 53
Goldby, F., 25
Goodchild, J., 268, 269
Gorton, J., 97
Gottschalk, A., 71, **72**, 120, 190, 331
Grace, Dr, 69
Grace, T. D. C., **72**
Grafe, Dr, 257, 261
Graham, D., **64**
Granit, R., 103
Grant, K., 239, 240, 291
Grasset, N., 151
Gray, H., 24, 26
Grech, A., 168
Greenland, R., **72**
Gregory, J. W., 270, 285, 287
Gregory, W. K., 302
Grimmett, C., 19

Hackett, B., 55, 57, 127, 130, 131, 183
Hackett, C. J., 55, *56-57*, 127, 130, 131, 183, 327
Haecker, Dr, 304
Haggerty, Dr, 126
Haile, Mrs, 318
Haldane, J. B. S., 74, 287
Hambidge, C. M., 275
Hancock, W. K., 116
Hargrave, N. C., **23**
Harris, S., 124, 133
Hartung, E. J., 239, 240, 241, 285, 290, 291
Harvey, M., 36
Hawkes, R. A., **72**
Hawson, Miss, 240
Haygarth, J., 140
Hearst, W. R., 300
Hele, I., 232, 269
Helmi, Indonesian Ambassador, 182
Helmi, K., 182
Hemer, W. H., 262, 321, 322
Henck, G., 244
Henderson, D. A., 142, *143*, 149, 151, 152, 154, 155, 164, 165, 166, 169, 172, 177, 178, 192, 329, 336
Hendrick, G., **64**
Herschel, F., 183
Heysen, H., 232, 275
Higgins, F., 250
Higginson, R. A., **28**
Hilary, E., 275
Hinks, A. R., 287
Hirohito, Emperor,
Hirsch, J., 65, 131
Hirst, G., 69
Hirt, A. (Anna), 16, 212
Hirt, C. (Crin), 16, 212
Hirt, E. (Elva), **213**
Hirt, E. L., 203, 211, **213**, 338
Hirt, J. G., 203, 211
Hirt, M. (*née* Kaiser), 211
Hirt, P. (Paula), 212, **213**
Hitler, A., 29
Hitti family, 31
Hogben, L., 286
Hohnen, R., 76
Holloway, B. W., 63, **64**, 71
Holmes, A., 287
Holmes, J. M., 270
Home, R. W., 117, 174
Hone, R., 275
Hoogstraal, H., 127
Horner, J., 275
Horsfall, F., 49, 53, 65
Horzinek, M., 173
Hotchkiss, R., 51, 65, 103
Howard, J., **189**
Howes, D., 71
Huntingdon, Dr, 288, 289
Hurst, E. W., 36
Hutton, J. T., **23**
Huxley, A. F., 183
Huxley, J., 286

Huxley, L. G. H., 98, 114, 275
Hyams, B. K., 16, 217, 218-231, 265

Ingram, J. C., 176
Irwin, W. M., **28**
Ishigami, Dr, 70
Ivanovic, K. I., 76

Jackson, A. V., 57, 328, 331
Jackson, M., 57, 328, 331
Jacoby, K., 92
Jacotot, H., 65
Jamieson, Mr and Mrs, 77
Jeans, J., 290, 292
Jeffreys, H., 288
Jeffries, S. W., 227, 234, 338
Jenner, E., 140, 163, 180
Jezek, Z., 164, 166, 168, 169, 336
Jiang, Y. T., 151, 177
Johns, B. L. G., **23**
Johnson, D.W., 302, 303
Johnson, J., 78
Johnson, R. T., 192
Johnson, S., 242
Johnson, T. H., 24
Johnson, W. W. S., 29
Joklik, Mrs, **64**
Joklik, W. K., 56, 63, **64, 72**
Jones, B., 176
Jones, H., 125, 130
Jose, I., 25
Joseph, M., 149
Judd, Prof., 245, 249

Kaiser family, 203
Kaiser, M., 212
Kamahora, J., 69
Kapsenberg, J. G., 167
Karmel, P. H., 275
Kawai, K., 246
Keay, R., 125, 129
Kelly, J. E., **23**
Kempe, C. H., 166
Keogh, E. V., 33, 34, 36, 39, *38-39,* 39, 40, 42, 49, 53, 57, 332, 339
Kew, O., 181
Key, Dr, 320
Khan, M. H., 257
Khodakevich, L. N., 169
King, E., 33
Kirk, R. L., 98, 100, 101, 174, 176
Kissing, Prof., 289
Kitaoka, Dr, 69
Kleeman, Dr, 21
Kleinig, T., 13,
Kleinigs, 17, 204
Knight, M. C., **23**
Konowalow, B., **64**
Kostrzewski, J., 152, 153
Krieger, Prof., 302

Lacroix, A., 273, 311
Ladnyi, I. D., 164, 165, 168
Laffery, K. J., 71, 332
Landsteiner, K., 51
Lange, Mr, 324
Larcia, B. D., 74
Larmor, J., 291
Laver, W. G., 71
Law, Prof., 302
Lawton, G., 274
Le Maître, L'Abbé, 288
Leach, R. H., 58, 81, 82
Lee, Dr, 250
Legge, M., 58
Lehmann-Grube, F., 71, **72**
Leonard, F., 243, 302
Lewis, B., 59
Lewis, J. B., 275
Lewis, M., 15,
Light, W., 268
Lipskaya, G., 76
Little, D., 176
Lobeck, Prof., 302
Lodge, O., 285, 288, 292
Logie, A., 63, **64, 72,** 127
Lord, C., 239
Love, J., **213**
Loveday, D., 164, 165, 168
Low, A., 130
Lowther, D. A., 71, **72**
Luria, S., 66, 90

Lush, D., 81
Luther, M., 203, 206
Lwoff, A., 65, 107
Lynn, M., 188

Macbride, E. W., 286
Macdonald, G. S., 232, 235
Mackaness, G. B., 59, 63, 127, 129, 332
Mackay, Dr, 260
Mackenzie, H. A., 98
Mackenzie, J. S., 189
Mackerras, I., 36
Mackerras, J., 40
Mackinder, H., 285, 286, 287, 289
Macnamara, J., 83, 85, 187
Madgwick, J., 98
Madigan, C., 22, 23
Magarey, J. R., **28**
Mahalanobis, P. C., 77
Mahler, H., 152, 168
Mahony, M., 126
Maitland, G. W. G., 38
Malherbe, Dr, 323
Manning, R. E., 164
Mao, T. T., 69
Marconi, M., 284, 285, 293
Marennikova, S. S., 147, 151, 167, 168
Marmion, B. P., 333
Marshall, A., 126
Marshall, I. D. 58, 59, 63, **64**, **72**, 85, *86*, 87, 95, 98, 192, 327
Martin, F. C., 271
Matsumoto, S., 69
Matsushito, K., 178
Matthews, R. E. F., 108, 128
Mawson, D., 21, 23, 269, 275
Mawson, P., 23
Maxwell, C., 290, 291
May, R., 184
McAuslan, B. R., **72**, 105
McCauley, D., 332
McCauley, J., 332
McClain, M., **72**
McCoy, F., 22
McCoy, W. T., 222, 223, 225, 269, 272
McDermott, W., 52
McDonald, R., 284
McFadden, G., 181
McGauran, P., 188
McGregor, Mr, 250
McPhie, J., **23**
McWilliam, Dr, 179
Mead, M., 126
Medawar, P., 49
Meiklejohn, G., 151, 154, 166
Melnick, J. L., 169, 171, 173
Menzies, R., 70
Mercer, E. H., 27, 70, 99, 129, 336
Meselson, M., 66
Michel, H., 304
Mill, H. R,. 286
Millar, S., 129
Miller, J. F. A. P., 75
Millikan, R. A., 288
Milne, E. A., 288
Mims, C. A., 63, **72**, 105, 333
Mitchell, T., 272
Mitchell, W., 270, 271
Montagnier, L., 170
Montagu, M. W., 140
Moore, E., **64**
Morris, B., 98, 100
Morrison, W., 115
Moss, B., 192
Motteram, W. M., **23**
Mountford, C. P., 24
Moyer, R., 181
Mugford, F. K., **28**
Mulvaney, J., 77
Munn, R. E., 126
Murphy, F. A., 172 , 173, 180, 192, 337
Murphy, I., 180
Murphy, W., **72**,
Murray, H., 239
Myers, K., 87
Mykytowycz, R., **64**, 87
Myles, M., 62

Nagler, F. P. O., 50
Nakano, J. H., 167, 177
Nasution, H. A., **67**
Needham, J., 131

Nel, L. T., 322
Newcombe, K., 129
Newland, M. C., **23**, 327
Newton, W., 176
Nichol, L. W., 101
Nicholson, A. J., 114
Nietz, Mr, 18
Nordén, Å., 82
Nourse, J. P., 245
Nuyts, P., 268

O'Connell, G., 50, 52, 65
O'Connell, J., 65
Ogilvie, Prof., 288
Ogston, A. G., 70, 99, 101

Pandit, C. G., 73
Paneth, F., 284, 319
Parker, F. L., 267
Parkin, L., 23
Partridge, P. H., 97
Patel. Dr, 74
Penkethman, A., **64**
Percy, E., 284
Perrier, Prof., 290
Piaget, M., 257
Pickering, Prof., 319
Pierce, C., 65
Planck, M., 291
Plantinga, A., 163
Poole, W., 87
Porter, R., 192
Poynton, A., 274
Price, A. G., 24, 270, 275
Price, J. R., 107
Prozesky, O. W., 149
Pryor, L. 19, 62

Qiu, S-B., 179

Radford, M., 179
Rankin, C. G., **28**
Rao, A. R., 166
Ratcliffe, F. N., 36, 42, *83*, 85, 88, 89, 91, 92, 327
Redman, S. J., 188
Reeds, C., 302
Rees, A. L. G., 116, 119, 125, 174
Reeves, W. C., 64, 86
Regnery, D. C., 66, 86
Renfree, B., **64**
Rhodes, A. J., 50
Richter, R., 320
Ride, L. T., 67
Ride, W. D. L., 67, 114
Riemann, J. A., 140
Rivers, T., 53
Rivett, D., 83
Roach, Mrs, 318
Roberts, E. M. (Bobbie), 34, 40 , **41**, 181, 339
Roberts, J. A., **75**,
Robertson, R. N., 116, 124
Robson, H. N., 66, **67**
Rogers, W. P., 114
Rolland, W. M., **23**
Rommel, E., 33
Rossiter, G. G., 333
Rossiter, M., 333
Rothenburg, Prof., 292
Rott, R., 173
Rous, P., 51
Rubin, H., 66
Rudall, R. J., 232
Russell, J., 248
Rutherford, E., 284, 285, 290, 291

Sabin, A., 66
Sachse, Mr, 274
Sambrook, J. F., 71, 107, 177
Sanarelli, G., 82
Sanders, A. G., 55
Sands, D. W., **28**
Sands, R., **28**
Sarkar, J. K., 166
Sawer, G., 98
Schacht, H., 252
Schaffer, W., **64**
Scholes, Mrs, **64**
Schonland, Prof., 291
Schreiner, K., 59, 61, **62**
Schreiner, Mrs, **62**

Schuster, A., 292
Scollay, E. J., **62**
Seagoe, Dr, 246
Sela, M., 126
Seton, E. T., 301
Sharp, Prof., 302
Shepherd, D., 22
Sheppard, J., 163
Shope, R. E., 65, 83
Shortridge, D., 327
Simpson, G. C., 288
Skeats, E. W., 210, 239, 285, 318
Slaney, B., 15, 58
Slatyer, R. O. 190
Slöss, Prof., 290
Small, P., 192
Smallwood, R. A., 170, 333
Smart, J., 13
Smeddle, Mr, 259, 260
Smirnyagin, V., 127
Smith, D. I., 124, 128, 133, 137
Smith, E., 288
Smith, H., 284
Smith, J. O., 328
Smith, J., 40
Smith, M., 130
Smith, S. H., 289
Smith, S. H., 291
Smuts, J., 285, 288
Sobey, W. R., 87
Soekarno, President, 67
Soper, F., 141
Southwood, R., **30**
Spencer, B., 210, 285
Spencer, L. J., 272
Spreckels, Mrs, 297
Sprigg, R., 23
Stamp, D., 251, 289, 317, 318
Stanhope, J., 188
Stein, W., 53
Steiner, G., 75
Stoker, M., 129
Stoward, P. M., 14,
Studdert, M. J., 172, 173, 334
Sturt, A., 269
Sturt, B. M., 269, 293
Sturt, C., 268, 269, 292, 293
Sturt, G. R. N., 292, 294
Sullivan, W. M., **213**
Sunderland, S., 31, 66, **67**
Sutton, K., 59
Svet-Moldavsky, G., 77
Swanage, M., 117, 175

Talbot, Colonel, 40
Tamiya, Dr, 69
Tate, R., 22
Taylor, H. C., 52
Taylor, T. G., 269, 270, 286, 288, 289, 338
Temin. H., 66
Terrel, D., 192
Terrell, R. D.,.178, 194
Terrell, T. 176
Theiler, M., 53, 65
Thiele, C., 234-236
Thomas, M., 197
Thompson, H. V., 87, *88*
Thomson, J. J., 284, 291
Thomson, L., 131
Tilley, Prof., 319
Tindale, N. B., 23, 24, 25
Tooy, W., **67**
Tregenza, J., 268
Trusedale, W., **64**
Turnbull, G. M., **23**

Vagg, F., **64**
Vallée, A., 65
Vallee, P., 117, 174
van Rooyen, C. E., 50
Verinskiöld, Prof., 286, 288, 289
Virat, B., 65

Walker, J. S., 233, 235, 338
Walker, R., 175
Wallnerova, Z., 128
Walsh, R. J., 36, 81, 114, 116, 334
Warburton, F., **72**
Ward, H. K., 34, 68, 85, 271, 334
Ward, L. K., 21, 269
Warner, R., 40
Wasasjerna, ., Prof., 290

Wavell, A., 30
Webb, C. (Kate), 62, 63
Webb, L. C., 62, 63
Webber, C., 332
Webster, R. G., 71, **72**, 192
Wegener, A., 288
Weir, R., **72**,
Wellington, Prof., 288, 289
Wells, H. G., 260
Whatson, W., 334
White, D. O., **64**, 105, 117, 125, 159, 171, 172, 173, 177, 192, 334
White, F., 189
White, G., 125, 129, 133
Whitworth, J., 188, 192
Whyte, H. M., 70, 99, 106, 110
Wickett, J., 165
Wilkins, H., 260
Wilkinson, J., 277
Wilkinson, R. S., **28**
Williams, Mr, 260
Williams, S., 15, 58
Wills, L. J., 319
Wilson, S. J., 260
Wilson, V., 299
Winter, A., **64**
Wittek, R., 168, 169
Wood Jones, F., 22, 23, *24-25*, 26, 27, 36, 268
Wood, H., 284
Wood, I. J., 43, 328
Wood, J. G., 114
Woodroofe, G. M., 58, 59, 63, **64**, 66, 71, **72**, 85, 87, 95, 327
Woods, D. D., 56
Wooldridge, M., 183, 184
Wright, J., 191
Wright, R. D., 99
Wynn, K., 165, 168

Yeltsin, B.N., 170
Young, P., 124, 126, 131, 133
Young, Prof., 325
Yu, H., 68

Zeeman, P., 284

Subject Index, Part I, Frank Fenner, Chapters 1–11, 18

Aboriginal Australians
expeditions to Central Australia, 23-24
prehistorians' pilgrimage, 77
research on skulls, 24-25
Australian Academy of Science, 113-120, *see also* Books, Donations
Australian Biological Resources Study, 115
Basser Library, personal archives, 3
Charles Fenner's archives, 3, 207
Botany Bay Project, 115-116
committees, 113, 115
election as Fellow, 75, 78
Fenner Conferences on the Environment, 117-119
Fenner medal for plant and animal sciences, 120
histories of the Academy, 116-117
Secretary, Biological Sciences, 114-115
video histories of scientists, 119

Books published
Burnet, F. M. and Fenner, F. (1949) *The Production of Antibodies*, 49
Fenner, F. and Ratcliffe, F. N. (1965) *Myxomatosis*, 88-89
Fenner, F. (1968) *The Biology of Animal Viruses*, 92
Fenner, F., McAuslan, B. R., Mims, C. A., Sambrook, J. F. and White, D. O. (1974). *The Biology of Animal Viruses*, 2nd Edition, 105-106
Fenner, F. and White, D. O. (1970) *Medical Virology*, 104
Fenner, F. and White, D. O. (1976). *Medical Virology*. 2nd Edition, 171
White, D. O. and Fenner, F. (1986) *Medical Virology*. 3rd Edition, 171
White, D. O. and Fenner, F. (1994) *Medical Virology*, 4th Edition, 172
Fenner, F. (1976) *Classification and Nomenclature of Viruses*, 128
Fenner, F. and Rees, A. L. G. (eds) (1980) *The Australian Academy of Science. The First Twentyfive Years*, 116
Fenner, F., Bachmann, P. A., Gibbs, E. P. G., Murphy, F. A., Studdert, M. J. and White, D. O. (1987) *Veterinary Virology*, 172
Fenner, F., Gibbs, E. P. G., Murphy, F. A., Rott, R., Studdert, M. J. and White, D. O. (1993) *Veterinary Virology*. 2nd Edition, 173
Fenner, F., Henderson, D. A., Arita, I., Jezek, Z. and Ladnyi, I. D. (1988) *Smallpox and its Eradication*, 163
Fenner, F. and Gibbs, A. J. (eds) (1988) *Portraits of Viruses. A History of Virology*, 171
Jezek, Z. and Fenner, F. (1988). *Human Monkeypox. Monographs in Virology*, 169
Fenner, F., Wittek, R. and Dumbell, K. R. (1989) *The Orthopoxviruses*, 169
Fenner, F. (ed) (1990). *History of Microbiology in Australia*, 173-174
Fenner. F. (ed) (1995) *The Australian Academy of Science. The First Forty Years*, 117
Fenner, F. and Fantini, B. (1999) *Biological Control of Vertebrate Pests. A History of Myxomatosis; an Experiment in Coevolution*, 175
Fenner, F. and Curtis, D. (2001) *The John Curtin School of Medical Research. The First Fifty Years, 1948-1998*, 175-176
Fenner. F. (ed.) (2005). *The Australian Academy of Science. The First Fifty Years*, 117, 175
Fenner, F. (2006) *Nature, Nurture*

and Chance: The Lives of Frank and Charles Fenner

Centre for Resource and Environmental Studies (CRES), 121-136
appointment as Director, 121-123
appointment of senior staff, 124
establishment, 106-107, 121
funding problems, 124, 133-134
lectures in Australia, 132-133
overseas travel, 126-131
overview, 133-136
SCOPE, 127-128
Chance, in my life and my father's, 338-339
Childhood, *see* Fenner family
China, visits, 1980, Australian Development Assistance Bureau, 176
1978, certification of smallpox eradication, 151
1988, Gweilin, as tourist, 178
1947, invitation of Chinese Academy of Medical Sciences, 67-69
1988, review of use of nematodes for biological control 178-179

Dalton Plan, 17
Director, JCSMR, 97-112, *see also* Books
appointment, 97
building developments, 101
changes in existing Departments and Units, Genetics, 98
Medical Chemistry, 98-99
Electron Microscopy, 99
Chairman, committee on CRES, 106-107
Chairman, committee on undergraduate medical school, 106
establishment of new Departments, Clinical Science, 99
Human Biology, 100
Immunology, 100
Pharmacology, 100
International Committee on Taxonomy of Viruses, 107-108
lectures in Australia, 103-104
overseas trips, 102-103
overview, role of Director, 109-111
report on CSIRO and medical research, 107
writing textbooks, 104-105
Donations, Australian Academy of Science, 190-191
Australian National University, CRES, 191
Fenner Hall, 191
JCSMR, 189-190
Nature and Society Forum, 191

Fenner family, 9-17
family life, 9-10, 16-17, 204
home in Adelaide, 10-11
Fenner, F. J., change of name, 9, religion, 18
Siblings' biographies, Lyell, 12-14, Thomas, 14-15, William, 15-16, Winifred, 14
Fenner Hall, ANU, 187-188, 207
Fogarty Fellowship, 103, 126, 177-178
Friendships, Australian friends, 327-335
overseas friends, 335-337

Honours, prizes and awards, 78, 108, 133, 185-187
House, in Adelaide, 11
Canberra, 59-62
Human monkeypox, 146-47, 155

Japan, 1957, interviews with virologists, 69-70
1991-96, Agency for Cooperation in International Health, 179-180
Japan Prize, 169-170, 178
John Curtin School of Medical Research, *see also* Professor of Microbiology, Director
appointment, as Director, 97
as Professor of Microbiology, 54
as Visiting Fellow, 159
establishment, 49, 55
governance, 72-73, 97-98
permanent building, occupation, 70, planning, 63-64
expansion of staff, 70-71, 100-103
temporary laboratories, 59, 60, 63

Marriage, 40-42
children, 58, 62-63
tribute to wife, 181-182
Mentors, Howard Florey, 182-183
centenary celebrations, 183-184
Macfarlane Burnet, 48-49
centenary celebrations, 184-185
René Dubos, 51-52
Myxomatosis, 82-89
history, 82-85
book, 88-89
research, Hall Institute, 85-86
JCSMR, 87-88

Nature, in my life and my father's, 337, 339
Nurture, in my life and my father's, 337-338

Papua-New Guinea Medical Research Advisory Committee, 77-78
Parental influence, father, 21, 24
mother, ,
Personal celebrations, 1994, Frank and Bobbie Fenner Conference, 192
1999, fifty years in the ANU, 192
Physical anthropology, *see* Aboriginal Australians
Professor of Microbiology, JCSMR, 55-95
academic staff of Department, 58, 63, 64, 71-72
accommodation in Melbourne, 57-58
appointment, 54
elections, Australian Academy of Science, 78
Royal Society, 78
US Academy of Science, 133
first meeting with Florey, 55-57
move to Canberra, 59
Overseas Fellow, Churchill College, 74-75
research, genetics of rabbitpox virus, 90
Mycobacteria, 81-82
myxomatosis, 82-89
pattern of work, 90-92
reactivation of animal viruses, 90
study leave, 1953, 64-65 1957, 65-66
visits, China, 1957, 67-69
India, 1960-61, 76-77
Indonesia, 1956, 66-67
Japan, 1957, 69-70
Visiting Fellow, Moscow State University, 76-77

Recognition other than by honours, prizes and awards, 187-189
Resident medical officer, 25-26
Research School of Biological Sciences, ANU, 75, 114
Rockefeller Institute of Medical Research, 49-53
accommodation in New York, 50
offer of Chair of Microbiology in ANU, 52
operation on wife, 52
research on tubercle bacilli, 50-51

Schools, 17-18
Adelaide High, 18
Rose Park Primary, 17
St Peters College, 15, 18
Thebarton Technical High, 17-18, 19
Scientific Committee on Problems of the Environment (SCOPE), 125-126
appointment as Editor-in-chief, 1976-80, 125
election as member, 1971, 125
Smallpox and its Eradication, 137-157,
Book: *Smallpox and its Eradication*, 163-166
archiving records, 163-164
early plans, 164
editorial board, 164-165
grants to JCSMR for my expenses, 166
launch, 168
related books, 168
Eradication, early attempts, 140-142

Intensified smallpox eradication program, 142-156
certification of eradication, 147-155
China, 151
global, 152-155
India, 148
Malawi, 149
South Africa, 149-150, 151
ensuring vaccine quality, 142, 144
declaration of eradication, 155-157
monkeypox and related viruses, 146-147, 155
priorities and strategies, 145
'whitepox viruses', 147, 155, 166-168
post-eradication activities, Committee on Orthopoxvirus Infections, 160-163
Committee on Variola Virus Research, 163
proposals for destruction of all stocks of variola virus, 162-163
Smallpox, the disease, 138-140
as bioterrorism weapon, 170
clinical features, 138
epidemiology, 139
history, 139-140
immunization, 140
Sport, 19, 23

University of Adelaide, 21-26
enrollment in Science, 21
M.D. degree, 25
poliomyelitis epidemic, 22
prizes, 22, 25, 26
students' affairs, 23-24
transfer to Medicine, 21
University of Sydney, Diploma of Tropical Medicine, 27

Walter and Eliza Hall Institute, 39-40, 49-51
appointment as Haley Research Fellow, 43
as ANU professor, 58
research, ectromelia, 47-48
myxomatosis, 85-86
Syme Prize, 52
Wife, née Bobbie Roberts, *see also* Marriage
2/2 Australian General Hospital, 40
Land Headquarters Medical Research Unit, 40
Award, Associate Royal Red Cross, 42
World War II, 27-45
Australia, Heidelberg Military Hospital, 44
2/2 Australian General Hospital, 33-34
Woodside, 2/6 Field Ambulance, 28
discharge, 44
enlistment, 27
malariologist, 34-44
Borneo and Moratai, 42
MBE award, 38
New Guinea, 34-38
Middle East, 29-32
2/1 Casualty Clearing Station, 31-32
Corps Headquarters, 29-31
2/6 Field Ambulance, 29
holiday trips, Egypt, 32, Petra, 31-32
malaria, 30, 31
return to Australia, 33
Research, New Guinea, 37, 38
Walter and Eliza Hall Institute, 39-40

Yellowstone National Park, 180-181

Subject Index, Part II, Charles Fenner, Chapters 12-17, 18

Apprentice printer, 209
Australites, 272-274, 302, 304, 311, 320
 Charles Darwin's, 272, 320
 lecture, BAAS 1937, 318-319
 publications, 278, 280
Australasian articles, *see* Science Notes

Bibliography, publications on education, 237-238
 publications on science, 276-280
Books published by Charles Fenner
 Bunyips and Billabongs (1933), 266, 279
  comment by F. Wood Jones, 266
 Fenner, C. et al., (1935) *The Centenary History of South Australia,* 279
 Mostly Australian (1944), 266
 Gathered Moss, 1946), 266-267
 South Australia—A Geographical Study, Structural, Regional and Human (1931), 272, 279
 An Intermediate Geography of South Australia (1934), 272, 279

Carnegie Foundation, 239
Chance, in Charles Fenner's life, 338
Childhood, 209-210
Children, *see also* Fenner Family, Part I, 9-17
 group photos with parents, 11, 12
 Charles Lyell, 212
 Frank Johannes, Part I, 212
 Winifred Joyce, 212
Clerk Maxwell centenary, 290-292

Dalton Plan, 222
Death of Charles Fenner, 215
 obituary, 215,
Director of Education, 226-236
 appointment, 226
 appreciation by J. S. Walker, 233
 comments by Thiele, 234-236
 extract from Hyams' article in *Biography, an Interdisciplinary Quarterly,* 226-231
 tribute by S. W. Jeffries, 234
Drehsa, 204-205

Faraday Centenary, 284
Fenner Coat of Arms, 205-207
Fenner family lineage, 197-207
 Fenner families in Germany, 198-203
Fenner family, England, 292
Fenner Hall, Canberra, 187-188, 207
Field Naturalists' Society, 267

Hirt family history, 203-205
 Emma Louise Hirt, 203, 212, 213
Historical Memorials Committee, 268-269
 Light, Colonel William, 268
 Nuyts, Peter, 268
 Sturt, Captain Charles, 268-269
  portrait by Ivor Hele, 269
  preservation of his house in Adelaide, 269
  visit to family in England, 292-293

Marriage, 211, 213
Melbourne Teachers' College, 210

Nature, in Frank's life and his father's, 337
Niedergrenzebach, 198-202
Newspaper, the *Talbot Leader,* 209
Nurture, in Frank's life and his father's, 337

Overseas trips, 1937, education, 239-263
 broadcasting for schools, 259-260
 cinema in schools, 261-262
 vocational education, England, 250-252, 257-262
  Germany, 252-256
  South Africa, 262-263

Switzerland, 257
United States of America,, 243-250
Overseas trip, 1931, scientific, 239, 317-320, 281-326
BAAS Centenary, 239, 284-290
BAAS, 1937, Nottingham, 317-320
Clerk Maxwell Centenary, 290-292
emotional reactions to overseas travel, 239-243
Faraday Centenary, 284
geographical/geological observations, at sea, 281-283
Canada, 303
continental Europe, 304-312
England, 283, 287
South Africa, 320-326
Spitzbergen, 312-317
United States of America, 294-303

Physiography, Werribee Gorge, 265
Prizes, David Syme Research Prize, 275
John Lewis Gold Medal, 275
Sachse Gold Medal, 274, 275
Pupil teacher, 209, 211

Royal Geographical Society of Australasia (South Australian Branch), 267
Royal Society of South Australia, 267

School, at Dunach, 209
School broadcasting, 232-233, 259-260
Science Notes, *The Australasian*, 257, 265-266
South Australian Museum, 215, 276
South Australian School of Arts and Crafts, 232
Superintendent of Technical Education, South Australia, 217-224
appointment, 214, 217
appreciation by Australian Broadcasting Commission, 232-233
article in *The Mail*, 231
retirement, 214
extract from *Australian Dictionary of Biography*, 217-218
extract from Hyams' article in *Biography, an Interdisciplinary Quarterly*, 218-225

Teaching positions, Charles Fenner
Ballarat School of Mines and Industries, 212, 214
Mansfield Agricultural High School, 212
pupil-teacher, 209, 211
Sale High School, 212
Teaching positions, Emma Louise Hirt, 212-213
Tektites, *see* australites
Thebarton Technical High School, 217, 222, 232

Wongana Circle, 275

University of Adelaide, Lecturer in Geography, 270-271
Charles Fenner Prize, Department of Geography, 271
University of Melbourne, Kernot Research Scholar, 210
BSc course, 210
DSc degree, 210, 265
Syme Prize, 275
University of Sydney, chair in Geography, 269-270